RANGE
MANAGEMENT

McGraw-Hill Series in Forest Resources

Henry J. Vaux, Consulting Editor

Walter Mulford was Consulting Editor of this series from its inception in 1931 until January 1, 1952.

RANGE MANAGEMENT

Third Edition

Laurence A. Stoddart
Late Professor and Head
Department of Range Management
Utah State University

Arthur D. Smith
Professor Emeritus
Department of Range Science
Utah State University

Thadis W. Box
Dean, College of Natural Resources
Professor of Range Management
Utah State University

McGraw-Hill Book Company

New York   St. Louis   San Francisco   Auckland
Düsseldorf   Johannesburg   Kuala Lumpur   London
Mexico   Montreal   New Delhi   Panama   Paris   Singapore
São Paulo   Sydney   Tokyo   Toronto

Library of Congress Cataloging in Publication Data

Stoddart, Laurence Alexander, date
  Range management.

  (McGraw-Hill series in forest resources)
  Includes index.
  1. Range management. I. Smith, Arthur D., joint
author. II. Box, Thadis W., joint author. III. Title.
SF85.S75 1975   636.08'4   74-26668
ISBN 0-07-061596-9

Range Management

    4 5 6 7 8 9 0 KPKP 7 9 8 7 6

This book was set in Baskerville by Creative Book Services,
division of McGregor & Werner, Inc. The editor was William
J. Willey; the designer was Creative Book Services; the production supervisor was Judi Frey.
Kingsport Press, Inc., was printer and binder.

To the memory of Dr. L. A. Stoddart who meant so much to us and to the profession of range management.

Contents

Preface

In the more than 30 years since the appearance of the first edition of *Range Management,* there have been many changes. New facts have been uncovered, basic concepts have been refined and tested by experience, and investigative techniques have been perfected. But even more important than the technical changes are the shifts in emphasis among the various rangeland products. Nonconsumptive uses, though not new, have become even more important. With increased human populations and greater demands for rangeland products, the need for clear understanding and greater knowledge of range ecosystems remains as vital as before. Nevertheless, no new conceptual framework differentiates the field of range management now from then. Basically, range management deals with the use of lands of low potential productivity maintained under extensive systems to produce water, red meat, wildlife, timber, and recreational opportunities in such a way that the basic resources, soil and vegetation, remain unimpaired.

Although an effort has been made to include all significant topics, space did not permit recognition of all literature pertinent to each. Documentation of literature citations includes those which seem to confirm most directly the basic principles presented. In the presentation of principles, an effort has been made to stimulate thinking and provide a basis for additional research and experimentation to validate or prove false views now held. In a rapidly changing world, range scientists must continue to test and discard outmoded concepts and replace them with sounder ones if range management is to remain viable. As was expressed in the preface to the second edition, "Growth rather than fixity must characterize a useful and vital science of range management."

Many of our colleagues offered suggestions and encouraged us in the venture. Those who reviewed sections of the manuscript appropriate to their specialties are Dr. George B. Coltharp, Dr. John O. Evans, Dr. John C. Malechek, Dr. C. M. McKell, Dr. Phil R. Ogden, Dr. D. I. Rasmussen, Dr. Lee A. Sharp, and Dr. Philip J. Urness. We are particularly indebted to Dr. Don D. Dwyer who reviewed many sections of the text and provided both encouragement and helpful criticism.

We are in an especial way grateful to Mrs. L. A. Stoddart who encouraged us in proceeding with the project and provided, as she did in earlier editions, valuable editorial assistance.

It is hoped that the third edition will continue to provide students of range management and related resources affected by range practices with an understanding of the principles involved in what must become, as demands for rangeland products increase, an even more important and challenging land-management profession.

ARTHUR D. SMITH
THADIS W. BOX

RANGE MANAGEMENT

Chapter 1

Characteristics of Rangelands

Range, rangeland, and *range management* are more easily described than defined. No single characteristic differentiates rangelands from croplands or forest lands. Rangelands are commonly arid and, hence, not suitable as croplands, but factors other than aridity may prohibit cultivation—rocks, shallow soils, rough topography, poor drainage, and cold temperatures. The term *range* connotes broad, open, unfenced areas over which grazing animals roam; but as rangelands are more intensively managed, fences—once useful for distinguishing range from pasture lands—are increasingly a part of the range scene. Once, with near universality, range could be defined as lands supporting stands of native plants, but the use of agronomic methods to alter native vegetation and the introduction of exotic species now make this criterion imprecise. Nevertheless, the term *range* probably evokes a quite similar image among professionals and laymen alike who have an acquaintance with lands and land use. Despite this fact, rangemen have been notably reluctant to define range; instead they have been content to describe and characterize it. (Blaisdell et al., 1970; Stoddart, 1967). (See Dyksterhuis, 1955, and Hedrick, 1966, for definitions of range and range management, respectively.)

WHAT RANGELANDS AND RANGE MANAGEMENT ARE

Rangelands are those areas of the world, which by reason of physical limitations—low and erratic precipitation, rough topography, poor drainage, or cold temperatures—are unsuited to cultivation and which are a source of forage for free-ranging native and domestic animals, as well as a source of wood products, water, and wildlife. These products may be found in all possible combinations, and at a given place only one may be of consequence. Many lands are too arid to significantly affect ground-water supplies or streamflow; at higher elevations water may be the principal contribution of rangelands. Wood products may be produced from lands which provide forage for native and domestic ungulates, although more commonly range areas do not support forest growth. Many species of wildlife are found largely and some solely on rangeland. By their nature rangelands are increasingly important as a place for man to engage in various outdoor recreational pursuits, and they provide environmental amenities—scenery and open space.

The importance of any particular product from, or use of, the range is not determined by physical factors alone. Culture and the stage of development of society are also important determinants. In primitive societies rangelands make their most important contribution by providing foods from wild plants and animals. In underdeveloped countries with pastoral economies, forage for domestic livestock is the primary contribution. In more developed countries where rangelands are associated with intensive agriculture and industrial development, as in the Western United States, water may be of greater value than forage. As economies become more complex and populations increase and become more highly urbanized, the importance of ranges as open space and places to seek relief in recreational pursuits will increase.

Despite these varied resources, the historical and most nearly universal characteristic of range at present is in the production of forage for wild and domestic animals. Though some may remain ungrazed because of physical limitations, or because other resources such as water may be deemed more important, it is difficult to conceive of a range without vegetation suitable as forage for animals. Forage is the central attribute of rangeland, as trees are of forests. In neither case are these attributes the sole determinants. They are, however, the pervasive ones. (See Fig. 1.1.)

Fig. 1.1 Livestock constitutes the major product of rangelands in developing countries. In developed countries its place among other rangeland products depends upon the particular range in question.

Range management is the science and art of optimizing the returns from rangelands in those combinations most desired by and suitable to society through the manipulation of range ecosystems. Range management is at once a biological, a physical, and a social science. It is *biological* because it deals with the response of vegetation to cropping and the response of the animals which harvest the crop; *physical* because climatic, topographic, and hydrological factors determine the kind and degree of use that can be made of range; and *social* because the needs of society determine the uses to which range resources are put.

At present there is a tendency to emphasize the scientific aspects of range management as is seen in the use of the term *range science* to cover the activities of both the practitioner and the researcher. Although the scientific aspects are being constantly expanded, range management is more than a science. Whether as a researcher expanding scientific knowledge or as a practitioner applying it, the distinctive attribute of the range manager has been, and is, the application of knowledge to the preservation of resources and the solution of social problems. This activity is best described as an art since it requires, by extrapolating from the pool of information available, the ability to synthesize a workable management plan for areas that differ in lesser or greater degree from those where the information was developed. It also requires the perception to detect early, evidence of changes in plants, soil, or animals, and the skill to alter plans to ensure a measure of stability to the ecosystem. This "feel" for the resource is the hallmark of the rangeman.

IMPORTANCE OF RANGELANDS

Rangeland ranks as a major land type whether measured by size, support for animal-based industries, or source of streamflow. Substantial portions of all major continents are rangeland. Worldwide, 30 percent of the world's land area is grassland; 27 percent is classed as forest; 10 percent is cropland (Semple, 1951). This does not, however, fully indicate the extent of the range resource. Considerable acreages of forests are grazed and many of the more arid portions of thè world, normally considered "deserts," contribute significantly to forage production in favorable years. An example of the former is the pine forests of the Southeastern United States where timber, wildlife, and domestic livestock are simultaneously produced from the same land. Vast areas around the Sahara and in central Australia are representative of the latter. Nor do these include the tundra which, in North America alone, occupies more than one million square miles and now is used only for native and semidomesticated animals (Klein, 1970), but which potentially may contribute significantly to range-forage production. It has been estimated that 47 percent of the earth's land surface is rangeland (Williams et al., 1968). (See also Table 1.1).

A more useful measure of the importance of rangelands is the contribution they make to animal production. In Australia, one-third of the cattle and sheep populations are supported by rangeland (Box and Perry, 1971). In the United States it has been estimated that 54 percent of the feed units consumed by all livestock and 76 percent of the feed units consumed by beef cattle come from forage (Hodgson, 1972). Unfortunately, available data do not separate world cattle populations by type, but the estimated populations of all cattle, sheep, and goats

Table 1.1 Distribution of Rangelands of the World by Continents exclusive of Forested Areas

Continent	Productive*	Potential†	Total
	(Percent)		
Africa	27	5	32
Australia	7	1	8
Europe and Asia	27	7	34
North America	9	5	14
South America	12	1	13

*Includes grasslands, tropical savanna, desert shrub, and woodland-shrub types.
†Tundra and dry deserts (see Fig. 2.1).

throughout the world are shown in Table 1.2. There are in addition 124 million horses (including mules and asses), 125 million buffalo, and 13 million camels.

These do not, however, indicate the most important countries from the standpoint of meat production. If one compares the data in Table 1.2 to those in Table 1.3 the disparity between animal populations and meat production is evident. Africa is credited with more cattle than the United States but produces only about one-third as much meat. Similarly, India has over 176 million cattle from which almost no beef is produced. These reflect cultural influences in which animals, particularly cattle, are valued for other than monetary returns. In some areas of Africa, cattle are evidence of wealth and are eaten only rarely at ceremonial functions or when they die naturally (Larson, 1967). The American Indians and particularly the Navaho in the Southwestern United States had similar notions of the horse as wealth, which led as it does in Africa to disregard for the range and widespread range abuse.

Table 1.2 World Population of Cattle, Sheep, and Goats by Principal Countries and Continents, 1969–1970

Data from U.N., F. A. O. (1970)

Continent and Country	Cattle	Sheep	Goats
	(Thousands)		
Europe	124,445	128,747	12,673
Spain	5,035	13,836	2,570
U.S.S.R.	95,000	130,665	5,148
North and Central America	168,367	28,035	14,431
Canada*	11,836	616	20
U.S.A.*	112,330	20,422	2,894
South America	197,664	123,101	30,122
Argentina*	48,000	43,900	5,380
Brazil*	95,008	24,333	14,744
Colombia*	19,576	1,850	900
Uruguay*	8,500	21,800	14
Asia	288,660	205,378	145,883
India*	176,450	42,600	67,500
Mainland China	63,100	70,600	57,000
Africa	150,979	149,107	118,959
Oceania	29,980	347,313	200
Australia*	20,700	176,000	85
World Totals	1,118,205	1,072,946	384,416

*Included in continental totals.

Table 1.3   Production of Meat Products from Range-type Animals by Selected Countries and Continents, 1970

Data from U.N., F. A. O. (1970)

Continent and Country	Beef	Mutton	Total
	(1,000 metric tons)		
Europe	8,630	1,026	9,656
Spain*	319	130	449
U.S.S.R.	4,800	800	5,600
North and Central America	12,560	320	12,880
Canada*	900	9	909
U.S.A.*	10,660	251	10,911
South America	6,135	413	6,548
Argentina*	2,930	200	3,130
Brazil*	1,720	56	1,776
Colombia*	435	3	438
Uruguay*	380	54	434
Asia	1,888	1,397	3,285
India*	172	357	529
Mainland China	2,200	600	2,800
Africa	2,407	995	3,402
Oceania	1,483	1,335	2,818
Australia*	1,078	760	1,838
World totals	40,103	6,886	46,989

*Included in continental totals.

Rouse (1970) suggests that a little more than half of the world's cattle exclusive of buffalo are grown for beef, milk, or both. Primarily, the meat-producing areas are Europe, the Western Hemisphere, Australia, and New Zealand. Slightly less than four-tenths of the cattle are in Asia and are kept principally for draft purposes, but they may be milked as well. The remaining tenth are in Africa where, except for those in South Africa, they contribute little other than milk, and even that is variable (Table 1.4).

RELATIONSHIP OF RANGE MANAGEMENT TO OTHER DISCIPLINES

Not all that comprises the body of scientific knowledge referred to as range science can be regarded as unique to the range field. Historically, ecology has had the greatest influence on the point of view and the methodology of the range student or practitioner. The dynamics of plant communities, first enunciated by ecologists, provided the foundation for range management. When a broader biological view was developed, in which plants and animals together were regarded as forming the ecological unit, the basis for today's concept of the ecosystem was laid. In his study and manipulation of range ecosystems, the range manager is a practicing ecologist, or if you prefer, an ecosystems engineer. His is the task of exploring, understanding, and manipulating elements of the whole biological system while recognizing the interrelationships among plant, animal, soil, water, and climate, thus expanding the science of range plants and animals to the whole resource complex—i.e., the ecosystem. Despite this broader concern, the plant is the foundation of the range ecosystem—the primary producer of foodstuff—and a knowledge of plants, and particularly plant physiology, is basic to the range manager, for unless the processes within the plant are known, the reactions of plants en masse cannot be understood. But plants are dependent upon soil; hence, the range manager must have a basic understanding of soils and soil formation. Knowledge of soils and plants together form the basis for the ecological principles fundamental to range management.

Plants are the source of food for countless animals which, in the process of food gathering, alter vegetation both individually and collectively. Accordingly, knowledge about food preference and behavior of animals is essential to and part of the body of knowledge of the range

Table 1.4 Average Meat Production per Animal and per Capita by Continent or Area, 1970

Data from Jasiorowski (1973)

Continent/area	Beef and veal, per animal	Mutton and goat, per animal	Meat production, per capita
	(Kilograms)		
Europe and U.S.S.R.	59.7	6.6	21.2
North America	87.7	11.1	49.1
Latin America	29.4	2.7	26.8
Near East	17.3	4.4	9.9
Far East and China	8.2	4.0	2.4
Africa	13.6	3.5	8.9
Oceania	45.8	5.7	145.8
World	31.2	4.7	12.3

scientist and practitioner. Many range areas produce concurrently or sequentially, forage and timber crops. Silvicultural principles must be known and adhered to, since grazing can have profound impacts on forest regeneration. Because ranges are predominantly arid by nature, the range manager must strive to achieve the most effective division among components of the hydrological cycle. In particular, maximum infiltration of water into the soil must be sought, bringing into play the fields of pedology, hydrology, and climatology. And since all products of the range are for the use of man, social, cultural, and economic considerations are vital to allocation of the range resource. These relationships are diagrammed in Fig. 1.2. Thus, the integration of knowledge from widely separate disciplines into a coherent system by the range manager is peculiar to range science.

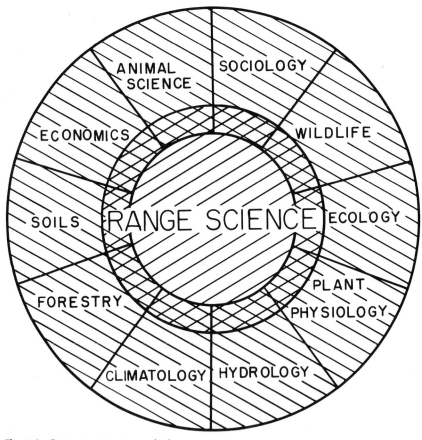

Fig. 1.2   Range management, whether as a practice or as a science, is intimately associated with other more or less closely allied disciplines.

PHYSICAL FEATURES AFFECTING USE OF RANGELANDS

One cannot understand the problems of managing ranges without a knowledge of their physical characteristics. Not only do these physical features determine the kind of vegetation available, but they also determine the manner and degree of use possible. These physical features include climate, soil, and topography. Together they cause grass to grow in the plains, forests to grow in the mountains, and shrubs to grow in the deserts.

Soil is produced by the action of climate and vegetation upon the parent-rock materials. Adequate precipitation makes luxuriant vegetation which, in turn, makes deep fertile soil. Conversely, under extreme aridity, the soil is poor. Climate, and more specifically precipitation, can be regarded as the most important single factor influencing forage production.

Topography not only influences the use to which ranges may be put, but affects plant growth also. On favorable exposures, conditions for plant growth are above the general average of the area. On unfavorable exposures, such as very steep slopes, erosion and runoff may be severe; hence, soil and moisture are unfavorable. Especially on hot south-facing slopes, vegetation may be very sparse.

Physical conditions on rangeland generally are unfavorable to luxuriant plant growth. Rangeland frequently occupies steep rocky hillsides, dry shallow plains, or cold windswept prairies, and for these reasons is inherently limited in productivity. A luxuriant growth cannot be expected even under the best management. The concept of site limitation is important in managing rangeland.

PRECIPITATION AS A FACTOR IN RANGE USE

The most constant characteristics of precipitation on rangelands are paucity and variability. Large areas receive less than 25 cm; few outside the tropical zone receive more than 65 cm per year. Even these amounts are not uniformly distributed and most of the annual rainfall may come within a few months. Annual variations are also common with precipitation in half or more years falling below the long-time average (Fig. 1.3).

Geographic Distribution of Precipitation

The geographic distribution of precipitation is influenced by the distance from oceans, varying inversely with distance. Thus, as one leaves the northern coastline of Australia at Darwin and travels southward

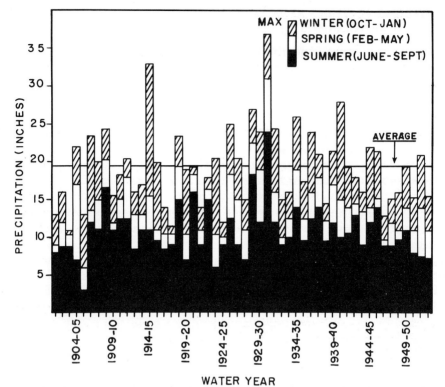

Fig. 1.3 Tremendous variations in precipitation, both annually and seasonally, make management of rangelands hazardous and complicated. . *(Data from Reynolds, 1959.)*

through 14 degrees of latitude (about 1,600 km), precipitation steadily declines to one-tenth that received on the coast (Perry, 1960):

	(cm/year)		(cm/year)
Darwin	149	Tennant Creek	35
Katherine	90	Alice Springs	25
Daily Waters	63	Charlotte Waters	13

Even greater differentials may exist. In the state of Washington parts of the Big Bend region in the center of the state receive 13 to 18 cm and parts of the Cascade Range, 160 km westward, receive over 300 cm of precipitation per year. Topographic influences, as well as distance inland, are involved here.

Topography causes wide differences in precipitation, often within a few miles. The average annual precipitation for Utah's weather stations below 1,200 meters (m) altitude is 25 cm; those between 1,200 and 1,500 m receive 32 cm; those between 1,500 and 1,800 m receive 34 cm; and those above 1,800 m receive 39 cm. The annual precipitation over the high westerly exposures of the Wasatch Range averages nearly 10 times that on certain deserts west of the Great Salt Lake.

Annual Variation in Precipitation

Low precipitation over the world's rangelands is made more serious by great variability. Monthly and yearly precipitation both deviate widely from their long-term mean, and prolonged periods of above or below normal rainfall are common.

When more than one year of below average precipitation occurs in succession, the effects are intensified (Fig. 1.4). Table 1.5 shows the precipitation averages of the Western United States during the historic drought period of 1933–1935. With the exception of Washington, all states had precipitation significantly below normal, with many having less than two-thirds of the normal rainfall. This moisture deficiency was

Fig. 1.4 Dead and dying animals around a water hole in Bechuanaland during the 1965 drought. Droughts are devastating to vegetation and range livestock, and to the people dependent upon them. *(Courtesy F. A. O.)*

Table 1.5   Average Precipitation for the 17 Western States during the Historic
1933–1935 Drought in Percentage of Long-time Normal

Data from U.S. Weather Bureau

State	1933	1934	1935
Arizona	86	76	112
California	84	75	92
Colorado	92	66	96
Idaho	103	89	68
Kansas	82	74	106
Montana	107	77	72
Nebraska	90	63	100
Nevada	74	79	96
New Mexico	89	70	102
North Dakota	79	56	105
Oklahoma	93	83	112
Oregon	107	98	78
South Dakota	91	65	81
Texas	84	86	121
Utah	79	71	81
Washington	137	111	84
Wyoming	85	76	86

unprecedented since man has studied weather in the West. This great
drought period, and the lesser droughts both preceding and following,
had profound influences upon range conditions. In many areas, established vegetation underwent complete change, even though grazing was
not heavy. So profound were the effects that recovery involved 5 to 15 or
more years.

The period from 1949 to 1954 was one of unprecedented drought
in Texas, and vegetation was affected greatly. Mortality of established
grass plants of curly mesquite *(Hilaria belangeri)* was 88 percent (Young,
1956).

Calculations of the number of years of precipitation significantly
below normal (less than 85 percent of normal annual precipitation) in
the Western United States show the greatest frequency of deficient precipitation in the Southwest, where over 40 percent of the years are
deficient. In the Northwest, only 10 to 20 percent are deficient. Over the
majority of the West, significantly subnormal precipitation occurs in 20
to 40 percent of the years. Further, these studies show that most years
are below average in precipitation; therefore, in studying charts showing
the average amounts of precipitation, one should not lose sight of the

fact that the amounts shown are usually available for plant development in fewer than half of the years (U.S. Department of Agriculture, 1922).

Seasonal Distribution and Character of Precipitation

In arid and semiarid areas the seasonal distribution of precipitation is exceedingly varied. Distribution is important because it determines whether vegetation receives moisture during its growing season or whether the moisture must be stored in the soil for use at some later period. This is more important in temperate climates, where favorable moisture and favorable temperatures may not coincide, than in tropical areas.

The distribution of precipitation on rangeland throughout the year is typically very unequal resulting in a "dry season" and a "wet season." For any location these wet and dry seasons occur during the same months of the year giving rise to characteristic climatic patterns. In some areas the disparity between monthly precipitation is small; in others it is very great. In interior Colombia 90 percent of the precipitation falls in the 7-month rainy season (Blydenstein, 1967). In temperate climates periods of high rainfall may come either during the warmest or the coldest months, giving rise to summer or winter wet seasons (Fig. 1.5). The summer rainy period may occur at the onset of the warm season, as in the northern Great Plains, or later in the summer, as in the Southwest. Where winter temperatures are sufficient for plant growth, winter rainy seasons may give rise to Mediterranean climates and vegetation, as in lands bordering the Mediterranean Sea, in California, and in parts of Australia. Annual plants are prominent in these climates. In high latitudes and altitudes the wettest months are also the coldest months when plant growth either cannot take place or is minimal. In consequence, moisture penetrates the soil to considerable depths and supports deep-rooted plants.

In tropical areas temperature usually does not limit plant growth and range productivity is controlled by the availability of moisture. Seasonal differences do not follow the typical summer-winter relationships of temperate climates, but wet and dry periods form the seasons. Plant growth is rapid during the wet season and plant production for the entire year may occur in a few weeks or months following the onset of rain. Plants are dormant during the dry season. Plant life usually consists of a mixture of herbaceous and shrubby vegetation adapted by dormancy or deep roots to survive long rainless periods.

The effectiveness of precipitation is not expressed by averages alone. The intensity of individual storms is of great importance. Very light rains are of no value to plants because moisture evaporates rapidly,

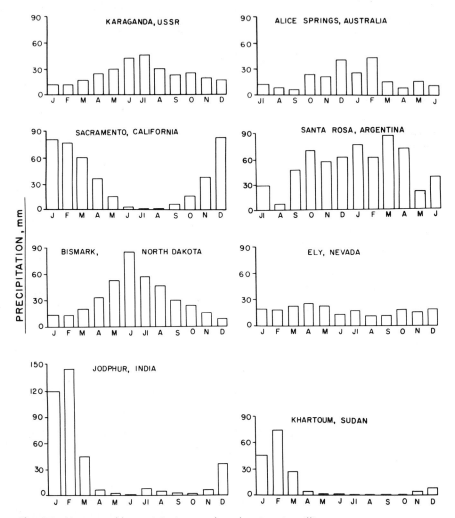

Fig. 1.5 Mean monthly precipitation at selected stations in millimeters. Even normal precipitation on rangelands is characterized by dry periods at some season of the year. Monthly sequences of stations in the Southern Hemisphere were offset 6 months to show the same summer-winter pattern.

especially if it falls on hot soils. Unless precipitation percolates into the soil to a depth where it is available to plant roots, it is of little importance. Under certain conditions, especially on clay soils, showers of less than 4 mm are of no value in increasing soil water (Shreve, 1914). In parts of the Southwest rains of less than 13 mm are insufficient to promote growth of range grasses unless preceded or followed by more effective

Table 1.6 Variation in Precipitation at the Desert Experimental Range Headquarters in Southwestern Utah

	Jan.	Feb.	Mar.	Apr.	May	June	July	Aug.	Sept.	Oct.	Nov.	Dec.	Total
					(Centimeters)								
1963	0.00	1.14	0.83	2.54	0.94	3.58	0.08	5.13	3.66	1.47	2.03	0.89	13.18
1964	0.41	0.18	0.89	3.43	0.99	0.84	1.45	0.86	0.36	0.38	1.55	0.81	12.14
1965	0.18	0.15	1.24	1.85	3.51	1.07	3.96	4.93	2.34	0.33	0.76	3.18	23.34
1966	0.20	0.53	0.15	0.46	0.10	0.58	0.76	1.57	2.06	0.41	0.20	3.07	10.11
1967	0.64	0.13	0.33	2.92	2.39	4.72	2.29	1.22	4.70	0.03	0.33	0.46	20.14
1968	0.05	1.27	0.86	2.57	1.27	2.03	2.39	3.05	0.33	0.97	0.38	0.23	15.39
1969	1.37	1.50	1.35	0.53	0.33	2.57	5.54	1.14	0.74	1.17	0.51	0.48	17.22
1970	0.08	0.61	2.62	1.75	0.00	0.99	4.55	1.17	0.38	0.08	1.07	1.14	14.43
1971	0.13	0.53	0.03	1.68	2.51	0.41	0.64	3.28	0.53	1.96	0.05	1.07	12.80
1972	0.00	0.00	0.00	0.86	0.00	1.63	0.10	3.28	1.75	1.93	0.61	1.85	12.01
Mean	0.31	0.60	0.83	1.86	1.15	1.84	2.18	2.57	1.69	0.87	0.75	1.32	15.07
Ratio max/min	∞	∞	∞	7.46	∞	11.51	6.93	5.97	14.24	6.53	4.06	13.83	23.09

rains (U.S. Department of Agriculture, 1941). Such ineffective storms are common in the more arid range areas.

Frequently, long periods may elapse without usable rainfall. In Reno, Nevada, during 1934 a period of 3 months contained only 2 days in which over 0.25 mm of precipitation was recorded. At Yuma, Arizona, only 12 days out of the year had over 0.25 mm of precipitation (U.S. Department of Agriculture, Weather Bureau, 1903–1937). At Kelton, Utah, records show 6 consecutive years without a trace of precipitation during August, and 7 consecutive months have occurred with no trace of precipitation. Typical of monthly patterns of precipitation in the salt desert in the Western United States are those from the Desert Experimental Range in Utah (Table 1.6).

At the other extreme are high-intensity storms in which most of the precipitation runs off not benefiting vegetation. Table 1.7 shows some such heavy rains in comparison with the annual total. Illustrative of the magnitude of some of these is one in 1938 in southern California (Burke, 1939). Precipitation had been about normal prior to February 27. Then rain began that, by March 3, had deposited 733 mm in the San Gabriel Range. In the flood that resulted, 87 lives were lost and property damage was placed at $79 million.

Table 1.7   Average Annual Precipitation Compared to Maximum Received during Short Periods of Time for Selected Western Stations

Data from U.S. Weather Bureau

Station	Number of days	Year	Precipitation, cm	Annual average precipitation, cm	Ratio, percent
Phoenix, Ariz.	1	1946	3.2	19.6	16.4
Phoenix, Ariz.	2	1946	5.8	19.6	29.7
Tucson, Ariz.	1	1943	6.5	29.2	21.7
Los Angeles, Calif.	1	1933	18.7	38.7	48.3
Los Angeles, Calif.	2	1933	25.6	38.7	66.1
Pueblo, Colo.	1	1946	5.9	29.6	19.8
Valentine, Neb.	1	1949	9.3	46.6	19.9
Del Rio, Tex.	1	1944	12.6	50.5	24.9
San Antonio, Tex.	1	1951	15.7	69.0	22.7
Modena, Utah	1	1943	4.9	26.8	19.0
Modena, Utah	4	1943	8.0	25.8	31.0
Lander, Wyo.	1	1951	5.0	32.1	15.4

Snowfall

Snowfall is a major determinant of plant communities and range use by animals at high elevations and high latitudes. As an example, in the United States, snowfall affects both the production of livestock and big game animals in about the northern two-thirds of the Western states. High mountainous ranges cannot be used at all by livestock in winter, and deer and elk are forced to seek ranges at lower elevations. Livestock either must be fed harvested forages or trucked or trailed over great distances to snow-free winter ranges. Periodically even these may be struck with an unusual combination of heavy snow, high winds, and cold temperatures causing substantial losses among domestic livestock and game animals alike.

The winter of 1948–1949 was an example (Table 1.8). That was the year of the sensational "aerial haylift," in which tons of hay and food concentrate were dropped from airplanes in emergency feeding operations over wide areas in the Western United States. Locally, snow accumulated to 4 to 10 times normal because of record low temperatures. At Elko, Nevada, daily mean temperature was below −17°C on 6 of the 31 days of January to give an average for the month of −15.2°C, a departure of −10.3°C from normal. A low of −38°C was set. It has been estimated that despite costly feeding by both air and ground crews, livestock losses reached 25 percent and their value reached $2.5 million (U.S. Department of Commerce, 1949) not counting reduced calf and lamb crops. Again in 1967 unprecedented snows on the high plateaus of northern Arizona and northern New Mexico caused losses among game

Table 1.8   Monthly Average Snowfall and Temperature in Nevada Winter Months of 1948–1949 Compared to Normal

Data from U.S. Weather Bureau

Month	Snowfall, cm	Temperature, °C
December 1948	29.0	−2.3
December average	16.3	−0.3
January 1949	50.0	−9.4
January average	22.9	−1.3
February 1949	27.9	−2.8
February average	19.1	+1.1

animals and livestock, and in 1973 tremendous livestock losses occurred in the plains areas of Colorado and New Mexico from April snowstorms.

WATER LOSS TO THE ATMOSPHERE

The low precipitation of most rangelands is made even more serious by high evaporation. Clear, hot days coupled with high winds cause a potential evaporation far in excess of precipitation. This high evaporation reduces soil moisture and increases water loss from plants, necessitating greater amounts of water if the plant is to function normally.

Humidity

Humidity is an important factor determining the effectiveness of moisture for plant growth. When humidity is high, greater plant growth can be expected from a unit of precipitation than when it is low. Humidity is generally low throughout range areas. The intermountain region, and especially the Southwest, has the lowest relative-humidity averages in the United States. Frequently, the minimum humidity falls below 30 percent, as contrasted with average minimums of 60 to 70 percent in the Eastern states. In midsummer, humidity minimums of 10 to 20 percent are common in the Western United States.

Evaporation

Evaporation determines to a large measure the effectiveness of precipitation. Evaporation data are remarkably few, but it is known that the Western United States has a high degree of water loss, especially in the Great Plains and the Southwest. In the Western United States, evaporation increases from north to south. Long-time data (U.S. Department of Agriculture, 1922) showing evaporation at two to five stations in each state during the 6-month period from April to September indicate rough averages of about 81 cm for North Dakota, 96 for South Dakota, 104 for Nebraska, 119 for Oklahoma, and 132 for Texas.

The Southwest has a high evaporation rate because of high temperature and low humidity. In 1948, Bartlett Dam, Arizona, had an evaporation rate of 338 cm. Total evaporation of 40 to 50 cm in each of the summer months is common in this area. Texas, Nevada, New Mexico, Arizona, and California all have stations which repeatedly report annual evaporation rates of over 250 cm (U.S. Department of Commerce, 1943-1949, 1952). Similarly, high evaporation totals occur elsewhere. At Barranquilla, Colombia, evaporation was 181.4 cm; at Jodhpur, India, it was 184.3 cm; and at Alice Springs, Australia, itwas 240 cm.

Table 1.9 Evaporation and Precipitation Records for Some Rangeland Weather Stations and Their Precipitation-to-Evaporation Ratios

U.S. Data from U.S. Weather Bureau

Station	Period involved	Precipitation, cm	Evaporation, cm	Precipitation-to-evaporation ratio
Lees Ferry, Ariz.	Long-time	12.4	224.5	0.055
Mesa, Ariz.	Long-time	18.1	202.3	0.089
Tucson, Ariz.	1930 only	31.2	219.8	0.142
Yuma, Ariz.	1924 only	1.3	338.9	0.004
Yuma, Ariz.	Long-time	11.0	280.1	0.039
Chula Vista, Calif.	Long-time	25.7	156.9	0.164
Oakdale, Calif.	Long-time	31.9	208.1	0.153
Fort Collins, Colo.	Long-time	38.3	110.5	0.346
Boulder City, Nev.	1938 only	14.3	305.2	0.047
Elephant Butte Dam, N. Mex.	1930 only	22.4	214.8	0.104
Las Cruces, N. Mex.	Long-time	21.5	219.5	0.098
Sante Fe, N. Mex.	1930 only	33.6	162.6	0.207
Austin, Tex.	1930 only	90.8	153.3	0.592
Tennant Creek, N. T. Australia*	Long-time	35.2	303.1	0.116
Jodhpur, India†	Long-time	38.0	184.3	0.206

*Data from Slatyer (1962).
†Report of the Central Arid Zone Research Institute.

Precipitation-to-Evaporation Ratio

The relationship between precipitation and evaporation is so important to plant growth that ecologists have adopted the precipitation-to-evaporation ratio as an index to water stress. Weaver and Clements (1938) ranked areas with a ratio below 0.20 as deserts, those between 0.20 and 0.60 as potential dry grasslands, those between 0.60 and 0.80 to 0.85 as true prairie, and those in excess of 1.00 as capable of supporting continuous forest.

Range areas commonly have precipitation-to-evaporation ratios below the 0.20 limit ascribed to true deserts (Table 1.9). Some are far below.

TEMPERATURE

Of the climatic factors, temperature probably is secondary only to precipitation in its influence upon vegetation. The range of temperature on high-latitude rangelands is great, though it is less variable from year to year than is precipitation. Temperature is of less importance in tropical latitudes where only periods of extreme heat inhibit plant and animal production.

Temperature belts are correlated closely with latitude but are influenced markedly by mountain ranges and proximity to oceans. In the Western United States where mountain ranges run north to south paralleling the coast, low-temperature belts extend far south along mountain ranges, and warm belts extend far north along the Pacific Coast. Altitudinal differences introduce marked variations in temperature within short distances. Four weather stations in Utah at altitudes of 1,700; 2,256; 2,652; and 3,100 m have mean July temperatures of 21°C, 19°C, 16°C, and 13°C, respectively. Their mean maximum temperatures for July are 30°C, 27°C, 23°C, and 18°C, respectively, or a decrease of 0.55°C for every 61-m increase in altitude. The normal expectation is an average annual temperature decrease of about 1.8°C per 300 m elevation increase (Trewartha, 1937). Because the Western United States is highly variable in altitude, the temperature may vary greatly within short distances.

The northern plains are the coldest portion of the Western United States range area, with the possible exception of the high mountains. Temperatures of 34°C to 40°C below zero are not at all uncommon in these areas, and extremes of 51°C below zero have been recorded.

At the other end of the spectrum are areas that have extremely high temperatures, as in tropical and subtropical rangelands at low elevation, the interior of Australia, and elsewhere. In the United States, average daily maximum temperatures in the central valley of California and the Southwest in summer months may exceed 61°C. In these areas, plant growth may be limited only by water availability, but animal use of rangelands may be restricted or modified by high temperatures.

Frost-free Period

The distribution and growth habits of native plants are influenced greatly by duration of the growing season. A plant must either be resistant to frost or need only a very short period for maturing its seed if it is to live in areas such as high mountains where the frost-free period may be less than two months.

Typical of mountains of temperate latitudes are those of the Western United States. Here the growing season is short, with the last killing frost generally occurring at lower elevations during May and at higher elevations after June (U.S. Department of Agriculture, 1922). The first killing frost in the fall may occur in September even at lower elevations.

Vast areas in the high mountains have a safe growing season of less than 90 days, though local areas may be subject to frost at any time. Coastal and midlatitude areas at lower elevations have a frost-free period of over 240 days.

WIND VELOCITY

Average wind velocities are high in flat, open plains where topographic barriers are at a minimum. In the United States, northern Texas and North Dakota are regions of maximum blowing. Also, Nevada is the center of a high wind-velocity area which extends north and east far into Idaho. Minimum wind-velocity areas center in southern Arizona and California, in northern Utah and western Wyoming, and in Oregon and Washington. Wind velocities are not of major importance in influencing vegetation development, but in association with extremes of temperature, effects may be severe. Hot winds sometimes sweep across the plains, generally from the south; and they sear vegetation rapidly, reducing soil moisture at the same time. During cold periods of the year, blizzards may cause winter kill of vegetation. Similar conditions exist in steppe and desert areas throughout Asia and Africa.

THE EFFECT OF CLIMATIC FACTORS
UPON RANGE ANIMALS

Not only do climatic conditions determine the amount and character of vegetation available to livestock grazing, but they have direct influences on livestock as well. Warm and wet climates are conducive to livestock parasites and diseases. Aridity and cool temperatures are factors in limiting the spread of disease among livestock. Warm foehn winds, and even cold winds, may serve a useful purpose in removing snow and thus exposing forage to livestock during the winter months. Such winds are called *chinooks* in the West and it is this type of wind that makes the northern Great Plains habitable by livestock in winter. Conversely, cold temperatures associated with blizzards, especially if accompanied by heavy precipitation, may freeze livestock. Big game animal populations in northern temperate zones are limited by winter conditions. Unseasonally wet weather, particularly if it occurs during lambing or calving periods, may have serious consequences for livestock. Cold and snowy winters make the use of supplemental feeds necessary. Annual variation in these requirements cannot be forecast; hence, wise stockmen in cold climates are prepared each winter to feed for extended periods.

Monsoonal climates with distinct wet and dry periods may result in

domestic animals breeding during the wet season with offspring dropped in the dry season. Poor reproduction, or heavy feeding in the dry
season, is often the result. Wet tropical areas may not be suitable to
European cattle, but support Zebu cattle, Banteng cattle, or buffalo.
Thus, not only may climate dictate management of livestock, but it may
also dictate the type of animals used (Fig. 1.6).

Animal production is correlated closely to weather. Generally, wet
years result in plentiful feed and good gains. Dry years may cause tremendous production decreases unless the stockmen resort to costly supplemental feeds.

Because of the many different sets of conditions that exist, range-
management practices must be varied accordingly. Necessarily, then, the
range manager must be alert to the infinite variation in conditions en-
countered upon the range, and he must plan his operation to meet the
restrictions imposed by nature.

Fig. 1.6 Selecting or breeding suitable types of cattle for each rangeland
area is important to successful livestock production. Brahman cattle like
these in Panama are more efficient in tropical and subtropical areas than
European types. *(Courtesy F. A. O.)*

BIBLIOGRAPHY

Blaisdell, J. P., Vinson L. Duvall, Robert W. Harris, R. Duane Lloyd, and Elbert H. Reid (1970): Range research to meet new challenges and goals, *Jour. Range Mgt.* **23:**227–234.

Blumenstock, George (1942): Drought in the United States analyzed by means of the theory of probability, *U.S. Dept. Agr. Tech. Bull.* **819.**

Blydenstein, John (1967): Tropical savannah vegetation of the Llanos of Colombia, *Ecology* **48:**1–15.

Box, Thadis W., and Rayden A. Perry (1971): Rangeland management in Australia, *Jour. Range Mgt.* **24:**167–171.

Burke, M. F. (1939): Rainfall on and runoff from San Gabriel Mountains during flood of March 1938, National Research Council, *Trans. Am. Geophys. Union* **vol. I,** pp. 8–15.

Dyksterhuis, E. J. (1955): What is range management?, *Jour. Range Mgt.* **8:**193–196.

Fretes, Ruben A., and Don D. Dwyer (1969): Range and livestock characteristics of Paraguay, *Jour. Range Mgt.* **22:**311–314.

Hedrick, Donald W. (1966): What is range management?, *Jour. Range Mgt.* **19:**111.

Hodgson, H. J. (1972): Forage and grassland progress, spring 1972, quoted in *Range Man's News* **4**(5):4.

Jasiorowski, H. A. (1973): Twenty years with no progress, *World Animal Rev.*, no. 5, 1–5.

Klein, David R. (1970): Tundra ranges north of the boreal forest, *Jour. Range Mgt.* **23:**8–14.

Larson, Floyd D. (1967): Cultural conflicts with the cattle business in Zambia, Africa, *Jour. Range Mgt.* **19:**367–370.

Perry, R. A. (1960): Pasture lands of the Northern Territory, Australia, *Land Research Series No. 5, Commonwealth Scientific and Industrial Research Organization.*

Reynolds, Hudson. G. (1954): Meeting drought on southern Arizona rangelands, *Jour. Range Mgt.* **7:**33–40.

Reynolds, Hudson G. (1959): Managing grass-shrub cattle ranges in the Southwest, *Agriculture Handbook No. 162,* U.S. Government Printing Office, Washington.

Rouse, John E. (1970): *World Cattle,* vol. 1, University of Oklahoma Press, Norman, p. xvi.

Semple, A. T. (1951): Improving the world's grasslands, *Food and Agricultural Organization of the United Nations Agricultural Studies No. 16,* Rome.

Shreve, F. (1914): Rainfall as a determinant of soil moisture, *Plant World* **17:**19–26.

Slatyer, R. O. (1962): Climate of the Alice Springs Area, in Lands of the Alice Springs Area, Northern Territory, 1956–57, *Land Research Series No. 6, Commonwealth Scientific and Industrial Research Organization,* pp. 109–128.

Squires, V. R., and N. L. Hindley (1970): Paddock size and location of watering points as factors in the drought survival of sheep on the central Riverine Plains, *Wool Tech. and Sheep Breeding* **17**:49–54.

Stoddart, L. A. (1967): What is range management?, *Jour. Range Mgt.* **20**:304–307.

Trewartha, Glenn T. (1937): *An Introduction to Weather and Climate*, McGraw-Hill, New York.

United Nations, Food and Agricultural Organization (1970): *Production Yearbook*, vol. 24, Rome.

U.S. Department of Agriculture, Office of Farm Management (1922): *Atlas of American Agriculture*, Part II, *Climate*. Advance sheets 2–7, U.S. Government Printing Office, Washington.

———, Weather Bureau (1903–1937): Report of the chief of the Weather Bureau, U.S. Government Printing Office, Washington.

——— (1941): *Climate and Man*, Yearbook of Agriculture, U.S. Government Printing Office, Washington.

U.S. Department of Commerce (1949, 1951, and 1952): Climatological data, and national summary, U.S. Government Printing Office, Washington.

———, Weather Bureau (1943–1949, 1952): *United States Meteorological Yearbook*, U.S. Government Printing Office, Washington.

———, ——— (1965): *World Weather Records*, U.S. Government Printing Office, Washington.

U.S. Department of the Treasury (1885): Report on the internal commerce of the United States: Part 3. The range and range cattle business of the United States House of Representatives Ex. Doc. 7, 48th Congress, 2d session, U.S. Government Printing Office, Washington, pp. 95-166.

Walker, A.L. and J. L. Lantow (1927): A preliminary study of 127 New Mexico ranches in 1925, New Mexico Agricultural Experiment Station Bulletin 159.

Weaver, J. E., and F. E. Clements (1938): *Plant Ecology*, 2d ed., McGraw-Hill, New York.

Williams, Robert E., B. W. Allred, Reginald M. DeNio, and Harold E. Paulsen, Jr. (1968): Conservation, development, and use of the world's rangelands, *Jour. Range Mgt.* **21**:355–360.

Young, Vernon A. (1956): The effect of the 1949–54 drought on the ranges of Texas, *Jour. Range Mgt.* **9**:139–142.

Chapter 2

Grazing Areas
of the World

The numerous combinations of factors affecting vegetation in the world
have resulted in correspondingly numerous vegetation combinations
typical of each set of conditions. Activities of man and his livestock, such
as grazing, cultivation, timber cutting, control of fire, and damming and
diverting of water, have qualitatively or quantitatively changed almost all
of the world's vegetation.

Several classifications of native vegetation have been made by
ecologists which in part are applicable to range usage. The broadest
classification normally used by ecologists to designate large geographical
areas with similar ecological characteristics is the *biome*. The major
biomes are grassland, desert, coniferous forest, deciduous forest,
tundra, and aquatic. Of these, grasslands and deserts provide the most
important range types. Forests may be important for grazing depending
upon the density of the tree stand and the management objectives
sought. Alpine tundra is locally important in mountainous areas of the
world, but polar regions have limited use other than for caribou, reindeer, and other wildlife.

Seven major categories of vegetation provide most of the world's grazing lands: grasslands, desert shrubs, woodland shrubs, tropical savannas, temperate forests, tropical forests, and tundra (Fig. 2.1). Each of these may be divided into smaller vegetation types.

The terms *grazing type* or *vegetation type* are not distinct in usage, and no distinction will be made here. They generally refer to the species or various combinations of species which have similar stature, morphology, and appearance and dominate or *appear to dominate* the range, thus giving it characteristic appearance. The term *vegetative region* will be applied to broad contiguous geographical areas of similar though not identical vegetative types. *Zone* will be used to characterize a vegetative region occupying a particular altitudinal life zone made up of one or more grazing types.

GRASSLANDS

Grasslands occur on every continent and many larger islands throughout the world. They represent some of the world's richest and most productive grazing types. Vegetation types as dissimilar as dense bamboo jungles, steppes, savannas, and arctic herblands, an area of over 17 million square miles, have been called *grasslands* by different authors (Barnard and Frankel, 1964). True natural grassland types, however, cover only about 3.57 million square miles (Shantz, 1954).

Grasslands, as used here, are made up of plant communities with a vegetative mixture of herbaceous plants dominated by the family Gramineae. These areas are typically free of trees, shrubs, or woody vegetation. They usually develop in areas with 25 to 75 cm of annual precipitation. Soils are deep, high in organic matter, and fertile, resulting in widespread conversion to croplands. Grasslands may occur from sea level to elevations of 5,000 m, but are best developed in the interior of great land masses and on plains (Polunin, 1960).

Grasslands of North America

The grasslands of the Great Plains form the largest single area of true grass range in the world, extending from southern Canada to central Mexico. They are bounded eastward by the deciduous forest and westward by the Rocky Mountains. They encompass the tall-grass, short-grass, and desert-grass regions. In addition, grasslands are represented by the intermountain bunchgrass and California annual grass regions.

The Tall-Grass Region Unfortunately for the range industry, the great tall-grass region of North America now forms the fertile farmlands

MAP LEGEND
of VEGETATIVE TYPES

	DESERT SHRUB
	GRASSLAND
	MEDITERRANEAN SHRUB
	TUNDRA
	WOODLAND
	TROPICAL SAVANAH
	FOREST of VARIED TYPES
DEF	DECIDUOUS, EVERGREEN FOREST
DF	DECIDUOUS FOREST
TF	TROPICAL FOREST
EF	EVERGREEN FOREST

Fig. 2.1 Vegetative types of the world. Major rangeland types are shown with symbols; types of little significance as ranges are identified only by letters.

of America's corn belt. Only the most xeric portions and those parts underlain by a soil too rocky for cultivation remain. Originally it occupied a strip 240 to 800 km wide from Canada to Mexico.

Physical conditions   Annual precipitation in the tall-grass prairie varies from about 100 cm in the east to 60 cm in the west. Annual evaporation increases from north to south with about 115 cm in North Dakota and 190 cm in Texas. High temperatures and heavy winds are common. Growth begins early in the spring and continues throughout the summer because of the usually reliable summer precipitation. The growing season varies from 4 to 9 months. The soils are deep, dark in color, and fertile. Some of the best cultivated pastures and farms in the world are to be found within this area.

Natural vegetation   Although many combinations of vegetation occur, by far the most important dominants are two species of the genus *Andropogon, A. scoparius* and *A. gerardi* (Fig. 2.2). The former is most abundant on uplands and the latter on lowlands. Together these two constitute about 72 percent of the vegetation on grazed climax prairie (Weaver and Tomanek, 1951; Weaver, 1954).

Grass types dominated by *Andropogon scoparius* probably exceed all other upland types combined. Uniform appearance, despite large numbers of flowering forbs, is usual in the prairie because the dominating grasses reach heights of 1.5 to 2 m, thus overshadowing associated plants.

Fig. 2.2 Tall-grass prairie in Oklahoma dominated by *Andropogon gerardi* and *A. scoparius*. This range has supported cattle at the rate of 1.5 to 3 ha per animal year for decades.

Important associations of *Stipa spartea* and *Sporobolus heterolepis* occur on uplands especially on shallow, rocky, or sandy soils, and *Spartina pectinata* occurs on bottomlands. In the vast and important sand-hills range area of northwestern Nebraska, *Calamovilfa longifolia* is the dominant species and *Bouteloua hirsuta, Sporobolus cryptandrus,* and *Muhlenbergia pungens* are important associates (Frolik and Shepherd, 1940).

Bordering the Gulf of Mexico, the tall-grass prairie is represented by the coastal prairie (Gould and Box, 1965). Here tropical genera such as *Elyonorus, Heteropogon, Trachypogon, Paspalum,* and *Panicum* join the bluestems as codominants (Box, 1961).

Throughout, drier portions of the tall-grass prairie are dominated by midgrasses such as *Agropyron smithii, Bouteloua curtipendula, Panicum virgatum, Sporobolus cryptandrus, Aristida* spp., and *Stipa* spp. (Costello, 1944; Hopkins, 1951). In periods of favorable precipitation, these grasses dominate better sites as far west as the Rocky Mountains; in unfavorable periods, they dominate poorer sites throughout the tall grass extending their dominance greatly during drought. During the drought from 1934 to 1941, *Andropogon scoparius, Poa pratensis,* and many other grasses decreased in density as much as 80 to 95 percent (Weaver and Hanson, 1941a). *Sporobolus cryptandrus* increased to a position of major importance, and from South Dakota to Kansas it became a dominant over thousands of acres. *Agropyron smithii* likewise increased rapidly. By 1945 parts of the prairie were 70 percent short grasses, the remainder being largely weeds (Albertson and Weaver, 1946).

Grazing value of tall grass   The tall-grass vegetation does not cure well on the range. Almost all the species leach badly and lose much of their nutritive value upon maturity. During the winter, these grasses are almost worthless; to ensure proper gains, animals must be fed some supplement high in protein. Because of their rankness of growth, the prairie grasses are seldom grazed by sheep; and older cattle, in general, do better than younger animals.

Heavy grazing on tall-grass vegetation may greatly change the composition. Though grasses form 97 percent of the normal composition, they may be reduced to less than 20 percent, with accompanying reduction in yield from 3,300 kg per hectare to less than 300 kg. Well-managed grazing seems to cause only little change, and deteriorated ranges will recover rapidly upon protection (Weaver and Hanson, 1941a; 1941b). Heavy grazing causes the invasion of *Poa pratensis, Bouteloua gracilis,* and *Buchloë dactyloides,* and continued use results in an increase in low-value weeds. In North Texas, *Stipa leucotricha* and *Andropogon saccharoides* are predominant in overgrazed prairies (Dyksterhuis, 1946), and especially in Oklahoma *Artemisia filifolia* is abundant on sandy soils.

The Short-Grass Region   The short-grass ranges are highly variable and have great geographic spans. They lie immediately east of the foothills of the Rocky Mountains and extend eastward to approximately the 100th meridian, where they give way to taller grasses. Southward, the true short grass reaches southern New Mexico and central Texas, where it merges with the desert-grass region. Large isolated tracts occur in northern Arizona. Northward it extends far into Canada, though its form in the northern parts is distinctly toward a midgrass type. Indeed, the term *mixed prairie* (Weaver and Albertson, 1956) is an appropriate one. Species composition at any one point is affected by two major environmental gradients—a temperature gradient that increases from north to south and a moisture gradient that increases from west to east. In northern areas cool-season plants dominate; in the south warm-season grasses make up most of the biomass. In western areas short grasses are the dominants; in eastern regions tall grasses are more important.

Physical conditions   Precipitation in the short-grass plains is between 25 and 65 cm with 70 to 80 percent received as rain between April and September during the growing season. The light rains do not percolate deeply and leached material is deposited to form a hardpan at the limit of moisture penetration. Vegetation is short-rooted and depends upon the moisture that falls during the growing season.

Natural vegetation   The short-grass plains, over vast stretches, are dominated almost exclusively by *Bouteloua gracilis* and, through much of its range, by a companion species, *Buchloë dactyloides*. Both are short in stature and form a dense sod (Fig. 2.3). Blue grama produces 50 to 95 percent of all forage on well-managed ranges and is the species upon which range management should be based (Costello, 1944). With these dominant grasses occur a variety of less important grasses, forbs, and low shrubs.

The grasses reproduce mainly by seed. Exceptions are buffalo grass (stolons) and western wheatgrass (*Agropyron smithii*) (rhizomes). Most of the grasses produce flower stalks at almost any time during the summer when rains come.

Grazing value of the short grass   The short-grass plains are noted for their excellent grasses which cure well on the range. Often forages may be in short supply in spring, for many of the species are warm-season plants.

Cattle are grazed on the plains yearlong in the southern parts; 8 to 10 months or, with supplemental winter feed, 12 months in the central

Fig. 2.3   The short grasses cover thousands of square miles of rolling
plains in North America, giving them a uniformly smooth appearance.

parts; and 4 to 8 months in the northern parts. In the central and
southern areas, if mountain summer ranges are available, the plains may
be grazed from fall to spring. Summer use is most common, for they
furnish the best forage then.

Despite the tremendously heavy grazing to which the plains have
been subjected, they have remained remarkably productive. The native
short grasses are outstandingly resistant to grazing, partly because of
their morphology. Upon too-heavy use, the grasses often give way to
weedy plants such as *Opuntia, Grindelia, Salsola, Yucca,* and *Gutierrezia*
(Fig. 2.4).

Unfortunately, dry-land cultivation has been attempted on vast
areas of the short-grass plains which are unsuited to cultivation without
irrigation. As a result, thousands of acres have been stripped of their
natural cover and later abandoned. Such lands produce annual weeds
and require many decades to return to normal. Wind erosion has been
extensive during drought periods on many of these abandoned lands.

Desert Grasslands   The desert grasslands extend from the Southwestern United States into northern Mexico, mainly at elevations of less
than 1,300 m. Despite being arid, it is a major range type.

Physical conditions   The desert grasslands are rough in topography,

Fig. 2.4 Snakeweed *(Gutierrezia sarothrae)* has invaded thousands of
acres of the short-grass plains upon misuse. The plant is little used by
livestock, although it is used by pronghorn.

with numerous hills and mesas. They are much the driest of the grasslands, having low precipitation and excessively high evaporation. Most
of the area receives between 25 and 50 cm of precipitation, of which 50
to 70 percent falls during the late summer. Rainfall does not percolate
deeply, and vegetation must secure the moisture rapidly or it is lost to
evaporation. Annual evaporation as high as 200 cm is common. Temperatures also are high.

Natural vegetation   The vegetation is highly variable, sometimes
being almost pure grass and sometimes grass with an open savanna of
desert shrubs forming an overstory (Fig. 2.5). Areas of pure grass are
now rare. Probably the most important single species is *Bouteloua
eriopoda,* because of its high palatability, vigor of growth, nutritiousness,
and ability to cure and form valuable dry-season forage. *Hilaria belangeri* dominates vast areas, especially in more eastern portions. Other
desert-grass species of importance are *Hilaria mutica, Aristida longiseta, A.
divaricata, A. purpurea, Bouteloua hirsuta, B. rothrockii, Buchloë dactyloides,
Sporobolus cryptandrus,* and, originally very important, *Muhlenbergia porteri.*

Chief among the associated shrubs are mesquite *(Prosopis juliflora),
Aplopappus tenuisectus, Larrea divaricata, Yucca* spp., *Quercus* spp., *Acacia*

greggii, Gutierrezia spp., *Flourensia cernua,* and *Opuntia* spp. Small coniferous trees, mostly juniper, may dominate rocky hills and washes.

Grazing value of the desert grass   The desert grasslands are valuable livestock range. Most of the grasses are highly preferred and nutritious; fortunately, they cure well on the ground and thus furnish feed through most of the year despite a short growing season. The range is used practically all year long. Sheep do well on these dry grasses, but cattle predominate and sheep are in the minority. Though some shrubs are used as forage, many are too thorny for ready consumption.

Heavy cattle grazing tends to increase the shrub growth at the expense of perennial grasses. Low-value shrubs invade the desert grasslands, upon destruction of the grass, until they become the dominant species. *Opuntia* spp., *Actinea* spp., *Gutierrezia,* and *Aplopappus heterophyllus* have increased. Likewise, many annual grasses, chiefly species of *Bouteloua* and *Aristida* become abundant. Rapid spreading of low-value mesquite, which collects shifting sand to form the familiar mesquite sand dunes, is also serious.

There are two views regarding desert grassland. Many investigators consider hot-desert shrub types to be degraded grassland; others believe that fire control by man has been a factor in increasing shrub growth in

Fig. 2.5 Desert grassland range in southern Arizona dominated by *Bouteloua eriopoda, Tridens pulchellus,* and *Bouteloua rothrockii.* Despite 18 years of protection, note mesquite invading.

the desert grasslands. According to this concept, the open grassland is really a subclimax to a true climax of low-growing trees and shrubs underlain by grasses. Substantiating this theory, shrubs, especially mesquite, have been shown to increase rapidly when fire is prevented, even when the range is completely protected from grazing (Humphrey, 1953). See also Wright (1972).

Intermountain Bunchgrass   The intermountain-bunchgrass type once occupied the area between the heavily timbered Cascade Range and the Rocky Mountains and extended from the Okanogan highlands in Washington south to about southern Idaho and Oregon, interrupted only by the Blue Mountains. Extensive areas also occur in western Montana east of the Bitterroot Range.

Great uncertainty exists as to exact original extent of the bunchgrasses since, with overgrazing, sagebrush has invaded the intermountain grasslands from the south. Some ecologists hold that much of southern Idaho, eastern Washington, eastern Oregon, western Montana, and large areas of northern Nevada and Utah are climax grassland (Stoddart, 1941). Research and early records tend to bear out this viewpoint in many respects. Climatically, however, this area differs from the major grasslands which characteristically receive most of their precipitation during the summer or growing season (Shantz and Zon, 1924; Smith, 1940). In this region of spring rainfall, grasses grow in spring and they dry up during the rainless summer. The soils under well-developed bunchgrass stands are black, deep, high in lime and organic matter, and resemble those of true grasslands.

Physical conditions   The precipitation of the intermountain- bunchgrass region is low, ranging from about 20 to 50 cm. Most of this falls from early September to early April, mostly as snow. Growth occurs mainly during the early spring months, with the vegetation remaining dormant during the dry summer. In the Palouse area of eastern Washington, the grasses are mature and dry by the first of July. A long frost period during the winter and a dry summer period force vegetation to remain dormant during the greater part of the year. Only during a brief spring period does frost-free temperature overlap the moist-soil period and provide green forage.

Natural vegetation   Clearly, the most important grasses of the intermountain-bunchgrass region are *Agropyron spicatum* and *Festuca idahoensis* (Fig. 2.6). *Stipa comata, Poa secunda, Koeleria cristata,* and *Poa ampla* are associated grasses. Bottomlands and clay soils often are dominated by *Elymus cinereus* and *Agropyron smithii;* sandy soils and rocky soils support *Oryzopsis hymenoides, Stipa comata,* and *Sporobolus cryptandrus.*

Fig. 2.6 In areas of abundant precipitation the Palouse area of eastern Washington is dominated by *Agropyron spicatum* and *Festuca idahoensis,* which form dense stands of high grazing capacity.

Bromus tectorum and other annuals have invaded the area almost everywhere as a result of fire and heavy grazing. In the more moist areas, broad-leaved herbs such as *Balsamorhiza* spp., *Lupinus* spp., *Achillea* spp., *Wyethia* spp., and *Helianthella* spp. are locally abundant.

California Annual Grasslands   The central valleys of California support dense stands of annual herbs. Most of these are introduced plants having their origin in the Mediterranean area.

The climate of the area is typically Mediterranean with cool, moist winters and hot, dry summers. Only about 10 percent of the rainfall comes during the summer season. Plant growth normally begins with the first rains in the autumn and dormancy occurs in the spring when the soil moisture is depleted. Some ecologists believe *Stipa cernua* to be the original dominant in the California valleys with associated species being *Poa scabrella, Stipa pulchra, Melica imperfecta, Sitanion hystrix,* and *Elymus triticoides.* Such great changes have taken place, however, that the conditions which originally obtained are only imperfectly known. At present less than 5 percent of the herbaceous cover is made up of perennials (Sampson et al., 1951).

Cultivation has preempted large acreages. Those remaining, be-

cause of fire and heavy grazing, are completely and probably permanently occupied by annuals and should be managed as such (Fig. 2.7).
This annual cover is dominated by *Avena fatua;* but it is rivaled closely by
several species of brome, chiefly *Bromus hordeaceous, B. mollis, B. rubens,
B. rigidus,* and *B. tectorum.* Other important grasses are *Hordeum
murinum, H. pusillum, Festuca myuros,* and *F. megalura* (Weaver and Clements, 1938). Native forbs are of secondary importance in this type,
though they are numerous. From a grazing viewpoint introduced forbs
such as *Medicago hispida* and *Erodium cicutarium* and *E. botrys* are important constituents and furnish good forage. On foothill ranges, a
savanna-like overstory type consisting of oak, pine, and *Ceanothus* is
common over this annualgrass type.

Green forage is available most years only from February to May.
Consequently, the area is grazed primarily by cattle and largely in fall
and winter, although frequently cattle are grazed yearlong on foothill
ranges (Bentley and Talbot, 1951).

World Grazing Types Homologous to American Grasslands

True tall-grass prairies are rare as world vegetation types. As in the
United States, they have been widely cultivated and little remains.
Where they still exist, they are invaded constantly by woody species, and
if fire is controlled, become savannas or woodlands (Fig. 2.8).

Fig. 2.7 Annuals have entirely replaced the former bunchgrass cover
over thousands of acres in California. These grasses furnish excellent forage from late winter until late spring or early summer.

Fig. 2.8 Wildebeests on tropical grasslands of East Africa. Small areas of tall grass occur throughout the tropics in association with the savanna type. *(Photo courtesy F. A. O.)*

The largest areas are the pampas of Argentina and Uruguay and the velds of South Africa. The pampas are similar to the coastal prairie extension of the North American prairie but with more large tussocks (bunchgrasses) and more species of *Stipa* and *Melica*. The South African veld, although similar in appearance to North American prairies, is dominated by *Themeda triandra* and other tall grasses primarily of the genera *Eragrostis, Aristida,* and *Andropogon* (Eyre, 1963).

The short-grass areas of the world have fared somewhat better as grazing types than have the tall-grass, but many have been cultivated, particularly in Eurasia. Although the area is marginal for crop production, food shortages and desire for national independence have caused periodic attempts to extend agriculture into the arid regions.

Vast areas of steppe remain in Western Europe and Asia. The grasslands are similar in appearance and function to those of the northern Great Plains. Indeed, most of the genera are the same and many species are similar. A typical *meadow steppe* in Russia contains a mixture of such sod-forming grasses as *Stipa pennata, S. lessingiana, S. joannis, Festuca sulcata,* and *Koeleria gracilis* (Keller, 1927). In the northern areas the grassland is dominated by tussock-forming grasses of the genus *Stipa;* in more southern areas sod-formers, mainly *Festuca,* become the dominant plants (Eyre, 1963). In the more arid areas, or where heavy grazing has occurred, woody plants, mostly *Artemisia,* replace the grasses.

Desert-grassland areas occur scattered throughout the desert-shrub types in every continent. In Australia the major desert-grassland areas are the xerophytic tussock grasslands dominated by species of *Astrebla* (Mitchell grass) and xerophytic hummock grassland dominated by spinifex (*Triodia* spp.) (Leigh and Noble, 1969). The Mitchell-grass type is valuable for grazing, making up about 6 percent of Australia; it is an important cattle range. It occurs on heavy clay soils and resembles the tobosa grass (*Hilaria mutica*) flats of North America (Fig. 2.9). The most important Mitchell-grass species are *Astrebla lappacea, A. pectinata, A. squarrosa,* and *A. elymoides.* Spinifex grasslands have little grazing value. The large hummock-forming *Triodias* are unpalatable, and water is scarce in the sand dunes where they occur (Fig. 2.10). Although they make up about 22 percent of Australia, they are largely devoid of livestock.

Scattered areas of desert grassland occur throughout the desert

Fig. 2.9 Mitchell-grass (*Astrebla* spp.) flats of northern Australia are similar to the tobosa flats in the Southwestern United States. *(Photo courtesy Comm. Sci. Ind. Res. Org.)*

Fig. 2.10   Large tussock grasses of low palatability form the spinifex type in central Australia. It is of low carrying capacity supporting only one to two animals per square kilometer per year.

scrub of Africa (McGinnies, 1968) from a few hectares to several square kilometers in size associated with desert-shrub types. Dominant grasses include *Cenchrus ciliares, Aristida, Sporobolus,* and other hot-desert species. Small "runon" (flood) areas in wadis support tall grasses such as *Chrysopogon aucheri, Bothriochloa insculpta,* and *Hyperrhenia hista.*

The tussock grasslands of the South Island of New Zealand, dominated by the genera *Poa* and *Festuca,* bear some resemblance to the intermountain-bunchgrass area of North America. Originally these grasslands occupied about 25,000 square miles (Eyre, 1963), but they have been largely cultivated. Some of the more northern plant communities in the Russian steppes also appear to have many of the same genera and general appearance as the Palouse prairie (Keller, 1927).

Annual grasslands occur around the Mediterranean Sea. Cool-season grasses of the genera *Bromus* and *Vulpia* combine with annual forbs to form a vegetation very similar to that of the California annual grasslands with most of the growth occurring in the winter months, and vegetation drying during the summer months.

Land Use Problems of Grasslands   Grasslands represent some of the most productive grazing regions of the world, yet their importance as a grazing type is being diminished constantly through cultivation, heavy grazing, and shrub encroachment. As the human populations grow, food chains shorten, and dry-land agricultural techniques improve, it is reasonable to expect that even more grassland will be lost to cultivation. Grasslands remaining as rangelands will be those too wet, too dry, too cold, or too high for crop production.

Grasslands are well suited for cattle use and if the preference for beef continues, grasslands will be used primarily for the production of cattle and buffalo. Ecologically, grasslands occur between the wetter forests and the drier desert shrub. With continued grazing by animals that prefer grass (cattle), grass competition will be reduced with trees invading from the forest and shrubs and succulents from the drier end of the gradient. The control of fire on most of the world's grasslands has favored the spread of brush. Vigorous programs of selective grazing by browsers (goats or wild ungulates) and a burning program may be necessary to maintain grasslands.

DESERT SHRUBLANDS

Low-growing shrubs cover the largest area of the world's rangelands variously referred to as desert scrub, arid bushland, or simply desert. Shrublands are characterized by an arid climate (usually less than 25 cm of rainfall), poorly developed soils, and a sparse vegetation dominated by woody plants usually less than 2 m in height. These "deserts" are widespread and each has unique features. Generally, they fall into two categories—hot-desert shrublands of tropical and subtropical areas and cold-desert shrublands of temperate latitudes.

North American Desert Shrublands

Both the hot- and cold-desert shrublands are present in North America. The intermountain desert shrub region is dominated by low shrubs mixed with cool-season grasses, and is used primarily in winter or spring. The hot-desert shrub region is represented by the yearlong ranges of the Chihuahuan, Sonoran, and Mojave Deserts.

The Intermountain Shrub   The intermountain region, which lies between the Rocky Mountains and the Sierra Nevada and the Cascade Range, is dominated over the majority of its area by deeply rooted, semidesert shrubs belonging mostly either to the Compositae or to the Chenopodiaceae families. It includes most of Nevada and Utah, and it extends into western Colorado, northern Arizona, eastern California, and the western half of Wyoming. Its original northern limit is much disputed, but its present range includes about one-fourth of Oregon (the southeast) and most of the southern half of Idaho.

Physical conditions   The range is irregular topographically and highly variable in soil and climate. Poor drainage conditions coupled with low precipitation result in concentrations of soil salts in the lowlands, and these greatly influence the vegetation.

Precipitation tends to be more abundant in the nongrowing season, and the summers are hot and dry. As a result, vegetation is deep-rooted or matures before summer droughts begin. Temperatures are low during the winter months; growth is confined to a brief spring period; and plant growth necessarily is limited.

Natural vegetation   Dominant shrubs are big sagebrush (Fig. 2.11) and black sage (*Artemisia nova*). Of secondary importance on drier and saltier soils is shadscale (*Atriplex confertifolia*). Additional important shrubs are various rabbitbrushes (*Chrysothamnus* spp.), *Sarcobatus vermiculatus, Gutierrezia sarothrae, Atriplex nuttallii, Eurotia lanata, Ephedra* spp., and in southern extremes, *Coleogyne ramosissima.* Sharp zonation of these species, a result of soil differences, is common, often resulting in almost pure stands of a single dominant species. The most important grass associates are *Agropyron spicatum* and *Sitanion hystrix.* Other important grasses are *Oryzopsis hymenoides, Hilaria jamesii, Stipa comata, Agropyron smithii, Elymus cinereus,* and various species of *Poa.* Forbs are of less importance, though local exceptions are common. On drier portions of the salt deserts, the most important forb associate is *Sphaeralcea* spp. On more moist foothill zones *Balsamorhiza, Lupinus, Wyethia, Aster,* and *Astragalus* are common. Throughout the region, early growing annual

Fig. 2.11   Sagebrush *(Artemisia tridentata)* dominates extensive areas in the intermountain region. Though sagebrush is of low palatability except in winter, associated plants make it a valuable grazing area.

grasses (mostly *Bromus tectorum*) which are adapted to producing seed in the brief period in the spring, when soil is moist, are common. Russian thistle is abundant under recurrent disturbance. The introduced poisonous weed *Halogeton glomeratus* has invaded much of Nevada, Utah, and southern Idaho.

Grazing value of the intermountain shrub   The big-sagebrush type and the salt-desert type are somewhat distinct in their grazing use, the former being essentially a spring-fall range and the latter a winter range. Since it occupies the moister areas, sagebrush is the more productive of the two. Many of the more productive sites have been cultivated, but the remaining area still is the most important spring and fall range in the intermountain region. Spring range is such a vitally important link between the vast winter and summer ranges of the West that the value of these sagebrush lands is far higher than their grazing capacity would merit. Both because of season of use and a shortage of spring-fall forage, heavy grazing has been widespread.

Shadscale and black-sage types reach greatest importance on lower elevations of Nevada and Utah where, despite the low productiveness of the soil and climate, these vigorous shrubs furnish winter grazing for

Fig. 2.12   Shadscale *(Atriplex confertifolia)* is abundant on dry and salty soils in the intermountain-shrub type, occurring often in pure stands.

thousands of sheep and cattle (Fig. 2.12). Almost all of this type has remained as rangeland because of low precipitation or heavy, alkaline soils which prohibit cultivation. Continued grazing after growing starts in the spring has been responsible for much damage. These winter ranges are the backbone of the intermountain sheep industry.

Hot-Desert Shrublands   Hot-desert shrublands extend from Texas, New Mexico, Arizona, and California south to central Mexico. This region has the most arid climate in North America, hence its vegetation has developed many adaptations to resist drought. Wide spacing of dominant shrubs is characteristic to accommodate the widely spreading root systems needed to secure moisture. Fleshy plants are common, and these draw enough water from temporary surface moisture following rains to store in their hydrophilic storage cells for use during dry periods. Perennial plants leaf out in the wet seasons and drop their leaves upon the inception of drought; annual plants survive by their ability to grow rapidly and set seed during the wet season.

Physical conditions   Precipitation in the hot-desert type is far below optimum for plant growth, the average being only 75 to 350 mm annually, with individual years producing less than 25 mm in some places. Moreover, evaporation rates are exceedingly high, 300 to 400 cm per year. Frequent and prolonged drought periods are common. Temperatures are high, maximums near 43°C occurring rather frequently over much of the area. The frost-free period is long, and temperatures below −4°C occur infrequently.

Natural vegetation   The vegetation of the Southern desert shrub is predominantly woody, though the region is characterized more than any other in North America by its large number of species and their variability. The most widespread dominant is *Larrea divaricata;* this frequently forms pure stands in which no other perennial grows. The exclusion of other species is achieved by a widely spreading root system, which occupies the area more thoroughly than the crowns indicate.

Secondary shrubs are *Franseria deltoidea* which often occur as an understory to the taller species and, in very dry and hot sections, *F. dumosa.* There are many species and forms of cacti, such as *Opuntia, Cereus, Ferocactus,* and *Echinocactus.* Bottomlands and moister flats often are characterized by dense stands of mesquite. Where alkali is moderately high, *Atriplex polycarpa* dominates wide areas (Nichol, 1937). Common upland-shrub species are *Acacia greggii, A. constricta, Calliandra eriophylla, Flourensia cernua, Cercidium microphyllum, Olneya tesota, Canotia holocantha,* and *Fouquieria splendens* (Fig. 2.13).

Although the original condition of the Southern desert is difficult to

Fig. 2.13 The Southern desert-shrub region is dominated by large and, frequently, deeply rooted woody and fleshy species. These are spaced far apart and form an exceptionally low cover.

reconstruct, there seems to be little question that perennial grasses were once more abundant, at least in the less arid portions, than they are today. In the eastern portions of the area, grasses become increasingly abundant with increasing precipitation until the desert-shrub region fades imperceptibly into the desert-grass region; and many of the dominants of the desert-grass region occur in the higher or eastern parts of the desert-shrub type.

Important perennial grasses that occur in the desert-shrub region are *Muhlenbergia porteri, Hilaria mutica,* and *H. rigida. Muhlenbergia porteri* was originally a much more important forage species in this type than it is today, having been reduced by grazing.

Grazing value of the hot-desert shrub   Few of the shrubs are important as forage. Much of the forage for livestock is furnished by ephemeral annuals which, in good years, provide a profuse growth of nutritious forage. Unfortunately annuals are undependable; and though forage is abundant in wet years, it may be almost entirely absent in dry years, which limits the value of the type for grazing. Because of the hot and dry condition of the desert range, much of it is used during only a short part of each year. Sheep and goats are grazed in some parts, especially during the winter; but they are far outnumbered over the region as a whole by cattle. Stock water is a critical problem, for surface water is scarce and wells must be unusually deep.

Homologues to North American Desert Shrublands

All arid regions of the world support some desert shrubland. Some are productive and are of major grazing value; others occur in areas of such low precipitation that they are little used. Many have no counterparts in North America.

Cold-Desert Shrublands   The cold-desert shrublands of Russia, Mongolia, and Southwest Asia are similar to those of the intermountain region (Petrov, 1972). Shrubs, mainly *Artemisia* and *Atriplex,* and grasses such as *Poa, Festuca,* and *Koeleria* grow in the valley and basin areas where they supply valuable forage. Sheep and goats are the most common livestock used on the browse ranges though camels are locally abundant. Cattle and buffalo are more rare; horses and asses fill the large-grazer niche.

A second large area of cold-desert shrubland exists in South America extending from the Rio Negro south to Tierra del Fuego (Soriano, 1956; 1972). This area is similar in physiognomy and production to the intermountain area. It is grazed primarily by merino sheep for fine wool production with large herds being run yearlong.

In Australia saltbush areas occur in the arid regions of the southern

Fig. 2.14   Fence line in a saltbush type in New South Wales. Prevailing winds from the right cause the merino sheep using the region to concentrate on the area to the left of the fence. *(Courtesy R. A. Perry.)*

half of the continent. The largest of these is an *Atriplex vesicara-Kochia aphylla* type (Fig. 2.14). On heavy soils, *A. nummularia* is the climax dominant shrub. As the shrublands are degraded with overgrazing, they become *Danthonia caespitosa-Stipa variabilis* grasslands. The area is grazed yearlong with merino sheep for wool production, but in recent years cattle have increased. The high salt content of the forage and the brackish water throughout the region dictate sheep use over most of the area which amounts to about 5.9 percent of Australia (Leigh and Noble, 1969).

Hot-Desert Shrublands   Australia also has large expanses of hot-desert shrublands dominated by *Acacia* which make up about 24.5 percent of the total land area. The type varies from a low open woodland-shrubland in the higher rainfall areas (43 cm) to low shrubs in tussock grasslands in the drier (13 cm) areas. It takes in the mulga *(A. aneura)* vegetative type. Mulga, or other *Acacia* species, usually occur in pure stands with open areas interspersed with shrubs such as *Eremophila, Dodonea, Hakea, Lycium, Myoporum, Exocarpus,* and *Cassia. Acacias* other than mulga may be *A. brachiptacha, A. pendula, A. sowdenii,* and *A. kempeana.* In the higher rainfall areas the main grasses are *Danthonia, Enneapogon,* and *Stipa. Eragrostis, Aristida,* and *Chrysopogon* are more common in the summer rainfall area. Spinifex is common in the drier desert-shrub areas. In the summer rainfall (250 to 740 cm) belt Eucalyptus species replace *Acacia* as the major woody component.

Australia's shrub ranges are used yearlong for livestock production. The southern shrublands are used mostly by merino sheep. The central and northern areas are used for beef production. Unlike other hot-desert-shrub areas, Australian shrubs are thornless and can be browsed readily.

South America has large areas of hot-desert-shrub ranges. The *Larrea* communities are the most nearly homologous North American grazing type. In aspect and dominant genera, the ranges of the dry regions of central Argentina are very similar to those of Arizona (Morello, 1958). The remainder of South America's shrublands resemble the cactus shrublands of the Sonoran and Chihuahuan Deserts. Low-growing, thorny shrubs and cacti are almost universally codominant. These plant communities have highly developed, drought-evading mechanisms and can withstand extreme climate conditions, but they do not produce much forage and are marginal for grazing. They are used yearlong by sheep, goats, donkeys, and some cattle. In addition to cacti, there are a number of leguminous shrubs (Eyre, 1963).

The world's largest hot-desert-shrub type extends in a belt from the west coast of Africa eastward across Africa, Arabia, and southwest Asia to the Thar Desert of India. This area is dominated by a sparse cover of shrubs usually less than a meter in height (Fig. 2.15). Rainfall is generally less than 30 cm which falls in the winter and early spring north of the Sahara and in the summer south of the Sahara. The dominant shrubby vegetation is a mixture of thorny shrubs of the genera *Acacia, Salvadora,* and *Gymnosporia* (McGinnies, 1968). *Euphorbias* replace the cacti of the American deserts. Grasses are mostly ephemeral and of tropical origin (LeHouérou, 1972). As could be expected from such a large area, vegetative composition is highly variable.

The entire area is used by nomadic herds and flocks of varied composition which shift with climatic conditions and availability of forage. Camels are universally present. Goats and sheep are normally herded together with goats outnumbering sheep in the more arid areas. Cattle and buffalo are raised only in the higher rainfall areas along the ends of savannas and woodlands.

An area of similar vegetation occurs in southwest Africa from central Angola to Karoo (Eyre, 1963). It has plants of the same life form and many of the same genera as the desert scrub of the Sahelian zone south of the Sahara.

Fig. 2.15 Hot-desert shrubland in Libya dominated by *Rhantherium suareo* and *Stipa lagascae* will support one sheep on 1 to 2 ha per year. *(Courtesy H. N. LeHouérou.)*

Problems in the Use of Arid Shrublands   Most of the problems associated with the grazing of arid shrublands are related to aridity. Water, whether for plant growth or for livestock, is in short supply (Fig. 2.16). Rainfall is low and sporadic, and drought common. More years of below-average rainfall occur than above-average, and, when rains come, they are usually intense, producing considerable runoff. Both extremely high and extremely low temperatures occur.

Effective use of the more arid shrublands can only be accomplished through a flexible system which permits their use when conditions are favorable and calls for the removal of livestock when conditions are unfavorable. It is virtually impossible to contain animals on a given area continuously without severe range deterioration and heavy livestock losses. Several systems of use have evolved to make efficient utilization of the sporadic forage supply.

In Western North America the cold-desert ranges are used mostly in winter. Livestock, usually sheep, are trailed or trucked from the high-elevation ranges to the deserts, the latter practice now being more common. Similar seasonal uses of shrub ranges occur in Asia, and gov-

Fig. 2.16 Camels watering in Algerian desert. Most of the problems related to the use of desert shrublands have to do with aridity; the sparse vegetation will support only animals with a high water efficiency. *(Courtesy F. A. O.)*

ernmental programs for seasonal use of sheep and goats have been proposed in Iran, Turkey, and Russia.

The hot-desert shrub ranges of the Americas and Australia are most often used as yearlong ranges, and ranch properties must be huge and embrace many soil and vegetation types so that alternate grazing areas are available. Livestock may be grazed on "runon" grass flats (Tobosa in North America, Mitchell grass in Australia) following a rain and moved to upland types where the grass cures better during the dry seasons.

In the desert shrublands of Africa and Asia, true nomadism is practiced. In most cases the use pattern follows that described for seasonal mountain ranges except that livestock movement is between wet- and dry-season ranges. Wet-season ranges usually are determined by three factors: (1) the availability of temporary drinking water for livestock; (2) the forage supplied by sporadic rains; and (3) the presence of muddy ranges, insects, or diseases of the higher rainfall areas.

Other than those problems incident to drought and aridity, desert ranges have many desirable features. Forage, though not abundant, is nutritious, diseases are uncommon, and a wide variety of livestock and game animals can be raised. Unfortunately, drought and overgrazing have depleted many of the fine shrub ranges, and desert encroachment into this valuable grazing type is widespread throughout the world.

SHRUB WOODLANDS

Closely associated with grasslands and desert shrubs are many highly variable woodlands. These usually occur in about the same annual rainfall belt as grasslands, but low-growing trees (usually less than 10 m) and dense shrubs are the dominant vegetation. The reasons that trees, rather than grasses, dominate are equally varied. Some of the major reasons are lack of periodic fire, poor rainfall distribution during climatic cycles, shallow and rocky soils, and heavy use by grass-eating animals. Many of these woodlands have a high potential for "range improvement" or vegetation conversion. Grasses can usually be seeded successfully following removal of the brush, but intensive management is necessary to prevent the return of the woody plants. Several types of woodlands occur in North America, but the major woodland-range types are in Africa, Australia, and South America.

Shrub Woodlands of North America

The woodland ranges of North America are mainly in Mexico, the Southwestern United States, California, Arizona, and the Rocky Mountains. They are commonly referred to as chaparral, a term of Spanish origin which originally meant an evergreen oak, though ecologists have come to apply it to any evergreen sclerophyllous shrub type of hot, dry regions. Chaparral is applied to any brushland in Mexico and in the Southwestern United States; it refers specifically to evergreen shrubs in California. Shrub types in North America can be divided into (1) California chaparral, (2) mesquite shrub, (3) oak brush, and (4) mountain brush. Intermediate between these and forests is the piñon-juniper woodland. Other woodlands occur associated with the forests of the Eastern and Southeastern United States, but they are not used extensively for grazing, and oak woodlands occur in California and Oregon.

The California Chaparral   The California chaparral occurs over large areas of southern California, extending south from Monterey and west from the Sierra Nevada into Mexico, covering 2½ million ha in California (Plummer, 1911). Some areas of central and southern Arizona are dominated by a similar type of chaparral which, though greatly reduced in species, is remarkably typical of the California chaparral in appearance (Nichol, 1937).

Physical conditions and vegetation   Climate in the California chaparral area is characterized by little summer rainfall and comparatively heavy winter rainfall. This, coupled with relatively high winter temperatures, results in most of the growth on the shrubs taking place in the spring, with plants appearing brown and lifeless during the summer (Plummer, 1911). Summer temperatures are high.

On drier areas at lower elevations, the chaparral growth is open, but the plants tend to have an even distribution (Fig. 2.17). At higher elevations, the shrubs grow very densely, especially on north slopes, forming an almost impenetrable tangle 3 m or more in height. At elevations above 2,400 m, the chaparral gives way to coniferous forests. Major species are *Ceanothus oliganthus, C. sorediatus, C. spinosus, Arctostaphylos pungens, A. gladulosa, A. viscida, A. canescens, A. tomentosa, Prunus ilicifolia, Rhamnus crocea,* and *R. californica* (Jepson, 1925). Open stands of brush support a valuable undercover of *Festuca megalura, Bromus rubens, Aira caryophyllea,* and *Stipa* spp. (Sampson et al., 1951).

Grazing value of the California chaparral   The chief function of the California chaparral is watershed protection. The region is subject to erosion because of steep slopes, torrential rains, and frequent fires. The chaparral-covered mountain slopes provide irrigation water to the farms on the coastal benchlands and culinary water to the coastal cities.

Livestock obtain some forage from the chaparral. During emergencies such as drought, some parts of the area are grazed by all classes of stock. Goats seem to do well on this ration; in some limited areas, goat

Fig. 2.17 The California chaparral is very dense and, in areas of higher precipitation, forms an almost impenetrable tangle. Similar types occur around the Mediterranean and the coast of southwest Africa.

grazing is common. Sheep and cattle can enter dense brush stands only with difficulty; consequently, large areas are unused.

The Oak Type Scrub oak occupies vast areas in the West, especially in the southern areas, sometimes almost to the exclusion of other shrubs. There are many species of oak, some evergreen and some deciduous, and they occupy a variety of habitats. The brush stand may be very dense, virtually excluding low-growing vegetation, or it may be an open savanna over grass or low shrubs (Fig. 2.18). The oak itself is seldom of great importance to grazing animals, with the exception of the acorns which are valuable as forage.

The oak types of the intermountain and Rocky Mountain foothills are composed of small and close-growing species, mainly *Quercus gambelii, Q. turbinella,* and *Q. undulata.* These areas generally are covered by a growth of shrubs so dense as to make grazing difficult but, usually, not impossible. In California, an open cover of oak trees, *Quercus wislizeni, Q. douglasii,* and *Q. dumosa,* form a savanna grassland. Savannas are found also in southern Arizona where oaks, mainly *Quercus emoryi* and *Q. oblongifolia,* grow in mixture with desert grassland (Shantz and Zon, 1924). Similar stands, mostly post oak *(Quercus stellata)* occur in central Texas.

Fig. 2.18 Dense stands of *Quercus gambelii* dominate extensive areas of the foothill ranges of the Rocky Mountains.

In southeastern New Mexico, a large oak type occurs, consisting of low-growing shin oak, mostly *Quercus havardii,* associated with grass species and a few desert shrubs. Similar shin-oak areas are found through the southern portion of the mixed-grass prairie on sandy soils. Many plant geographers rank these types as grass rather than oak.

In grazing use, the dense scrub-oak types of the Rocky Mountain and intermountain regions are analogous to the mountain brush and piñon-juniper types, being used to best advantage as spring and fall range. Not infrequently, however, they are grazed yearlong. The oak-grass savannas of California and the Southwest, generally, are used yearlong, as are the adjacent grasslands.

Mesquite Shrubland   Mesquite *(Prosopis)* shrublands occur in the 50 to 75 cm rainfall belt of southern Texas and northern Mexico and, in drier areas, along watercourses there and elsewhere in the Southwest. A great many combinations of vegetative communities occur from dense mesquite forests in the wetter regions of northeastern Mexico (Rojas-Mendosa, 1965) through mesquite savannas to dense and complex mixtures of many brush species (Box, 1961).

In southern Texas and northeastern Mexico, mesquite is associated with from 5 to 15 other species of woody plants such as *Acacia rigidula, A. farnesiana, A. tortuosa, Celtis pallida, Xanthoxylum fagara,* and *Diosperos tex-*

ana. Grasses associated with the type typically are those of the coastal prairie portion of the American tall grass (Gould and Box, 1965). These woodlands become productive grass ranges when brush is controlled, but revert quickly to woodlands if not managed intensively (Powell and Box, 1967). Apparently, much of the mesquite type was once grassland, and brush has increased since the control of fire (Box et al., 1967).

Mesquite also occurs in areas of less than 50 cm precipitation as part of a shrubland-desert grassland mosaic and is mainly located in Mexico west of the Sierra Madre range and in West Texas, New Mexico, and Arizona. Associated grasses are mostly the same species as occur in the desert grasslands.

The mesquite type is used primarily for beef production under yearlong grazing, though in Mexico, goats and sheep may be grazed also.

The Mountain-Brush Type   The mountain-brush type usually exists as a discontinuous transition zone between coniferous forest and grassland or sagebrush land. Like the California chaparral, it does not extend far northward but reaches its best development in Utah, Arizona, New Mexico, and Colorado. It also occurs commonly in scattered areas largely as a transition between the California chaparral and the coniferous forest, called *soft chaparral* by Jepson (1925).

No single species can be said to dominate the type, for the great variety of soil and climate causes first one and then another of the many constituents to exert local dominance. Important genera are *Cercocarpus, Amelanchier, Symphoricarpos,* and *Ceanothus.* These four genera are represented by several important species, and many other genera are locally important. Among these are *Acer, Artemisia, Holodiscus, Prunus, Physocarpus,* and *Rubus.*

Usually, the mountain-brush type is open, with considerable grass and forb growth as an understory. This, together with the high palatability and nutritive value generally characteristic of the shrubs, makes the type valuable for grazing despite the drought and rough topography that typify it. This area is grazed to best advantage during the spring and fall, and it forms an important link in the seasonal movement of stock in many parts of the West. Only its small area prevents this type from holding a position of foremost importance as a source of range forage.

The Piñon-Juniper Woodlands   The piñon-juniper type is characterized by low-growing conifers that are not lumber-producing species, and grazing thus becomes the chief use of this land. The common species of piñon pine are *Pinus edulis* and *P. monophylla;* the junipers are chiefly

Juniperus monosperma, J. osteosperma, J. pachyphloea, and *J. scopolorum.* These small trees, either alone or in mixture, form an open overstory on vast areas of the Western foothill lands and give to these hills a characteristic woodland aspect mainly south of the 42d latitude (Fig. 2.19).

Physical conditions and vegetation The piñon-juniper lands, generally, are intermediate in elevation, lying between the desert shrubs or grasses and the coniferous forests, at elevations seldom above 2,000 m. This intermediate foothill area usually is steep, and erosion removes the soil rapidly. It is characterized by less than 40 cm of precipitation per year and high temperatures.

Often in the northwest portion of its range, the piñon-juniper region seems to be close to sagebrush in its ecological requirements, the two being sometimes mixed and sometimes separated into small alternes, the sagebrush usually occurring on the moister, deeper soils, and the piñon-juniper on the drier and more rocky sites. Juniper often invades adjacent types of vegetation, possibly as a result of decreased fires and increased grazing.

Unfavorable growing conditions make the piñon-juniper region one of the lowest producing of all Western range areas. Major plant species are *Cercocarpus* spp., *Cowania stansburiana, Agropyron spicatum, A.*

Fig. 2.19 A typical open stand of juniper *(Juniperus osteosperma).* Species of piñon pine often are intermixed with juniper. Note the young trees invading adjacent sagebrush land.

smithii, Bouteloua gracilis, Stipa spp., *Hilaria jamesii, Oryzopsis hymenoides, Bouteloua curtipendula,* and *Muhlenbergia. Sasola kali* var. *tenuifolia* and *Bromus tectorum* also have become an important part of the forage in the northern and central portions of the piñon-juniper region.

Grazing value of piñon-juniper Piñon-juniper is used for spring range. Since the soils are shallow and rocky, it has escaped cultivation and now is often made to support the livestock that once grazed a much vaster acreage during the spring grazing season. Extensive overgrazing was once estimated to have caused 60 percent depletion (U.S. Department of Agriculture, 1936). Almost no forage remains in many areas except that provided by the trees themselves, which has led to tree removal and seeding. The consumption of some of this coniferous foliage appears to be normal during cold periods on winter ranges, but domestic stock could be expected to consume little or none of it on correctly used spring ranges. It is more heavily used by deer in winter.

World Homologues to North American Woodlands

Large areas of sclerophyllous shrublands similar in their aspect to those of California occur around the Mediterranean Sea, in Australia, South Africa, and Chile (Eyre, 1963).

In Australia, this chaparral-like vegetation, called *malee,* occurs in the winter rainfall area of 20 to 42 cm (Leigh and Noble, 1969) and covers a broad area from western Australia eastward around the Bight into Victoria. The main malee species are *Eucalyptus oleosa, E. dumosa, E. socialis,* and *E. viridis.* Much of the higher rainfall area has been plowed and put into wheat. The drier areas are used for sheep range with a stocking rate of only about 1 sheep to 15 ha yearlong.

Other major areas of sclerophyllous scrub are in South Africa and Chile. Like those of the Mediterranean, they frequently occupy areas that formerly were forested. Each has shrub species peculiar to them. The South African area is dominated by the Cruciferae, Rosaceae, and Leguminosae families. *Quillaja saponaria, Kagenackia oblonga,* and *Rhus caustica* occur in most of the Chilean shrub communities (Eyre, 1963). These chaparral types are typically so dense as to inhibit substantial production of herbaceous vegetation, and the grazing value of the type depends primarily on the forage value of the shrubs themselves. In general, the type is best suited to wildlife, sheep, and goats.

Woodlands Not Represented in North America The tropical woodlands that occur in many areas do not have a counterpart in North America. The "thorn forests" of South America, Asia, and Africa are somewhat similar to the mesquite type of North America. These types are found

where the forests are flanked by dry regions, usually expressing themselves as broad ectones of low, gnarled, thorny trees and dense under-
brush. In Africa they may give way to tropical savannas in more favorable situations or where fire has held the woodland in check.

Thorn forests occur on the Indian subcontinent in the drier monsoonal area. Although used by cattle and buffalo, they do not appreci-
ably add to the world meat supply.

Vast areas of thornless woodland occur in Australia and form a
fairly significant grazing type (Fig. 2.20). These woodlands have been
classified into a number of vegetation types by Australian ecologists
(Leigh and Noble, 1969; Moore and Perry, 1970), but they can be
grouped into tropical eucalypt woodlands and temperate eucalypt woodlands. Tropical woodlands occur in the northern portion of the country
where a summer rainy period produces 42 to 75 cm of rain followed by a
dry period of 3 to 9 months. Woody species are mainly *Eucalyptus* with a
mixture of other shrubs, such as *Acacia, Hakea,* and *Terminalia.* Grasses
include tall tropical grasses *(Sorghum, Themeda)* and midgrasses. The
major use is for cattle production.

In temperate climates, woodlands occur in the winter rainfall area
of 25 to 50 cm and are dominated by eucalypts. Grass genera include

Fig. 2.20 Mulga in central Australia used primarily by cattle. The typically bare areas in this type yield runoff water to the adjoining grasses and
shrubs.

Danthonia, Stipa, Chloris, and *Poa.* Sheep grazing forms the major livestock use, though cattle production is replacing sheep in some areas. Improved pastures are common. Where woodland is cleared and phosphate fertilizer added, clover and cultivated grasses produce high yields.

Vast areas of tropical woodland and scrubland remain in Brazil. These areas are used for cattle grazing, but livestock operations are primitive and production from them is low.

Problems in Grazing Woodlands Tropical woodlands present special problems that are usually not associated with those of temperate areas. Insects, particularly the tsetse fly of Africa, may thrive in the woody habitat and transmit disease to livestock. Although total rainfall may be moderate to low, the monsoonal nature of the climate of most tropical woodlands causes high amounts of rainfall in a relatively short time followed by a predictable drought. In consequence, there are floods and temporary swamps, soils are impoverished, and vegetation is leached making nutritional problems inevitable.

Most woodlands occur in areas of sufficient precipitation to grow grass. If the land-use objective is to produce beef or wool and economics are favorable, conversion to grassland is a valid alternative. Although the chances of establishment of grasses in woodlands are high, keeping the grasses productive is a continuing problem. The manager must work against natural successional forces rather than with them, requiring higher inputs of labor and management than where one works with nature.

The diverse vegetation makes woodlands almost ideal for growing several kinds of animals; however, in America and Australia a single kind of animal often is used—cattle or sheep. In Africa sheep, goats, camels, cattle, and donkeys may use the same range. In Asia and the Northern Territory of Australia where it is a feral animal, the buffalo makes efficient use of the browse range.

Woodlands are highly productive for wildlife (Fig. 2.21). In Africa where wild animals use woodlands in conjunction with grassland and savanna types, as many as 10 or more species of ungulates may graze a particular area.

TROPICAL SAVANNA

One of the major grazing types in the world, the tropical savanna, does not occur in the United States. It is the major grazing type between the Tropic of Cancer and the Tropic of Capricorn. It is characterized by a grassland with an open stand of trees spaced approximately as far apart

Fig. 2.21   Herbs, shrubs, and trees all produce forage in the mixed woodlands, such as this one in Botswana, and support a wide variety of animals.
(Courtesy F. A. O.)

as their height. Many ecologists consider it to be an artificial type because
it grades into a forest on its more humid boundary and into a grassland
on its arid boundary.

The climate of savannas is largely monsoonal with distinct wet and
dry seasons. Plants make rapid growth at the onset of rains with lush
vegetation occurring within days of the first monsoonal rain. During the
dry season, only the woody plants escape dormancy and even some of
them shed their leaves. Rapid animal growth takes place during the wet
season, but severe nutritional problems are faced during the dry season.

The largest savanna type is the *Acacia*–tall-grass savanna of Africa.
It stretches in a broad belt from east to west across the center of Africa
between the desert scrub to the north and the tropical forests to the
south. Similar savannas occur in eastern and southern Africa. Flat-
topped *Acacias* and *Combretum* trees 8 to 15 m in height cover the landscape in orchardlike stands (Fig. 2.22). Herbaceous vegetation, mainly
grasses, provides a dense ground cover and most of the annual dry-
matter production. In the high-rainfall areas, dominant grasses are
Andropogon, Imperata, Hyparrhenia, and *Panicum.* In the drier climates

Aristida, Cenchrus, and other desert grasses provide the major part of the forage.

Savannas are difficult to identify. Some ecologists describe much of the woodland area in Australia, Asia, and South America as savannas. Indeed, many areas between forests and grasslands exhibit the savanna structure. Some are held in the open, parklike condition by recurring fires. So long as fires persist, they remain savannas; when fire is controlled, they revert to woodlands. Even in what is regarded as the true savanna belt of Africa, the area is changing constantly. Its humid boundary is threatened by woodland encroachment as grazing intensity increases and fire is controlled. Its xeric boundary recedes as the desert encroaches following drought, heavy grazing, and removal of trees for charcoal.

Although threatened as a type, the savanna ranges are extremely important for both wildlife and domestic livestock. Most of the game animals of Africa use the savanna at some time of the year. Many species of large ungulates have evolved highly specialized feeding habits —giraffes eat the trees, eland the browse, and zebra the grasses. Domestic livestock also are varied. Camels have sufficient "top feed," goats have adequate browse, and sheep and cattle have grasses and herbs.

The savanna ranges of Africa are almost all used by nomadic pas-

Fig. 2.22 Savanna grassland in East Africa. The savanna type occurs between woodlands and grasslands throughout the tropics.

toralists. In some areas, savannas are used during the wet period with herds taken to the forest boundary during the dry period. In others, the flocks and herds graze desert and desert-shrub vegetation during the "wet" and return to the savanna for the "dry." Except where European-type ranches have been developed, the savanna is seldom used yearlong.

Problem of Managing Savanna Range   The major problem of managing savanna ranges is in maintaining a proper balance between grass and woody plants. Heavy use by cattle and control of fire almost always lead to a dense woodland and lowered forage production. Use by browsing animals in proper balance with cattle can hold the woody plants in check maintaining a proper balance among forage plants (Skovlin, 1971). This balance is difficult to achieve without wildlife. Ecologically, the native animals are best suited to harvest the savanna forage, but from an economic standpoint, it is often desirable to increase domestic livestock at the expense of the game herds. Hunting clubs and game cropping are used to some extent to keep a place for wildlife in the face of a rapidly developing livestock industry.

Nutritional and disease problems of livestock are those typical to monsoonal tropical areas. Nutrition can be vastly improved by seeded pastures (Fig. 2.23). Tsetse-fly populations occur along watercourses and in the dense bush in most of the African savanna making use of the areas impractical during the wet season. Tropical tall grasses usually do not cure well, and the nutritive value of the forage is low during the dry season. Water for domestic livestock may be ephemeral over much of the area, dictating the areas where stock can be grazed during the dry period.

TEMPERATE FORESTS

Both coniferous and deciduous forests occur throughout the temperate regions of both hemispheres. Where they occur in dense stands and are managed primarily for timber production, little grazing is produced other than for wildlife. In naturally open stands and where they are made open by timber cutting, considerable grazing may be provided. The forested grazing regions are represented by several distinct types which vary with respect to species, density, and grazing value.

The Coniferous Forests of North America

The coniferous forests, occupying higher and, hence, moister mountainous lands throughout the West and the Northwestern coastal area of the United States are characterized by evergreen trees mostly of the

Fig. 2.23 Improved pasture in tropical savanna in northern Australia. Annual sorghums have been replaced by Townsville lucerne which cures well for dry-season forage.

genera *Pseudotsuga, Pinus, Picea,* and *Abies.* Since variations include precipitation from 38 to 254 cm per year, altitude from sea level to some 3,658 m, and temperature from almost frost-free to the extremely cold winter and short growing season of the high mountains, there is a great vegetal variety within the region. Open stands of small trees of the Southwest and Rocky Mountains support a dense undergrowth of forage plants. Many of the dense and tall-growing forests of the Northwest support little undergrowth and are of almost no value to livestock. Of the vast open coniferous forest in the West, some 28 million ha cannot be grazed because of steep topography, insufficient forage, and dense tree growth (U.S. Department of Agriculture, 1936).

Many forest types have large open parks supporting excellent stands of desirable grasses and forbs which, because of high precipitation, produce a lush crop. These ranges are generally in much better condition than other Western grazing types.

The Ponderosa Pine Forests The Ponderosa pine forest type occurs from northern Arizona and New Mexico northward into northeastern California, eastern Oregon and Washington, southern Idaho and east to Colorado and western South Dakota. It is not continuous within this area but occurs in mountains where conditions are favorable. It has its greatest development on the plateaus of Arizona and New Mexico. It does not occur to any extent in the Great Basin. Because of the wide spacing of trees in mature stands and the mosaic of different-aged stands, a wide variety of herbaceous and shrubby plants provide a mixture of forage. Open parks, which provide most of the forage, are common (Fig. 2.24).

Grazing value varies greatly. The major grasses in the southern parts are *Festuca arizonica, Blepharoneuron tricholepis, Muhlenbergia montana, Sporobolus confusus,* and *Poa* spp. Forbs important as forage include species of *Lupinus, Aster, Senecio, Delphinium, Agoseris, Achillea, Erigeron,* and *Arnica.*

Northward the dominant grasses are *Festuca arizonica* and *Muhlenbergia montana. Festuca idahoensis* replaces *Festuca arizonica* in the northern part of Colorado and Wyoming. Other important grasses are *Andropogon scoparius, Blepharoneuron tricholepis, Agropyron trachycaulum, Koeleria cristata, Danthonia intermedia,* and *D. parryi. Bouteloua gracilis* is of

Fig. 2.24 Under protection from grazing for over 20 years, this stand of *Pinus ponderosa* has produced an undercover of valuable forage plants dominated by *Purshia tridentata* and *Stipa comata.*

importance only in the lower edge of the ponderosa pine zone where it comes in contact with piñon-juniper and oak.

In the far north and west, *Festuca idahoensis, Agropyron spicatum, A. smithii, Poa canbyi, Calamagrostis rubescens,* and *Carex* spp., are most common. Shrubs, which sometimes occupy an important position under the most open stands of ponderosa pine, are *Quercus gambelii, Cowania stansburiana, Symphoricarpos* spp., *Amelanchier* spp., and *Cercocarpus montanus, Prunus virginiana, Purshia tridentata, Artemisia* spp., and *Physocarpus malvaceous.*

Grazing value   Most of the ponderosa pine type is open and supports an abundant stand of herbaceous plants. In addition, the many dry parks and meadows of the higher forests provide excellent forage.

Both cattle and sheep are grazed for 4 to 6 months during the summer. Some forests to the south are grazed during the mild winter months, and lower grasslands are grazed during the summer, when the forage species make their best growth. The grazing capacity of the forest is dependent upon the elevation and openness of the tree cover. It varies from practically valueless stands of dense trees to open park and meadow grasslands.

In many places the topography makes uniform livestock distribution difficult; hence, many accessible valley bottoms, hilltops or mesas, and open level parks are grazed heavily.

Heavy grazing over large areas has caused the replacement of bunch grasses by *Poa pratensis, Stipa robusta, Muhlenbergia filiformis, Sitanion hystrix, Chrysopsis villosa, Actinea richardsonii,* and *Artemisia frigida.* Other less desirable species, such as *Bromus tectorum* and *Gutierrezia sarothrae,* are locally abundant.

The Douglas Fir-Aspen Zone   The Douglas fir-aspen zone is a highly important forest-grazing zone throughout the central Rocky Mountains. The vegetation is varied, changing with soil moisture and aspect. The principal vegetation types are Douglas fir (*Pseudotsuga menziesii*), lodgepole pine *(Pinus contorta),* and aspen *(Populus tremuloides).*

The forage value of the Douglas fir area varies considerably. At lower altitudes, the vegetation and grazing values are similar to those of the ponderosa pine type. At higher elevations and more favorable sites, dense stands prevent undergrowth allowing only secondary forage species such as *Calamagrostis rubescens, Arnica cordifolia,* and *Vaccinium* spp.

Most of the grazing in this zone is to be found in the treeless parks and in the areas dominated by aspen (Fig. 2.25). Its sparse foliage allows considerable sunshine penetration and hence a valuable undergrowth of

Fig. 2.25 Aspen *(Populus tremuloides)* dominates thousands of acres in mountain ranges of the Western United States and is underlain by valuable forage plants which make the type one of the highest-producing in the West.

forage plants. The zone receives 50 to 62 cm of precipitation, and the soil generally is highly productive.

Among the many important forage plants that occur in the aspen type are *Bromus* spp., *Festuca idahoensis* and *F. thurberi*, *Carex* spp., *Poa* spp., *Agropyron* spp., *Vicia* spp., *Geranium* spp., *Mertensia* spp., *Lathyrus* spp., *Achillea lanulosa*, *Senecio serra*, *Wyethia amplexicaulis*, *Amelanchier* spp., and *Symphoricarpos* spp.

On moist sites the understory in aspen stands is predominantly forbs including the following genera: *Heracleum, Ligusticum, Osmorhiza, Angelica, Delphinium, Rudbeckia,* and *Thermopsis.*

Because of the wide variety of forage plants and the more favorable moisture that characterize the aspen type, it is one of the most productive grazing types in the mountain west.

The Spruce-Fir Zone In the spruce-fir zone, dominated by *Picea engelmannii, P. pungens,* and *Abies lasiocarpa,* the trees usually grow in clumps, interspersed with open parks and meadows (Fig. 2.26). The climate is severe in winter, and snow may cover the ground until the middle of June. The grazing season is short, but high precipitation results in excellent forage for a limited time. Both sheep and cattle thrive on these high ranges, and much of the area will support an animal unit for the summer on each 1.5 to 2 ha. The forage plants are similar to those of the Douglas fir-aspen zone.

Many grassland parks originally were dominated by *Festuca thurberi* but now many of these areas are characterized by *Stipa lettermanii, Poa reflexa, Phleum alpinum, Trisetum spicatum, Potentilla filipes, Helenium hoopesii, Collomia linearis,* and *Madia glomerata.*

Sagebrush extensions up to elevations of 3,000 m are common in which *Artemisia tridentata* is dominant. Other shrubs common at middle and lower elevations in the spruce-fir zone are *Symphoricarpos oreophilus, Purshia tridentata,* and *Chrysothamnus lanceolatus.*

Wet meadows that occur in this zone support dense stands of *Agrostis* spp., *Bromus* spp., *Poa* spp., *Phleum* spp., *Deschampsia* spp., *Carex* spp., *Juncus* spp., *Trisetum, Senecio* spp., *Achillea lanulosa,* and *Taraxacum officinale* which furnish a wealth of forage to many kinds of animals.

The Northern and Northwestern Coniferous Forests The northern Rocky Mountains of Wyoming, Montana, Idaho, and Canada support a dense stand of trees of commercial size, and, through most of this area, timber production, and not grazing, is the most important land use.

Fig. 2.26 Open parks and meadows between clumps of spruce and fir at high elevations in the Rocky Mountains furnish excellent grazing.

Within these forests numerous dry meadows dominated by *Stipa, Agrostis, Carex, Danthonia,* and *Agropyron* occur where most of the forage is produced. Lodgepole pine forests, especially, are lacking in forage. These northern forests are used by both cattle and sheep and usually are grazed during the summer only.

Farther north, in the Canadian taiga, the coniferous-forest type is not heavily used by domestic livestock. Its grazing use is primarily for caribou and other wildlife.

The coniferous-forest region of the Pacific Coast—from California through Canada to Alaska—is the most important timber area of America and its value as a grazing type is minor. Aside from certain logged and burned areas that have been seeded to pastures and certain incidental uses, mostly as summer sheep range, they are not grazed.

Many logged and burned forests in the Northwest have reverted to chaparral, which occasionally furnishes grazing where the brush is not too dense. These areas are dominated in California and parts of Oregon by species of *Ceanothus, Arctostaphylos, Myrica, Quercus, Arbutus, Prunus,* and *Rhamnus. Ceanothus integerrimus* is by far the most important forage species. Farther north, important species are *Salix, Alnus,* and *Acer* and occasionally *Sambucus, Corylus,* and *Betula.* Many grasses and forbs occur as an understory to the chaparral where it is not too dense.

Under the more open-growing coniferous trees of the Northwest occur various forage plants, such as *Poa, Carex, Festuca, Bromus, Melica, Deschampsia, Agrostis, Lupinus, Vicia, Lathyrus,* that furnish valuable feed to livestock.

Southeastern Pine Forests Most of the Eastern and Southeastern United States is covered with forests of varying density and composition. Only the pine forests of the Southeast are important for grazing and they constitute a major cattle-producing area. Livestock are produced on a combination of farmland, permanent pastures, and forested ranges. The forested ranges are highly variable, but they are similar with regard to grazing use. This area extends from the Atlantic Ocean to about 96° west longitude and from the Gulf of Mexico to about 37° north latitude. With an average of over 125 cm of rainfall and with 200 to 365 days of frost-free weather, vegetation-growing potential is high. Heavily leached acid soils are a limiting factor. Nearly 81 million ha of forested range is grazed, but nutritious forage is available from native grasses for only about three months in spring and a few weeks in autumn. Protein and phosphorus are especially deficient (Campbell, 1951). Heavy dependence upon improved pasture and supplemental feed is, therefore, necessary.

Forest types are dominated by longleaf pine *(Pinus palustris)*, slash pine *(P. elliottii)*, shortleaf pine *(P. echinata)*, loblolly pine *(P. taeda)*, and mixed hardwoods. Grazing has limited place in upland hardwoods and in shortleaf-loblolly pine types where the low grazing value does not justify the risks of damage to tree reproduction (Shepherd, 1950). The lower Coastal Plain, mostly longleaf and slash pine, is by far the most important of the Southern areas for grazing. Open and cutover forest produce abundant forage (Fig. 2.27) (See Chapter 12). Many coastal-plain areas are unforested and produce a wealth of grasses which with good management are productive ranges. Grasses include *Andropogon, Panicum, Axonopus affinis, Sporobolus curtissii, Aristida, Paspalum,* and *Cynodon dactylon.* Periodic prescribed burning generally is recommended. This appears to be necessary to make efficient grazing use of the range. Winter burning increases spring cattle gains twofold to threefold (Campbell, 1947).

Bottomlands of the Mississippi Delta and swamps of Virginia and the Carolinas contain numerous openings in the hardwood forests which are dominated by *Arundinaria tecta* (Biswell and Foster, 1942).

Throughout this region, cattle is the most important range animal. Hogs are locally important, but they are serious range pests. Sheep are

Fig. 2.27 Grade cattle grazing slash-pine plantation in Louisiana. *(Courtesy U.S. Forest Service.)*

kept only as farm flocks and are seldom used as range animals. Grazing is second only to timber production as a land use in Southeastern forests, but perhaps the fastest growing demand for forested lands is for recreational areas.

Forest Ranges of the World

Most of the temperate mountain ranges of the Northern Hemisphere have vegetation similar to that of the Rocky Mountains. Species of *Pinus, Picea,* and *Abies* are common and widespread, and *Larix, Populus,* and *Betula* species are intermixed with them. Herbaceous species are similar to those of North America in that most are of cool-temperate origin (Fig. 2.28).

Temperate forests are not widespread in the Southern Hemisphere. The mountains of South America have formations similar to those of North America, but their size is small and their distribution limited to the southern Andes. Australia has a coastal fringe of temperate Eucalyptus forests. None of these, however, form major grazing types. Grazing, where it exists, is done in association with other uses or on previously forested lands converted to pastures (Fig. 2.29).

Problems of Using Temperate Forest Grazing Types   Problems in the use of temperate forests arise from their limited grazing season and their wide

Fig. 2.28 High-elevation herbland parks similar to those in North America occur throughout the Northern Hemisphere. Here yaks graze a park in Mongolia. *(Courtesy F. A. O.)*

Fig. 2.29 Temperate forests often are cleared and converted to tame pastures as in the case of this *Eucalyptus* forest in New South Wales which now supports sheep on a mixture of clover and grass. *(Courtesy C. S. I. R. O.)*

use for other purposes. Climatic conditions dictate that much of the mountainous rangeland can be used only for a short time in summer. Their use thus depends upon other sources of forage throughout the remainder of the year. Where none such occur, they remain unused.

Most temperate forests are not primarily range types. Grazing exists in conjunction with other uses. In dry mountainous areas, water may be the most important range product. In scenic areas, recreation may be of prime importance. Timber production is always a major objective of managing forest land. When grazing has a secondary role in land use, special concessions must be made to the other uses. These tradeoffs must be of major concern to any range manager who uses forested lands (U.S. Department of Agriculture, 1972).

TROPICAL FORESTS

Dense, complex mixtures of trees, vines, and epiphytes are found in tropical areas of all continents and most major tropical islands. Although the area is large and is of great interest to ecologists, it is of little importance for grazing. Hot, wet conditions endanger the health of domestic animals, and diseases and parasites are common. Little herbaceous vegetation is produced under the dense canopy and it is usually of poor forage quality.

Where commercial cattle operations exist in the tropical-forest type, they are dependent upon improved pastures from which the forest has been removed and grass established. Under careful management they are highly productive (Fig. 2.30).

European cattle have not produced well in tropical areas. Zebu have a much better record of heat tolerance and disease resistance. Banteng cattle and buffalo cope successfully with forest-edge situations in the tropical areas of Asia and the Pacific islands. If these adapted species were managed properly, tropical-forest areas could be made a major grazing region (Macfarlane, 1968). In their present condition, tropical forests are scarsely a range type producing forage only for wildlife.

TUNDRA

The low-growing tundra vegetation of polar areas and high elevations is a unique grazing type. The northern tundra is used mostly by wild animals, but the alpine tundra of the high mountains is used by livestock in conjunction with the lower-elevation mountain ranges.

Fig. 2.30 Tropical woodland converted into clover pasture in Ceylon without removing trees. High rainfall makes such pastures productive. (Courtesy F. A. O.)

Fig. 2.31 Llamas, alpacas, and sheep graze alpine tundra in Peru. Grazing is diminishing in alpine areas of North America which now are prized as wilderness. *(Courtesy F. A. O.)*

The Alpine Tundra The alpine tundra occurs at elevations above timberline, generally upward of 3,000 m. Steep slopes and rough, rocky ledges make grazing difficult, but much of it is accessible to stock. The grazing season is short, with much of the area covered by snow for 9 to 10 months of the year.

In the United States the dominant climax species is *Kobresia bellardii* with *Poa alpina, Trisetum spicatum, Deschampsia caespitosa, Poa lettermanii, P. pattersonii, P. rupicola, Carex drummondii, C. chimaphila, Draba oligosperma, Sibbaldia procumbens, Salix petrophila, S. rivalis,* and *S. saximontana* common on favorable sites.

Three alpine communities are common: meadow communities on more sheltered benches, slopes, and level areas where soils are well developed; willow communities of the taller species, especially near the lower borders of the alpine-tundra zone; and alpine marshes where snowbanks contribute to a continuously moist habitat. Similar high-mountain tundras occur in Europe, South America, and New Zealand (Fig. 2.31).

Alpine-tundra areas in North America have a high aesthetic appeal and are in much demand by recreationists. Backpackers and wilderness seekers use most alpine areas during summer months and land-use conflicts between them and grazers have arisen. Their highest value is as watersheds, and because of the generally low production of forage and the short growing season, there is little reason to graze these areas extensively.

The Arctic Tundra   Arctic tundra covers almost 5 percent of the world's surface, forming a treeless cap on the Northern Hemisphere. It goes south to about 60° north latitude in America, Greenland, and Europe but retreats northward past the Arctic circle in central Asia.

The vegetation of tundra consists of low, cushionlike tufted perennial herbaceous plants and lichens. Total precipitation is usually less than for forests, and the moisture that falls remains frozen most of the year. This frozen state coupled with strong winds creates a physiologically dry environment, which ecologists have referred to as a wet desert.

Tundra vegetation is not widely used by traditional domestic livestock. However, herds of domestic reindeer use the European and Asian portions of the Arctic tundra (Fig. 2.32). Attempts are being made to use reindeer and the native musk-oxen as domestic livestock in North

Fig. 2.32 Peary caribou *(Rangifer tarandus)* on arctic tundra, Axel Heiberg Island, N.W.T., Canada. By using animals adapted to the conditions there, the arctic tundra could become a significant source of red meat. *(Courtesy G. R. Parker, Canadian Wildlife Service.)*

America. The grazing value of the tundra type is not likely to be great if emphasis is put on use of traditional meat-producing animals. Even when used by native animals like caribou, the lichens, which make up a large part of the diet, are affected greatly by trampling and 10 years may be required for recovery (Pegau, 1970). However, like the tropical forests, tundra occupies a large portion of the earth's surface and with special management has a potential largely untapped.

BIBLIOGRAPHY

Albertson, F. W., and J. E. Weaver (1946): Reduction of ungrazed mixed prairie to short grass as a result of drought and dust, *Ecol. Monogr.* **16:**449–463.

Barnard, C. and O. H. Frankel (1964): Grass, grazing animals, and man, in historic perspective, in C. Barnard (ed.), *Grasses and Grasslands,* Macmillan, London, pp. 1–12.

Bentley, J. R., and M. W. Talbot (1951): Efficient use of annual plants on cattle ranges of the California foothills, *U.S. Dept. Agr. Circ.* **870.**

Biswell, H. H., and J. E. Foster (1942): Forest grazing and beef cattle production in the coastal plain of North Carolina, *N.C. Agr. Expt. Sta. Bull.* **334.**

Box, Thadis W. (1961): Relationships between plants and soils of four range plant communities in South Texas, *Ecology* **42:**794–810.

———, Jeff Powell, and D. Lynne Drawe (1967): Influence of fire on South Texas chaparral communities, *Ecology* **48:**955–961.

Campbell, Robert S. (1947): Forest grazing in southern Coastal Plain, *Proc. Soc. Am. Foresters,* 262–270.

——— (1951): Extension of the range front to the south, *Jour. Forestry* **49:**787–789.

Costello, David F. (1944): Important species of the major forage types in Colorado and Wyoming, *Ecol. Monogr.* **14:**107–134.

Dyksterhuis, E. J. (1946): The vegetation of the Fort Worth prairie, *Ecol. Monogr.* **16:**1–29.

Eyre, S. R. (1963): *Vegetation and Soils—a World Picture,* Aldine, Chicago.

Frolik, A. L., and W. O. Shepherd (1940): Vegetative composition and grazing capacity of a typical area of Nebraska sandhill range land, *Neb. Agr. Expt. Sta. Res. Bull.* **117.**

Gould, Frank W. (1962): Texas plants—a checklist and ecological summary, *Tex. Agr. Expt. Sta. Misc. Pub.* **585.**

———, and Thadis W. Box (1963): *Grasses of the Texas Coastal Bend,* Texas A & M University Press, College Station, Tex.

Hanson, Herbert C. (1950): Ecology of the grassland II, *Botan. Rev.* **16:**283–360.

Hopkins, H. H. (1951): Ecology of the native vegetation of the Loess Hills in central Nebraska, *Ecol. Monogr.* **21:**125–147.

Humphrey, R. R. (1953): The desert grassland, past and present, *Jour. Range Mgt.* **6:**159–164.

Jepson, W. L. (1925): *A Manual of the Flowering Plants of California,* Associated Student's Store, Berkeley, Calif.

Keller, Boris (1927): Distribution of vegetation of the plains of European Russia, *Jour. Ecology,* **15:**189–233.

LeHouérou, H. N. (1972): Continental aspects of shrub distribution, utilization, and potential—Africa, the Mediterranean Region, in Wildland shrubs—their biology and utilization, *U.S. Forest Service General Technical Report,* INT-1, 1972, pp. 26–36.

Leigh, John H., and John C. Noble (1969): Vegetation resources, in R. O. Slatyer and R. A. Perry (eds.), *Arid Lands of Australia,* Australia National University Press, Canberra, pp. 73–92.

Macfarlane, Victor W. (1968): Protein from the wasteland, *Austral. Jour. Sci.* **31:**20–30.

McGinnies, William G. (1968): Vegetation of desert environments, in W. G. McGinnies, B. J. Goldman, and P. Paylore (eds.), *Deserts of the World,* University of Arizona Press, Tucson, pp. 381.

Moore, Milton, and Rayden Perry (1970): Arid rangelands, in R. Milton Moore (ed.), *Australian Grasslands,* Australia National University Press, Canberra.

Morello, Jorge (1955): Estudios botanicos en los regiones aridas de Argentina, *Rev. Agron. Noreste Argentino* **1:**385–542.

——— (1958): La provencia fitographica del Monte; Opera Lilloaderia y el bosque en el criente de Salta, *Rev. Agron. Noreste Argentino* **3:**209–258.

Nichol, A. A. (1937): The natural vegetation of Arizona, *Ariz. Agr. Expt. Sta. Tech. Bull.* **68.**

Pegau, Robert E. (1970): Effect of reindeer trampling and grazing on lichens, *Jour. Range Mgt.* **23:**95–97.

Petrov, M. P. (1972): Continental aspects of shrub distribution, utilization, and potentials—Asia, in Wildland shrubs—their biology and utilization, *U.S. Forest Service General Technical Report,* INT-1, 1972, pp. 37–50.

Plummer, F. G. (1911): Chaparral, *U.S. Dept. Agr. Forest Service Bull.* **85.**

Polunin, Nicholas (1960): *Introduction to Plant Geography and Some Related Sciences,* McGraw-Hill, New York.

Powell, Jeff, and Thadis W. Box (1967): Mechanical control and fertilization as brush management practices affect forage production in south Texas, *Jour. Range Mgt.* **20:**227–236.

Rojas-Mendosa, Paulino (1965): Generalidades sobre la vegetacion del Estado de Nuevo Leon y Datos Acerca de su Flora, Tesis Doctoral, Universidad Nacional Autonoma de Mexico, Mexico, D. F., 124 pp.

Sampson, A. W., Agnes Chase, and D. W. Hedrick (1951): California grasslands and range forage grasses, *Univ. Calif. Agr. Expt. Sta. Bull.* **724.**

Shantz, H. L. (1954): The place of grasslands in the earth's vegetative cover, *Ecology* **35:**143–151.

———, and R. Zon (1924): Atlas of American agriculture, Part I, The physical basis of agriculture, Advance Sheet 6, Section E, *Natural vegetation,* U.S. Department of Agriculture, Bureau of Agricultural Economics, Washington.

Shepherd, W. O. (1950): The forest range in southern agriculture, *Jour. Range Mgt.* **3**:42–45.

Skovlin, Jon (1971): Ranching in East Africa: a case study, *Jour. Range Mgt.* **24**:263–270.

Smith, A. D. (1940): A discussion of the application of a climatological diagram, the hythergraph, to the distribution of natural vegetation types, *Ecology* **21**:184–194.

Soriano, Alberto (1956): Los distrites floristicas de la Provincia Patagonica, *Rev. Inv. Agr.* **10**:323–347.

———— (1972): Continental aspects of shrub distribution, utilization, and potentials—South America, in Wildland shrubs—their biology and utilization, *U.S. Forest Service General Technical Report,* INT-1, 1972, pp. 51–54.

Stoddart, L. A. (1941): The Palouse grassland association in northern Utah, *Ecology,* **22**:158–163.

U.S. Department of Agriculture, Forest Service (1936): The western Range, 74th Cong., 2d Sess., Senate Doc. 199.

————, ———— (1972): The nation's range resources—a forest-range environmental study, *U.S. Forest Service Res. Report* **19.**

Weaver, John E. (1954): *The North American Prairie,* Johnsen Publishing Co., Lincoln, Neb.

————, and F. W. Albertson (1956): *Grasslands of the Great Plains,* Johnsen Publishing Co., Lincoln, Neb.

————, and F. E. Clements (1938): *Plant Ecology,* 2nd ed., McGraw-Hill, New York.

————, and W. W. Hensen (1941a): Native midwestern pastures; their origin, composition, and degeneration, *Univ. Neb. Conserv. Bull.* **22.**

————, and ———— (1941b): Regeneration of native midwestern pastures under protection, *Univ. Neb. Conserv. Bull.* **23.**

————, and G. W. Tomanek (1951): Ecological studies in a midwestern range; the vegetation and effects of cattle on its composition and distribution, *Univ. Neb. Conserv. Bull.* **31.**

Whyte, R. O., Jr., G. Moin, and J. P. Cooper (1959): Grasses in agriculture, *FAO Agr. Studies No. 42,* Food and Agricultural Organization of the United Nations, Rome.

Wright, Henry A. (1972): Shrub response to fire, in Wildland shrubs—their biology and utilization, *U.S. Forest Service General Technical Report,* INT–1, 1972, pp. 204–217.

Chapter 3

Development of
Range Management

Range management is regarded by many as an American science of rather recent origin. Exactly when range management as a profession began is debatable, but in the 1890s personnel of the Division of Agrostology, later part of the Bureau of Plant Industry, were formulating the foundations of range management. The writings of Jared Smith (1895) and H. L. Bentley (1898) in West Texas, and of Frederick Colville (1898) in the Northwest were among the first to outline clearly and concisely the problems of grazing livestock on open rangelands. Concurrently, others were making surveys of range conditions in the Great Basin (Kennedy and Doten, 1901) and the Red Desert of Wyoming (Nelson, 1898).

From these early explorations came recognition of some of the problems unrestricted grazing use had created (Shear, 1901). Recommendations for solving these problems took the form of proposals for seeding denuded ranges and controlling livestock use through management systems which look very much like grazing systems now being used. A permit system for livestock grazing on Indian lands was begun in 1891 and in 1894 regulations were formulated regarding grazing use of the forest reserves (Stoddart, 1950).

Search of records indicates that range research may have begun at Abilene and Chandler, Texas, with the work of Bentley and his associates in the late nineteenth century. Shortly after the turn of the century several Western state agricultural experiment stations and federal agencies began range research. In 1903 the Santa Rita Range Reserve was established in southern Arizona. By 1907 this had become an important experimental area (Stoddart, 1950). By 1910, eight state stations and the U.S. Forest Service were working on range problems (Talbot and Crafts, 1936).

Some Western universities began to teach courses in range management in the early 1900s, and in 1916 a curriculum in range management was established at Montana State University. The American Society of Range Management, which was established in 1948, has steadily grown in the United States and expanded abroad. This evidence seems to point to the fact that range management, as a profession and as a science, is indeed American.

But, although Americans can lay clear claim to the name range management, and to the establishment of range management as a science, if we take a careful look at the history of mankind we should not be so presumptuous as to assume that Americans invented range management. The roots of the profession are lost somewhere in the prehistory of *Homo sapiens* (Box, 1968). One has only to work among primitive pastoral people to realize that, although we in America can rightly lay claim to establishing range management as a science and to coining the word by which our profession is known, the elements of range management in which man, plants, land, and livestock were bound together on an ecological basis is not original with us.

MAN-PLANT-ANIMAL SYSTEMS IN HISTORICAL PERSPECTIVE

Many anthropologists believe that the early humanoids were roving, carnivorous apes (Ardrey, 1961), and that the early evolution of man began in heavily wooded savanna grasslands at the moist end of a gradient of grassland ecosystems (Ripley, 1966). As he learned to use tools and fire, man cleared the forests, thus extending grasslands into the humid regions. At the same time he moved into and adapted himself to the xeric conditions of desert areas. As man thus extended into new areas, he became part of the ecosystems he inhabited and they in turn changed biotically and environmentally. In part this was due simply to his presence, but, as a thinking, rational animal, he acted in ways that intensified his impacts upon the environment. By capturing animals and domesticating them he developed a symbiotic relationship in which the animals provided him with food, transportation, and clothing, and he, in turn, provided them with protection from predators, and water during dry periods. We have no written record of this early relationship, but it was in this era that grazing, the antecedent to range management, began.

In the earliest written record of man we find strong evidence of a close, almost spiritual, relationship between man and beast. The early Zoroastrian hymns and proverbs give specific instructions on the care of animals. The bull and the ram are central in the Mithraic tradition. The Bible, the Koran, and other holy books all deal with this close relationship. Many of the African religions, though having no written scriptures, have traditions of man and cattle being together in the creation (Mbiti, 1969). In their religion, it is generally thought that God gave man cattle, and that man is therefore responsible for them.

Effects of Primitive Man on the Ecosystem

Under the conditions of primitive man, natural biotic communities evolved with plants, animals, soils, and climates developing concomitantly. Although primitive man was mentally capable of altering ecosystems, his low population density and his lack of land-modifying tools and domestic animals left his ultimate influence on the landscape minor. Fire was his only major land-modification tool, and it was a natural ecological force that predated man himself. Plant communities evolved primarily under the grazing pressures of large herbivores, and the type of plants and their densities largely reflected their grazing habits and food preferences.

Effects of Pastoral Man

We cannot be certain when pastoral cultures came into being. There is evidence that grazing of domesticated animals occurred as early as 9000 B.C. Goats were well established in the Middle East in 7000 B.C.; sheep about 1,000 years later. Cattle were present in Greece by 6200 B.C. and horses and asses probably somewhat later. By 5,000 years ago these animals were widespread from the Nile to the Indus (Pearse, 1971).

At some point within this time span pastoral cultures were developed. Although different patterns emerged from place to place, there was a common element best described as that of man in nature in which man strives to live in harmony with his surroundings not for the sake of conservation, but for his very survival. A balance is struck among man, livestock, and the environment (Fig. 3.1).

Pastoral cultures are today most commonly nomadic with individual families or groups shifting from place to place with the season in response to availability of forage and water (Lewis, 1961) or, in some cases, to avoid insects. A major part of the world's rangelands are used by pastoral tribes, most notably in Asia and Africa. In Africa, there are 50,000 pastoral families south of the Sahara alone (Brown, 1971). For them, livestock is the sole means of support, and wild species are ignored almost completely.

Although the principles of range management are largely unknown among pastoralists, there is some degree of ecological soundness and viability in pastoral systems (Mahoney, 1966). But, if the culture is changed through an input in technology and new wants develop, environmental degradation follows (Box, 1971). Control of disease either in humans or livestock removes the natural factors limiting population and puts new pressures on the environment. The customarily large herds become even larger.

Fig. 3.1 Nomadic family and livestock at a water hole in Mauritania. Camels, cattle, sheep, goats, and asses commonly form nomadic pastoral herds, often walking many miles for water. *(Courtesy F. A. O.)*

There are two principal reasons why large herds of animals are kept: prestige and survival. In the view of some anthropologists, large herds in developing countries are kept primarily for prestige and status, which leads to range deterioration (Konczacki, 1967). Without question many more animals are kept than are needed to meet cultural obligations such as bride or blood payments; the need for subsistence of the nomadic family in the face of drought is a major factor in the maximization of animal numbers (Corti, 1970).

The average pastoral family of five to six in East Africa needs about 14,500 calories per day to remain alive. These calories come directly from their animals in the form of milk, blood, and meat (Brown, 1971). Diets vary regionally and seasonally. Milk forms the basis of the diet and comes from all kinds of animals. Meat usually comes from small stock—sheep and goats. In order to meet the daily requirement of 14,500 calories per family, 21 liters of milk, or its equivalent in meat or blood, must be produced requiring herds of 40 to 50 cattle per family. Any increase in human population results in a direct increase in grazing animal populations. This dependence on products from the top of the food chain—meat and milk—has resulted in broad-scale range deterioration as populations increased.

Effects of Technological Man

With the advent of modern man, greatly intensified pressures were exerted on rangelands. New species of both plants and animals were introduced and populations of native fauna were much reduced. In consequence, plant populations changed radically. Those plants that were most palatable to domestic livestock declined in vigor and abundance and became minor elements in the community; less palatable plants became more abundant. Whole plant communities were changed in composition or were replaced entirely by others. Brush and woody plants increased on many rangelands, often creating more favorable habitat for deer and other wildlife (Leopold, 1939). It was the effect of technological man that necessitated the development of range management as a profession.

DEVELOPMENT OF RANGE MANAGEMENT IN AMERICA

Early American history is interspersed with events growing out of the development of the range-livestock industry. The Spanish introduced the animals into America, the pioneer Texans cared for them, and there developed a new pattern of living upon which Western American life was built. Out of this tradition, modern range management was born.

Original Grazing Animals of the American Continent

A summary of the conditions existing at the time America first was inhabited is illuminating. No entirely new influence was exerted on the rangelands with the coming of domestic livestock, for, previously, countless numbers of wild mammals grazed the entire continent. Although these animals generally were not so abundant as domestic livestock, it is not unlikely that excessive numbers of the native fauna caused local or perhaps general misuse and overuse of the range. Over 67 million animal units (Table 3.1) were believed to be present. This is almost as much as the estimated animal units of livestock using the area today. The white man as a new ecological force merely realigned the influences already present and intensified the environmental complex.

Early Introduction of Livestock into America

The beginnings of the livestock industry in America are associated with exploration and colonization. Columbus is reported to have landed domestic stock, including horses and sheep, in the West Indies in 1493 (Anon., 1939); but it was not until 1515 that Cortez, landing in Mexico, brought livestock to the continent of North America. It is believed that cattle were not included but, rather, that only horses were brought. It is known that cattle were introduced in the year 1521 by Gregario Villalobos (Barnes, 1926), who landed them in eastern Mexico. These, however, did not reach the territory of the United States.

De Soto is reported to have brought horses to Florida in 1539

Table 3.1   Numbers of Big-Game Animals Estimated to Have Been Present Originally in the United States

Data from Seton (1927)

Animal	Numbers	Number per animal unit (455-kg base)	Animal units
Bison	50,000,000	1.0	50,000,000
White-tailed deer	40,000,000	7.7	5,195,000
Pronghorn antelope	40,000,000	9.6	4,167,000
Elk	10,000,000	1.9	5,263,000
Mule deer	10,000,000	5.8	1,724,000
Black-tailed deer	3,000,000	8.0	375,000
Bighorn sheep	1,500,000	5.6	268,000
Mountain goat	1,000,000	7.0	143,000
Total	155,500,000	.....	67,135,000

(Anon., 1939); but the first record of the importation of cattle into the territory properly a part of the United States is that of Coronado, who in 1540 traveled from western Mexico northward through Arizona, New Mexico, and Colorado, reaching as far north and east as Kansas (Sampson, 1928). Considerable numbers of stock were included in the expedition—1,000 horses, 500 cows, and 5,000 sheep according to Stewart (1936). From these herds, escaped and abandoned animals began stocking the range area. During the next century, considerable numbers of animals were brought to the numerous Spanish settlements established in New Mexico, Texas, Arizona, and California. Finding ample forage and favorable climate, the early stock increased abundantly, though it was not until many years later that production of range livestock was to become a national industry.

Little can be learned about the quality and kinds of livestock of these early days, but from what is known of the stock in European countries at that time it must be conjectured that these represented a motley array of kinds, shapes, and breeds. Certainly these introductions took place before the development of the improved breeds of today.

Beginnings of Range-Livestock Production

When the settlers from the East came to the Mississippi Valley about 1830 and merged with livestock men moving northward from Texas, the range-livestock industry can be said to have begun. Broad expanses of grassland gave an impetus to the livestock industry, and though for some years there was a lack of transportation and market facilities, there began a general increase in the Western livestock population. This was aided by the gradual diminution of the danger from marauding Indians as forts and settlements were built in the wake of trappers and adventurers.

This growth resulted from nothing but the naturally favorable conditions for livestock. Because of unlimited forage, the animals did well without conscious aid and husbanding by the ranchers.

There were periodic booms, which greatly stimulated the ranching enterprises of the West. The first of these occurred during the Civil War. The need for supplies for Confederate armies provided a ready market for the cattle of Texas. This was only temporary, for the Union blockade soon closed these avenues of commerce. At the close of the war, because of few cattle in the East and monetary inflation, prices remained high, which stimulated efforts to get animals to market (Osgood, 1929). This led to a most interesting stage in the developing livestock industry.

Texas Trail Herds   Though the Texas Longhorn (Fig. 3.2) had been known on the markets before the Civil War, the number of these cattle was not great. The first account of a northward drive was that of Edward Piper, who, in 1846, took a herd from Texas to Ohio (Wellman, 1939). The year 1866 marks the beginning of this movement in considerable proportions. Soldiers returning from the Civil War saw an opportunity to make money by marketing cattle, which had increased remarkably on the ranges, and there was a great rush to this business.

The trails going north from Texas were not single trails but merely a general direction of travel made up of numerous small trails converging at river fords and mountain passes. These trails varied in distance according to the routes followed (Fig. 3.3).

Movements of livestock often took the form of a series of drives that lasted years; animals leaving Texas as 2- to 4-year-old stock often were sold as 4- to 6-year-olds or more. Similarly, from the northern areas came herds of steers on the way to Eastern markets over long and hazardous trails.

These first great movements of livestock were accomplished only with great difficulty. The cattle were wild and unmanageable, especially

Fig. 3.2   The Texas Longhorn steer with long legs and lack of body depth was well suited to trailing the long distances encompassed by early cattle trails. The cattle owned by nomadic tribes are similar to these.

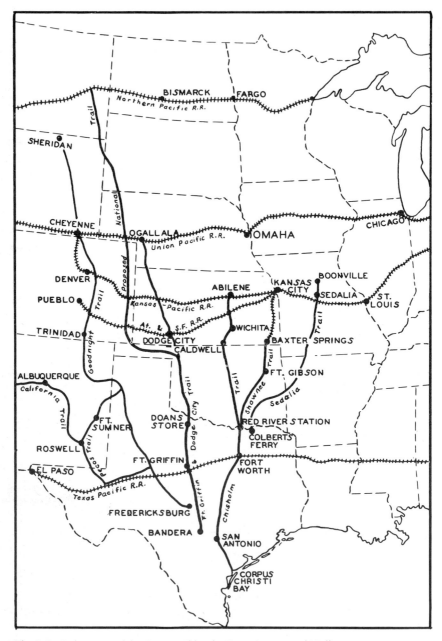

Fig. 3.3 Early routes of the Texas trail herds. *(From Lantow and Walker, New Mex. Agr. Exp. Sta. Bull. 59, and U.S. Dept. of Commerce, House of Executive Documents, 45th Cong., 2d Session, Vol. 20, 1884.)*

while they were being moved through unfamiliar country. The numerous rivers took their toll; many cattle were drowned while crossing. Even more serious than these natural hindrances were those associated with Indians and with white settlers, who had strong feelings against the entrance of Texas livestock (McCoy, 1932). The boon that offset these difficulties was the purchase of animals for breeding purposes by the ever-widening circle of new ranchers in the plains area.

Rise of the United States Sheep Industry   Although sheep were among the animals first introduced into the New World, they were, up to the time of the Civil War, relatively unimportant on the Western ranges. There were at various times influences that stimulated the raising of sheep, such as the American Revolution and the War of 1812, but these had resulted mainly in an increase in the numbers of sheep on farms.

The development of markets for wool and mutton during the eighteenth century made little progress in America, but, about the time of the War of 1812, decreases in foreign wool and the resultant increase in woolen mills in America caused rapid price increases. Mutton prices likewise increased and gave an impetus to the development of the industry. There were few sheep west of the Mississippi River in 1840. Though the industry was stimulated by the Civil War, there was no material increase in range-sheep production until there were railroads to carry the wool crop. Following this, there began a rapid development of the sheep industry. Sheepmen found much range that was admirably adapted to their herds and were attracted by the money to be made from the grazing of free lands.

During the years from 1865 to 1901 the sheep industry saw a period of trail herding almost as spectacular as that of the cattle-trail herds (Fig. 3.4). Sheep were trailed in large numbers from California, and later from Oregon, to stock ranges farther east and to fattening and marketing points in the Midwest. During 1880, nearly 600,000 head are reported to have been trailed over well-established trail routes (Wentworth, 1948).

By 1880, there were considerable numbers of sheep, with some states possessing greater numbers than they later maintained. By 1910, nearly all the West had reached maximum numbers. Early-day sheepmen soon found that they could increase their grazing land by seasonal migrations, sometimes covering vast areas. They grazed their animals yearlong and moved to the high and cool mountains during the summer months and to the low and less snowy deserts during winter months.

The effect of increasing sheep numbers upon the cattle industry was severe. As compared with sheep, cattle were more difficult to move.

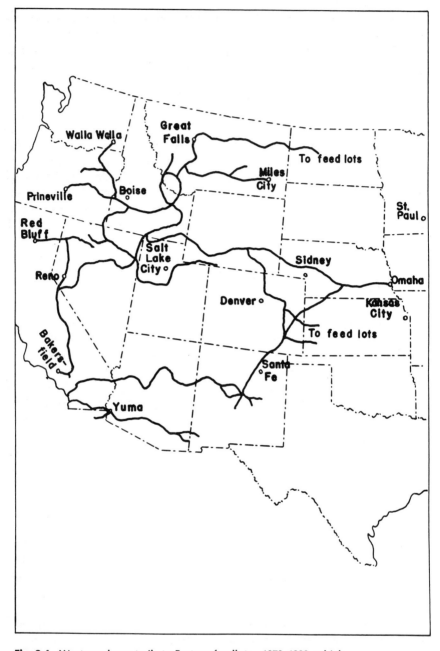

Fig. 3.4 Western sheep trails to Eastern feedlots, 1870–1900, which rivaled the early-day cattle trails in adventure and tradition. *(Reprinted by permission from* America's Sheep Trails—History, Personalities, *by Edward N. Wentworth,* © *1948 by the Iowa State University Press, Ames, Iowa.)*

Ranch headquarters had to be maintained, and animals became accustomed to occupying the ranges in the vicinity. This was not true in many instances of the early-day sheep operator. With headquarters consisting of a sheep wagon, a veritable home on wheels, he was prepared to move as he wished, seeking the best grass from one range to another. Although the cattlemen already were using these ranges, it did not seem to the sheepman that he was trespassing, for the land was not held by the cattlemen except by squatter's right. The cattleman felt that, by reason of previous use, the land was his. When the forage became scarce, the sheepman could move, whereas the cattleman was compelled to remain. The result was an intensely bitter feeling between the two classes of range operators which is only now passing. During the early period, illegal slaughter of cattle and sheep was common, and at times even human lives were taken in the intensely bitter contest for supremacy of the range.

Gradually, the sheepman and his herds became a part of the American West, and the cattlemen realized that sheep production was a permanent industry.

Westward Spread of Livestock

As miners, soldiers, trappers, and others pushed into the West, it became possible for the central and northern plains to be occupied. Cowmen came in increasing numbers, and farmers were not far behind. These settlers, aided by the westward extension of the railroads, offered an excellent market for livestock. The Western cattle industry boomed in the inflationary period following the Civil War, and, except for a brief period of panic in 1873, phenomenal increases in cattle numbers occurred until 1885.

Cattle were selling at high prices in the Eastern markets, and, as a result, speculators were encouraged by the hope of almost fabulous returns to be made from the livestock business. Immense inflow of capital took place from the Eastern states as well as from abroad. This was encouraged by investment syndicates, which emphasized the great returns from the cattle industry.

As financial conditions improved, the industry expanded rapidly. The handling of finances was often lax, as was the managing of ranch properties. This laxity was often the result of the owner's living in some distant city or even in the British Isles.

This early development of the cow business was described by the late H. L. Bentley (1898) of Abilene, Texas:

> Men of every rank were eager to get into the cow business. In a short time every acre of grass was stocked beyond its fullest capacity. Thousands of cattle and sheep were crowded on the ranges when half the number was too

many. The grasses were entirely consumed; their very roots were trampled into the dust and destroyed. In their eagerness to get something for nothing, speculators did not hesitate at the permanent injury, if not the total ruin, of the finest grazing country in America.

In a few short years the number of cattle built up to alarming proportions in West Texas. The grass resource was considered to be almost limitless in its capacity.

Though these conditions would soon have resulted in retrenchment of the range-cattle industry, an immediate collapse was precipitated by the winter of 1885–1886—a winter of unprecedented severity. The effects of this harsh weather were intensified by concentrations of cattle that had been removed from Indian territories. The result was a decimation of cattle that has been estimated to have left 85 percent of the animals dead over wide areas on the range. The situation was well described by the colorful Don Biggers (1901, pp. 22, 24–25) as follows:

> In the winter of 1884 began a series of most disastrous years ever known to the cattle industry. . . . When a blizzard would sweep over the country the cattle would drift before it, and it was then no uncommon sight to see great herds of cattle rolling southward, nothing to eat, nothing to drink, pelted by sleet and covered with snow; while around them the pittiless [sic] blizzard seemed to howl in fiendish glee. . . .
>
> The winter of 1886 was very severe and in the spring of 1887 occurred, beyond a doubt, the awfullest die-up ever known in the United States. From the Canadian borders to the Rio Grande the range country was covered with carcasses. . . . When the blizzards came the cattle would drift south until they came to the southern line of fence. Unable to go further they would move back and forth, pressing close to the fence or stand in clusters suffering from cold, hunger and thirst, and trampling out every vestige of grass. One would fall or lie down and others would tumble over it, and soon there would be a heap of dead along the line of fence. I saw one instance and heard of many others, where, for a distance of two and three hundred yards [sic], the heaps of dead bodies were higher than the fence. Over these bodies the snow drifted and sifted between them, soon forming a solid, frozen mass, over which hundreds of living cattle walked, tumbled over the fence and drifted away. This awful spectacle was to be repeated again in 1894.

During this period many herds were almost entirely wiped out, and others suffered tremendous losses. As a consequence of these disasters, much of the money which had supported the boom was withdrawn. However, there did remain some individuals with a true regard for cattle raising. These were the real pioneers of the present cattle industry.

From this time it was apparent that immense profits were a myth and that the future industry must settle itself on a more stable basis.

Similar situations occurred in other continents where modern civilizations occupied new lands. In Australia sheep numbers built up rapidly following settlement and exceeded the carrying capacity of the range. Large die-offs occurred and numbers have since stabilized at a relatively low level (Fig. 3.5).

Development of Scientific Range Management

In America the tragedy of livestock buildup and die-off spawned a new scientific field—range management. Spurred by disasters and observing the near-desert condition of the range, ranchers asked the Department of Agriculture to conduct research on rangelands. In 1894 more than 2,000 circular letters were sent to ranchers in West Texas. More than 1,000 livestock men answered these letters. They indicated that their range carrying capacity was declining and that research should be initiated on ways and means of improving their rangelands (Box, 1967). In consequence, Jared G. Smith, Assistant Agrostologist, U.S. Department of Agriculture, examined the Western ranges and wrote:

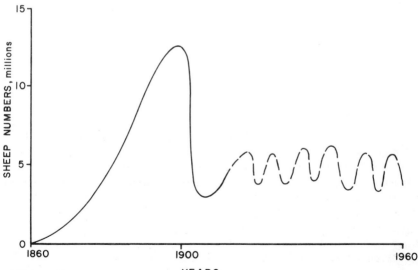

Fig. 3.5   Trend of sheep numbers in rangeland areas of Australia. After the initial rise and fall, numbers remain fairly stable but the number of watering points, and thus the area actually grazed, has steadily increased. *(From Perry, 1957.)*

There has been much written during the past 10 years about the deterioration of the ranges. Cattlemen say that grasses are not what they used to be; that the valuable perennial species are disappearing, and that their place is being taken by less nutritious annuals. This is true to a very marked degree in many sections of the grazing country.

The one great mistake in the treatment of cattle ranges, the one which always proves most disastrous from a financial standpoint, is over-stocking. It is something which must always be guarded against. The maximum number of cattle that can safely be carried on any square mile of territory is the number that the land will support during a poor season. Whenever this rule is ignored there is bound to be loss. The present shortage of cattle all through the West is due to the fact that the ranges were stocked up to the limit that they would carry during the series of exceptionally favorable grass years preceding the years of drought. Then followed a series of bad years, when the native perennial grasses did not get rain enough to more than keep them alive. The cattle on the breeding grounds of the West and Southwest died by thousands from thirst and starvation. It may seem like throwing away money not to have all the grass eaten down, but in the long run there will be more profit if there are fewer head carried per square mile (Smith, 1895, p. 322).

The following year (1896) C. C. Georgeson was sent to Texas to organize a series of experiment stations. One station was located at Abilene to serve western and central Texas and another at Channing, Texas, to serve the Texas Panhandle, Kansas, Oklahoma, Colorado, and New Mexico. H. L. Bentley was hired as a special agent to conduct range experiments at Abilene. His studies on correct stocking rates, deferred rotation grazing, range pitting, and range seeding are the first recorded efforts in range-management research.

Range research was established by most state and federal agencies in the United States soon after the turn of the twentieth century, and by 1910 eight state agricultural stations had one or two projects each (Talbot and Crafts, 1936). Organized range research developed through several distinct stages—the exploratory period prior to 1905, limited intensive studies between 1905 and 1909, organized experiments in the mountain West and Great Plains from 1910 to 1927, and an expansion phase from 1928 on (Chapline et al., 1944). Prior to the 1960s range research was designed primarily to maximize forage production for domestic livestock. Current trends in range research are geared to optimize the functioning of the entire range ecosystem.

Higher education contributed immensely to the growth of the range-management profession. The existence in each state of land-grant colleges dedicated to teach practical aspects of agriculture provided the setting for development of range courses based upon ecological theory. Range-management courses were begun in most Western land-grant colleges during the first two decades of the twentieth century. These could be found in botany, animal husbandry, forestry, and agronomy departments. Publication of textbooks on range (Sampson, 1923 and 1928; Stoddart and Smith, 1943 and 1955; Saunderson, 1950; Clawson, 1951) set the pattern for an organized body of knowledge that could be taught and formed the basis of the range profession.

SPREAD OF RANGE MANAGEMENT WORLDWIDE

Ecological management of grazing lands has been practiced by many nationalities and for varying lengths of time. The terms *range management* and *rangeland* have only recently spread from their American origin to other nations and cultures. For instance, the term *range* first appeared in widespread use in East Africa in about 1958 when the Kenya Ministry of Agriculture established a range program (Brown, 1958). Following the publication of a book on range management in East Africa by Heady (1960), the term became well established. The continent with the largest percentage of rangelands, Australia, has only recently established a section of rangeland research in its governmental research division (Perry, 1967).

Wherever there are vast areas of wild lands, some form of range management is practiced. It is called by different names in different places—*manejo de pastizales* in Mexico, *range management* in the United States and Canada, *pastoral industry* in Australia (Heathcote, 1969), or *the nomadic sector* in East Africa (Konczacki, 1967).

The form and substance of range management varies with the ecological resources available in a given area. Equally important in determining the kind of range management that is practiced are the level of economic development of the country, the political and economic institutions serving the industry, and the culture of the people involved.

The extremes vary from comprehensive studies of entire ecosystems using mathematical models and systems analysis and including literally dozens of basic disciplines to work that is more like anthropology than biological science. However, in all cases, a common set of biological principles applies. The central discipline is ecology. In some instances, range management may involve basic plant ecology, animal-population analysis, or other strictly biological sciences. In most, range management can be considered as human ecology, involving not only the biological principles, but man's institutions, his technology, and his culture as well.

LAND POLICIES AND RANGE MANAGEMENT IN THE UNITED STATES

Land policies greatly affected the course of range management in the United States. A case study can show the evolution of range management with land policies.

America's first settlers found a wealth of land of apparently limitless extent. This land was unowned and unsurveyed. The government soon faced the problem of surveying and developing a land policy for disposing of this vast resource.

The student of land management in the United States will note that the government has been at fault in its land policy and is not a little to blame for many present-day problems in conservation. A difficulty with these policies in the Western United States has been that they were designed for farmland and made no provision for a settler to obtain sufficient acreage for range-livestock production. This, together with the belief of many early settlers that land was inexhaustible, resulted in untold damage to the land and economic loss to the nation.

Acquisition of the Public Domain

At the time of its inception, the federal government had no land. During the period from 1781 to 1802, seven states turned their unowned lands over to the government, and this land became the first public domain (Hibbard, 1924).

Though these original lands were the basis for development of land policy, later acquisitions hold special interest. Most of the Western lands were acquired not for the want of more land, but for indirect considerations. For instance, the Louisiana Purchase of 1803 was consummated to gain control of the port of New Orleans. The Northwest Territory was desirable to protect the valuable fur trade and for strategic reasons. The Southwest Territory, wrested from Mexico in 1846, was wanted largely because of the necessity for providing good harbors on the West Coast. The purchase of the Gadsden Territory was made to establish a more satisfactory boundary and to facilitate the building of a railroad to the Pacific. Since there was no great value attached to the land itself, there was little concern for its judicious administration.

Land-Disposal Measures

The land-disposal measures enacted by the United States government changed as time went on, as new circumstances arose, and as different economic segments were able to influence the Congress. Gates (1968, p. 765) identifies four objectives:

> American land policy from Independence to the end of the 19th century had four objectives inherited from the colonial period: (1) to produce revenue for the government; (2) to facilitate the settlement and growth of new communities; (3) to reward veterans of wars; and (4) to promote education, the establishment of eleemosynary institutions, and the construction of

internal improvements by grants of land. Spokesmen for all four of these objectives were to clash over the relative importance of each and were to cause the adoption of measures that were inharmonious and incongruous with others.

Two phases of public-land-disposal policy can be recognized in the period to which Gates refers. A third, retention, developed at the close of the nineteenth century and remains viable today.

Sale of Lands   The first phase in disposal was marked by emphasis on the revenue to be secured; hence, early disposal acts were sales measures. The infant government possessed a great excess of land but a serious shortage of cash (Clawson, 1951). Throughout this period the settler was dissatisfied because of both inability to pay and stubborn belief in the free rights of the squatting days. The government was dissatisfied because of the slowness of sales and the increasing laxity in payment.

The failure of the many sales laws enacted during the period from 1785 to 1891 (Senzel, 1950) may be attributed to (1) the government's failure to limit the maximum size of purchases, (2) the government's failure to establish a high enough minimum price, and (3) the government's adoption of credit sales.

Homestead Era   During the second phase, settlement of the country was an important motive, and the homestead acts endeavored to make acquisition of land easy and inexpensive. Even during the first period, when land was regarded chiefly as a source of revenue, there was leniency toward the squatter because of the government's desire to settle the West. Beginning about 1841, the squatter was given ever-increasing rights. Many attempts were made to obtain free-land laws, and these attempts led to the passage of the 1862 Homestead Act. This law allowed the settler to acquire 160-acre tracts after living on the land for 5 years and making certain improvements. Unfortunately the act contained a commutation clause which enabled the settler to purchase land for $1.25 per acre after a residence of 6 months. As a result of this, many of the tracts were not occupied after commutation (Van Hise, 1910), but were sold to large lumber companies or ranchers and became part of vast enterprises.

By 1870, most of the highly productive lands of the Middle West had passed to private ownership. This left only the less productive semiarid lands to be settled. The allotted 160 acres were pitifully inadequate to support a family in these areas. The Enlarged Homestead Act of 1909 increased the homestead size to 320 acres in nine Western states

(Hibbard, 1939) and provided that one-fourth of the land should be cultivated. The Three-year Homestead Act of 1912 shortened the residence period required to 3 years but missed the real issue by its failure to increase acreage.

In 1916, the Stock-raising Homestead Act was designed for the Far Western lands not adapted to cultivation. This act gave stockmen 640 acres (260 ha), an area supposedly large enough to carry 50 head of cattle. The size of these range homesteads was grossly inadequate, and the provisions of the act requiring improvements and reserving water holes were unsatisfactory (Hibbard, 1939). As a result, less than half of the stock-raising homesteads were patented during the first 12 years that the act was in effect.

The failure of the government to determine and specify proper land use in the West has resulted in the plowing of thousands of acres of land unsuitable to cultivation. For many years, almost every Western state had tracts of once-plowed, abandoned land of little use for grazing. The federal government is not alone responsible for these improper land policies; state and county governments likewise made many serious errors in administration and taxation.

Federal Grants   The second phase in land disposal also was marked by numerous grants to railroads and other agencies for roads, canals, and similar developments. Since land sales had fallen short of expectation as a source of national income, the government began to look upon its land as a means to national development and internal improvements (Senzel, 1950).

Overshadowing all internal-improvement grants were the railroad grants. Grants of alternate sections of land for several miles on each side of the line were considered sound investments. Since the presence of railroads would, presumably, double land prices, in reality the government would gain by the grant.

Of tremendous importance in land-disposition history were the educational grants. These were given upon admission of states to the Union and over the years varied greatly as to form. Some were general grants designating certain sections within each township, usually in support of common schools. Later grants of specified acreages were made for specific educational purposes—university, land-grant colleges, normal colleges, school of mines, and other special educational institutions. Approximately 86 million acres were granted to states under various educational-grant acts (Public Land Law Review Commission, 1969).

Reservations   Finally, there was a third phase during which conservation of important resources was dominant and the withdrawal of lands was emphasized. Toward the close of the nineteenth century, it became apparent that there were areas in the West that would attain greatest usefulness if maintained in public ownership. Although a considerable area had been set aside for Indians, there were few specific reservations for the public. The realization that resources were limited and that land was being misused resulted in appreciation of the need for conservation.

In 1891, the first forest reserves were set aside by President Harrison. During the administrations of Cleveland, McKinley, and Theodore Roosevelt, further additions were made to this rapidly growing area. Since the time of Theodore Roosevelt, numerous areas have been added to the national forests and some areas have been returned to the public domain. Extensive areas have been acquired by purchase in the eastern half of the United States. Much of this land is grazed by livestock, and it forms an important part of the range resource.

Public lands were reserved for various purposes until, in 1934, there remained less than 200 million acres of unappropriated and unreserved lands, exclusive of Alaska. With the passage of the Taylor Grazing Act in that year the remainder of these lands, with a few exceptions, was withdrawn from entry. Thus culminated a long struggle to secure the adoption of measures that would provide for some form of control over the federal lands.

As early as 1899, efforts had been made in Congress to secure legislation permitting the leasing of public domain to stockmen. Between 1906 and 1920 many other measures with similar objectives were introduced in Congress (Barnes, 1920).

Initially, the Taylor Grazing Act provided for 80 million acres to be included in grazing districts. In 1967 there were over 138 million acres included in organized grazing districts. Federal agencies in 1968 held some 755.4 million acres of which 700 million acres were from the original public domain, the remainder having been regained from private ownership (Public Land Law Review Commission, 1970). The total federal holding is one-third the area of the United States; the percentage varies from 29 to 86 percent of individual Western states, exclusive of Alaska (Table 3.2). Although in the past there have been occasional vigorous movements to convert part of this land into private ownership, there is at present widespread acceptance of the idea that, for the most part, these areas should remain a permanent possession of the federal government. The disposition and present administration of the public lands and the acreages are shown in Table 3.3.

In all phases of land disposal there was evidenced only a striving toward accommodation to existing conditions. There was no forward-looking policy designed to anticipate and meet future situations. Thoughtful planning and scientific outlook resulted only as time brought to light the errors of earlier policies.

Not until 1964, when Congress authorized the creation of the Public

Table 3.2   Percent of Land in 12 Western States under Federal Ownership*

Data from Public Land Law Review Commission (1970)

State	Percentage	State	Percentage	State	Percentage
Alaska	95.3	Idaho	63.9	Oregon	52.2
Arizona	44.6	Montana	29.6	Utah	66.5
California	44.3	Nevada	86.4	Washington	29.4
Colorado	36.3	New Mexico	33.9	Wyoming	48.2

*Federal lands comprise about 33 percent of the total area of the United States.

Land Law Review Commission, was there an attempt to look at public-land laws in their entirety and define a national policy respecting the remaining public lands. The Commission was directed to identify all land laws, examine administrative practices of all federal land-managing agencies, and to make recommendations as to whether public lands

Table 3.3   Disposition and Status of the Federal Lands of the United States

Federal lands	Thousands of acres
Lands, titles of which have passed from the United States, to 1967*	
Homesteads	287,300
Cash sales and miscellaneous disposals	432,200
Grants to railroads corporations	94,300
Grants to states	229,300
Total area disposed	1,043,100
Public lands in the United States to 1968†	
Bureau of Land Management	470,383
Forest Service	186,893
Fish and Wildlife Service	26,559
National Park Service	23,300
Department of the Army	11,400
Bureau of Reclamation	8,686
Department of the Air Force	8,564
Corps of Engineers (Civil Works)	7,148
Bureau of Indian Affairs	4,947
Department of the Navy	3,601
Atomic Energy Commission	2,140
Agricultural Research Service	435
Department of Transportation	167
Department of the Interior	26
Other Agencies	1,119
Total	755,368

*U.S. Department of the Interior (1967).
†Public Land Law Review Commission (1970).

should be "(a) retained and managed or (b) disposed of, all in a manner to provide the maximum benefit for the general public" (78 Stat. 982). Over 2,600 different statutes were identified. The Commission (1970) recommended that:

> The policy of large-scale disposal of public lands reflected by the majority of statutes in force today be revised and that future disposal should be of only those lands that will achieve maximum benefit for the general public in non-Federal ownership, while retaining in Federal ownership those whose values must be preserved so that they may be used and enjoyed by all Americans.

Thus, there is recognition of the likelihood that most if not all the present acreages will remain in public ownership. It is not yet certain what legislation will result from the studies or what policies will finally be enunciated by Congress.

Results of Federal Land Policies

One result of the government's failure to regulate use of the public land was a serious reduction of land values. Over the period from about 1860, when cattle grazing on the public domain became general, until 1934, when the Taylor Grazing Act was passed, little provision was made for administering grazing on public land. Notable exceptions were certain reserves such as the national forests. The result of this general lack of supervision was an intense competition among users of public land to secure as much as possible from them. Since one individual had as much right on the area as another, there was nothing to prevent him from placing stock on a range, even though it might already be supporting all the animals the forage justified. Further, there was always a race to get to the range first and thus to graze the choice forage. This resulted in complete disregard of wise seasonal grazing practices. Such conditions could only result in extensive damage to the ranges.

The land laws also contributed to range deterioration, since the amount of grazing land that a person could acquire legally was insufficient to care for a herd large enough to support a family. This encouraged heavy use of the owned range and made necessary heavy grazing of public ranges in the vicinity. Likewise, permitting land to be taken up for crop production in areas that were unsuited for the purpose led to much damage. In most cases, it was not discovered, or not admitted, that the land was not sufficiently productive for cropping until after several years of cultivation. During that time, the forage was destroyed and the land subjected to erosion, which was sometimes so severe as to interfere seriously with revegetation after abandonment.

Similarly, the method of dispensing lands to the states, the public-school lands particularly, complicated any effort at management. These tracts, scattered as they were throughout the state in 640-acre units (260 ha), were impossible to administer. Had they been economic grazing units, they undoubtedly would have been used more effectively.

The Bureau of Land Management and the Forest Service, likewise, are handicapped by the existence of small tracts of patented land throughout areas that they administer. These areas are owned by individuals or corporations which in many instances have too little land to justify close supervision. Obviously such conditions make any efforts at control of land use difficult (Fig. 3.6).

FUTURE OF RANGE MANAGEMENT

The future of range management in the world depends primarily upon human population growth. If population growth is stabilized, we can expect range management to develop worldwide in much the same form as it exists in America today. There will probably be a shift from consumptive to nonconsumptive uses. Rangelands will be used more for purposes other than the production of meat; factors such as recreation, watershed, and aesthetics will become increasingly important to the range manager.

But the human population probably will not be stabilized in the world community in the near future; consequently, some major changes in range management will take place. In the short run we may see croplands put back to grass. However, if the human population continues to grow, it will be imperative to use all land that can grow crops for the production of human food. Some areas now in rangeland will be converted to crop production. New dry-land techniques will be employed to make many of these areas productive and some will be brought under irrigation. Although the rangeland area will decrease, the majority will, because of physical and economic limitations, remain rangeland.

Animal protein will be produced from those areas unsuitable for crop production. This meat production may not come from the traditional livestock such as sheep and cattle. Differential efficiency of animals may dictate that we raise goats, camels, donkeys, eland, rabbits, or kangaroo rats on the rangelands to achieve maximum protein production. This may in turn call for changes in people's customs, their dietary habits, and their attitudes toward food—for red meat will be a luxury crop.

Food chains will be shortened and more vegetable crops will be eaten. New crops, such as high-lysine corn, will be developed, and world cultures will gradually shift to eating plant products rather than meat.

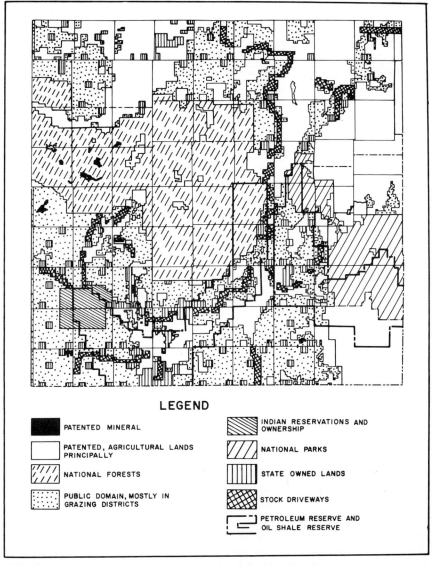

LEGEND

■ PATENTED MINERAL	INDIAN RESERVATIONS AND OWNERSHIP
□ PATENTED, AGRICULTURAL LANDS PRINCIPALLY	NATIONAL PARKS
NATIONAL FORESTS	STATE OWNED LANDS
PUBLIC DOMAIN, MOSTLY IN GRAZING DISTRICTS	STOCK DRIVEWAYS
	PETROLEUM RESERVE AND OIL SHALE RESERVE

Fig. 3.6 The land ownership patterns of an area in southern Utah, show-ing the great complexity of ownership and the minute subdivisions of land which face the administrator of rangeland in the West.

Rangelands will serve increasingly as major recreational and waste-disposal areas (Fig. 3.7). Whether or not human population is control-led, multiple use of rangeland resources for services other than the production of red meat will be a necessity.

Fig. 3.7 Recreation is becoming an increasingly important use of both public and private rangeland. Here a party is enroute to an area once used only as a wintering area for range cattle. The spare tire is being removed to permit passage through the narrow point on the livestock trail. *(Courtesy S. M. Clark.)*

The basic use will still be determined by the ecological factors of the environment. Social, political, and cultural factors will modify ecological constraints. Tradition largely dictates land use whether in Australia, East Africa, or elsewhere. The major difference is that in the developing countries of the world, the attitude of the producer is paramount; in the developed and industrialized nations of the world, the opinions, attitudes, and social customs of the consumer may matter more than those of the people actually on the land.

The social structures of man must and will change if the earth becomes crowded. In the meantime, however, severe misuse of the resource may occur because traditional behavioral patterns are not changed as rapidly as biological considerations would require. The challenge to the range manager, both now and in the future, is to bring a proper balance between the biological realities of the site and the demands of the life-styles of the people he serves.

BIBLIOGRAPHY

Anonymous (1939): America's first pigs landed 400 years ago, *Am. Cattle Producer* **21**:14.

Ardrey, R. (1961): *African Genesis,* Dell, New York.

Barnes, W. C. (1920): The vanishing public domain, *Producer* **2**(6):5–9.

—— (1926): The story of the range (reprinted from Part 6 of the hearings before a subcommittee of the Committee on Public Lands and Surveys), U.S. Senate, 69th Cong., 1st Sess.

Bentley, H. L. (1898): Cattle ranges of the southwest: A history of exhaustion of pasturage and suggestions for its restoration, *U.S. Dept. Agr., Farmers Bull.* **72.**

Biggers, Don Hampton (1901): *History That Will Never Be Repeated,* by Lan Franks (pseud.), High-Grade Printing Office of Biggers', Ennis, Tex.

Box, Thadis W. (1967): Range deterioration in west Texas, *Southwestern Historical Quart.* **9**:37–45.

—— (1971): Nomadism and land use in Somalia, *Econ. Dev. and Cultural Change* **19**:222–228.

—— (1972): Factors affecting the management of the world's rangelands, Proceedings of the 25th Anniversary of the Range Science Department, Texas A & M University, College Station, Tex.

Brown, Leslie H. (1958): Range management in the United States of America, Report to the Government of Kenya, Cyclostyled, 41 pp.

—— (1971): The biology of pastoral man as a factor in conservation, *Biol. Conserv.* **3**:93–100.

Chapline, W. R., Robert S. Campbell, Raymond Price, and George Stewart (1944): The history of western range research, *Agr. History* **18**:127.

Clawson, Marion (1951): *Uncle Sam's Acres,* Dodd, Mead, New York.

Colville, Frederick J. (1898): Forest growth and sheep grazing in the Cascade Mountains of Oregon, *U.S. Dept. Agr. For. Div. Bull.* **15.**

Committee on the Conservation and Administration of the Public Domain (1931): A report, Washington.

Corti, Linneo N. (1970): Range management in the developing countries, *Jour. Range Mgt.* **23**:322–324.

Gates, Paul W. (1968): *History of Public Land Law development,* A Report for the Public Land Law Review Commission, U.S. Government Printing Office, Washington.

Heady, Harold F. (1960): Range management in East Africa, Government Printer, Nairobi, Kenya.

Heathcote, R. L. (1969): Land tenure systems: Past and present, in R. O. Slayter and R. A. Perry (eds.), *Arid Lands of Australia,* Australia National University Press, Canberra, pp. 185–208.

Hibbard, B. H. (1924): *A History of the Public Land Policies,* Macmillan, New York. Reprinted by Peter Smith, New York (1939).

Kennedy, P. Beveridge, and Samuel B. Doten (1901): A preliminary report on
    the summer ranges of western Nevada sheep, *Nev. State Univ. Agri. Expt.*
    Sta. Bull. **51.**

Konczacki, Z. A. (1967): Nomadism and economic development of Somalia, *Can.*
    Jour. of African Studies **2:**162.

Leopold, Aldo (1939): A biotic view of land, *Jour. Forestry* **37:**727–730.

Lewis, I. M. (1961): *A Pastoral Democracy*, Oxford University Press, London.

Mahoney, Frank (1966): Range management in the Somali Republic, in Arthur
    H. Niehoff (ed.), *A Casebook of Social Change*, Aldine, Chicago, pp. 155–164.

Mbiti, John S. (1969): *African Religions and Philosophy*, Heinimann, Nairobi,
    Kenya, p. 289.

McCoy, J. G. (1932): *Historical Sketches of the Cattle Trade* (reprinted), The Rare
    Book Shop, Washington.

Nelson, Aven (1898): The red desert of Wyoming and its forage resources, *U.S.*
    Dept. Agr., Div. of Agrost. Bull. **13.**

Osgood, E. S. (1929): *The Day of the Cattleman*, University of Minnesota Press,
    Minneapolis.

Pearse, C. Kenneth (1971): Grazing in the middle east: past, present, and future,
    Jour. Range Mgt. **24:**13–16.

Perry, Rayden A. (1967): The need for rangelands research in Australia, *Proc.*
    Ecol. Soc. Austr. **2:**1–14.

Public Land Law Review Commission (1969): Land grants to states: A study by
    the staff of PLLRC (multil.).

―――― (1970): One-third of the nation's land, U.S. Government Printing Office,
    Washington.

Ripley, S. Dillon (1966): The challenge of adapting human populations to arid
    environments, *Proceedings of the ICASALS Symposium I*, Texas Technological
    University Press, Lubbock, Tex., pp. 23–33.

Sampson, Arthur W. (1923): *Range and Pasture Management*, Wiley, New York.

―――― (1928): *Livestock Husbandry on Range and Pasture*, Wiley, New York.

Saunderson, Mont H. (1950): *Western Stock Ranching*, University of Minnesota
    Press, Minneapolis.

Senzel, Irving (1950): Brief notes on the public domain, U.S. Department of the
    Interior, Bureau of Land Management, mimeographed.

Seton, E. T. (1927): *Lives of Game Animals*, Vol. III, *Hoofed Animals*, Doubleday,
    New York.

Shear, Cornelius L. (1901): Field work of the Division of Agrostology: A review
    and summary of the work done since the organization of the division, July
    1, 1895, *U.S. Dept. Agr. Bull.* **25.**

Smith, Jared G. (1895): Forage conditions of the prairie region, in *U.S.*
    Department of Agriculture Yearbook, U.S. Government Printing Office,
    Washington, pp. 309–324.

Stewart, George (1936): History of range use, in The Western Range, Senate
    Doc. 199, 74th Cong., 2d Sess., pp. 119–133.

Stoddart, Laurence A. (1950): Range management, in Robert K. Winters (ed.),
    Fifty Years of Forestry in the U.S.A., Society of American Foresters,
    Washington, pp. 113–135.

————, and Arthur D. Smith (1943): *Range Management,* McGraw-Hill, New York.

————, ———— (1955): *Range Management,* 2d ed., McGraw-Hill, New York.

Talbot, M. W., and E. C. Crafts (1936): The lag in research and extension, in The Western Range, Senate Doc. 199, 74th Cong., 2d Sess., pp. 185–192.

U.S. Department of Agriculture, Forest Service (1936): The Western Range, Senate Doc. 199, 74th Cong., 2d Sess.

U.S. Department of the Interior, Bureau of Land Management (1967): Public land statistics, U.S. Government Printing Office, Washington.

Van Hise, C. R. (1910): *The Conservation of Natural Resources in the United States,* Macmillan, New York.

Wellman, P. I. (1939): *The Trampling Herd,* Carrick and Evans, New York.

Wentworth, Edward N. (1948): *America's Sheep Trails,* Iowa State College Press, Ames, Iowa.

Chapter 4

Plant Morphology and Physiology in Relation to Range Management

It is fortunate for humanity that plants are so efficient and versatile. One cannot study plants without being impressed by the precision of their physiological activity, the versatility of their adjustment to new conditions, their adaptation to withstand adversities, and the tenacity with which they cling to life, struggling against unfavorable climate, invading and competing plants, and animals that consume their foliage. The plant is a truly wonderful machine, and given an opportunity, it serves as an unsurpassed and untiring benefactor to humanity.

Plants capture and convert the sun's energy into food substances for themselves and animals by the process of photosynthesis. Alone in nature they *produce* food. This food provides the energy for maintaining the life processes of the plant, provides structural materials for its growth, and supports all animal life either directly or indirectly. In addition, plants return oxygen to the atmosphere while removing carbon dioxide, thus negating the consumption of oxygen and release of carbon dioxide by animals and contributing to the maintenance of an oxygen/carbon-dioxide ratio in the atmosphere acceptable to animal life. Were plants removed from the earth, animal life could not exist.

Cropping of vegetation by animals upsets normal plant functions. The reduction of leaf area lessens the capacity of the plant to manufacture foods. Food storage is diminished and the sequence of growth stages disrupted. Fortunately, plants can withstand grazing up to a point without being adversely affected. Beyond this point, which differs among species with the stage of growth and the rigors of the habitat, plants may be weakened and even killed. The physiological processes by which the individual plant grows and the morphological changes that take place during growth are thus basic to range management. A plant's survival depends upon its ability to: (1) synthesize and store food for maintaining plant functions, (2) form vegetative structures for renewal of top growth, (3) maintain a healthy root system, and (4) produce reproductive organs. These are all interrelated and depend upon there being sufficient foliar tissue to synthesize the energy and plant structures that are required.

FOOD SYNTHESIS BY PLANTS

Food synthesis takes place primarily in the leaves of green plants where chlorophyll-bearing cells, in the presence of sunlight, combine the carbon from the air with water and mineral nutrients from the soil to form carbohydrates. The precise manner in which this takes place and the nature of the intermediate and final products vary among plants.

Most grasses fall into one of two groups with respect to the kind of nonstructural carbohydrate present in vegetative parts—those in which starches predominate and those in which fructosans predominate. The former are of tropical or subtropical origin; the latter are of temperate origin. Some contain neither starch nor fructosan in any quantity (Table 4.1). Starch is the most abundant carbohydrate found in trees (Kramer and Kozlowski, 1960).

Factors Influencing Synthesis

The manufacturing operations of the plant depend upon many factors, including (1) physiological efficiency of the plant, (2) amount of carbon dioxide in the air and the freedom with which it enters the leaf, (3) area of leaf surface, (4) intensity and quality of light, (5) water supply, (6) temperature, and (7) soil nutrients. The most important of these are water and leaf surface.

The water available to a plant determines photosynthetic activity in two ways: it is a necessary chemical constituent for the reaction; and it

Table 4.1   Grass Genera Grouped According to the Most
Abundant Kind of Nonstructural Carbohydrate Stored in
Stem Bases

Taken from Smith (1972) and Ojima and Isawa (1968)

Fructosans	Sucrose	Starches
Agropyron	Avena	Andropogon
Agrostis	Sorghum	Bouteloua
Alopecurus	Zea	Buchloë
Arrhenatherum		Cynodon
Bromus		Distichlis
Calamagrostis		Echinochloa
Dactylis		Eragrostis
Elymus		Leptoloma
Festuca		Muhlenbergia
Hordeum		Oryza
Lolium		Oryzopsis
Phalaris		Panicum
Phleum		Paspalum
Poa		Phragmites
Triticum		Sorghastrum
		Spartina
		Sporobolus
		Stipa

serves to keep the plant turgid and the stomata open and functioning thus permitting ready entry of carbon dioxide. Reduction in the rate of photosynthesis to 50 or even 25 percent of normal, and virtually complete cessation results from continued drought. In addition, water carries mineral nutrients taken from the soil.

The area of leaf surface exposed, other factors being equal, determines the rate of food manufacture. The average rate of photosynthesis is between 0.8 and 1.8 gm of sugar per hour for each square meter of leaf, depending upon the species (Miller, 1938). Thus the more foliage, the more food manufactured. Leaf area is the factor of most practical significance to the range manager, since it is the one most subject to alteration.

Except for soil nutrients, which may be manipulated by means of fertilization, the remainder are little subject to modification at present. Plant breeding, however, offers possibilities of developing more efficient plants as well as plants with wider ranges of tolerance to stress.

Uses of Products of Photosynthesis by Plants

Foods manufactured by the plant are used for *growth, respiration* (energy for physiological processes), or are *stored* for later use. The percentages of manufactured foods required by functions in an apple tree are 57, 38, and 5 percent for growth, respiration, and storage, respectively (Kramer and Kozlowski, 1960).

Food storage takes place in the roots and stem bases of perennial herbaceous plants, in the stems and roots of woody plants, and in the seeds of annual plants. These food reserves are used (1) in respiration during dormancy, (2) in growth following winter or drought dormancy, and (3) in reproduction. These functions are very important to the welfare of the plant, and it is essential that range plants be allowed to manufacture and store sufficient food during the period of active growth to fulfill these needs.

Carbohydrate Storage Cycle in Grasses   After winter or drought dormancy, plants grow at a rapid rate when temperatures and soil moisture are favorable. Such growth draws heavily upon stored food reserves in roots and stem bases which exist as nonstructural carbohydrates—sugars and starches as opposed to cellulose and like materials. Although roots are commonly thought of as being the principal place for storage of carbohydrates, White (1973) concluded that in grasses the stem bases and modified underground stems were the most important sinks. However, Coyne and Cook (1970) report almost identical carbohydrate concentrations in the roots and crowns of *Sitanion hystrix*. It seems probable that the lack of uniformity in partitioning plants into parts and analytical differences may give rise to apparent contradictions. Further, reserves usually are expressed on a percentage basis and consequently are affected by the amount of structural material in the particular plant part being analyzed. Moisture stress, too, ever an important factor on arid rangelands, may affect both the amount and the form of carbohydrates in plants (Dina and Klikoff, 1973). Although the exact area of storage varies among plants, heavy grazing tends to reduce reserves and slow growth during the following season.

Despite these irregularities, the general pattern of food storage and use can be stated. Coincident with the beginning of growth at the end of the dormant season, there is a decline in stored carbohydrates. This occurs in spring in cool climate and with the onset of the rainy season in warm climates (Fig. 4.1). The length of time during which stored foods are being depleted with the onset of growth may be as little as a few days in grasses (White, 1973) or as much as months in some desert shrubs (Coyne and Cook, 1970). The period ends when food manufacture by the newly formed leaves exceeds the needs for metabolism and growth. During the initial growth "draw down," reductions in stored carbohydrates up to 75 percent have been reported (McCarty and Price, 1942), though others have found less marked reductions.

As vegetative growth proceeds, there is a gradual replenishment of

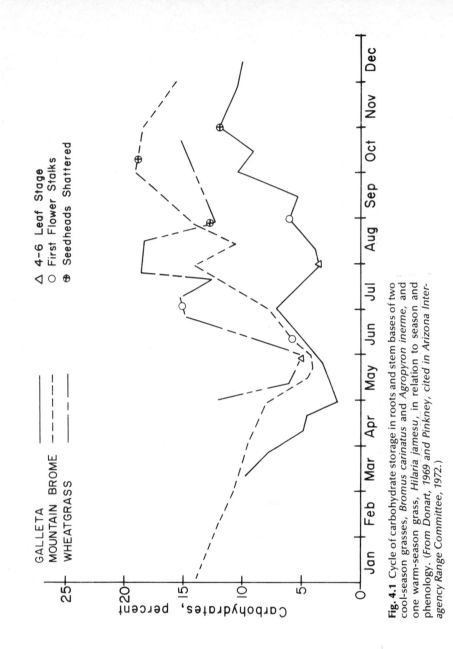

Fig. 4.1 Cycle of carbohydrate storage in roots and stem bases of two cool-season grasses, *Bromus carinatus* and *Agropyron inerme*, and one warm-season grass, *Hilaria jamesu*, in relation to season and phenology. (*From Donart, 1969 and Pinkney, cited in Arizona Inter-agency Range Committee, 1972.*)

stored carbohydrates even though a considerable amount is required for metabolism and growth. Reduced carbohydrates are found again at the time of flowering and seed maturity, although species differ in the amount of this reduction. Reserve carbohydrate reductions at these periods generally are supposed to be due to high food requirements of flowers and fruits, especially the latter. Hyder and Sneva (1959), however, reported similar reductions in headed and unclipped crested wheatgrass (*Agropyron desertorum*) plants. Following seed production, there is a period during which the food reserves rise. This increase may occur even after the leaves are seemingly inactive. If fall regrowth occurs, it may again cause a reduction in the carbohydrate reserves. Once active growth ceases, food reserves gradually decline through the dormant period, being used in respiration. That which remains at the end of the dormant season provides the material with which growth begins. Heavy grazing, by limiting food accumulation, thus affects growth and subsequent plant production.

Carbohydrate-Storage Cycle in Shrubs   Carbohydrate storage in shrubs has been studied less than in grasses. Shrubs are more diverse than grasses, varying greatly among species with respect to phenology, woodiness, and permanence of aerial portions, and retention of leaves. These differences can be expected to give rise to different food-storage patterns. Evergreen trees, for example, have different carbohydrate cycles than do deciduous trees. Food storage in evergreens is more erratic and lacks the marked reductions which accompany the beginning of growth in the spring characteristic of deciduous species (Kramer and Kozlowski, 1960).

It is likely that deciduous and evergreen shrubs have different carbohydrate storage, patterns, as do trees, although this has not been well documented. As compared to grasses, shrubs have more places for storage—i.e., roots, root crowns, and stems. This multiplicity of storage structures greatly complicates determinations of available carbohydrate reserves.

Moreover, neither relative nor total nonstructural carbohydrate concentrations follow the same seasonal pattern among these different parts of the shrubs. Generally, percentage of carbohydrate concentrations in the roots of shrubs are greater than in the stems (McConnell and Garrison, 1966), though this is apparently not true for big sagebrush (*Artemisia tridentata*) and, in certain seasons, for winterfat (*Eurotia lanata*) (Coyne and Cook, 1970). This may not mean, however, that there are more reserve foods present there. For example, despite the fact that higher percentages of carbohydrates are found in the roots of trees, total

amounts of carbohydrates in the stems are greater than in the roots due
to the greater volume of top material. Relative amounts of reserve carbohydrates in tops and roots of shrubs may depend upon the season,
greater quantities being present in roots of bitterbrush (*Purshia
tridentata*) except in midsummer (Fig. 4.2).

Perhaps more importantly, the primary storage structures for
shrubs have not been fully identified, whether in the stem or roots. A
priori, it seems reasonable to expect that the reserves nearest to the place
of use would be most readily available and important. For shoot growth,
these would be most recently produced twigs; for roots, those in the
roots or root crown. Further, it is important to know the secondary
sources: stem material, root crowns, or roots. As with one's personal
finances, knowledge of the total amount of bank deposits may be less
useful than knowledge of where these are and how readily they can be
drawn upon when the need arises.

Different phenological sequences are exhibited by different shrubs

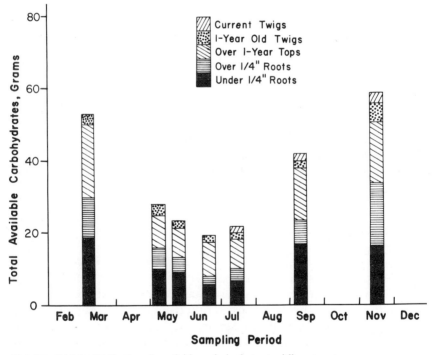

Fig. 4.2 Weight distribution of available carbohydrates in different parts
of plants of *Purshia tridentata* throughout the year. Note that these are
actual amounts, not concentrations per unit weight. *(From McConnell and
Garrison, 1966.)*

which affect carbohydrate-storage cycles. In bitterbrush, flowering and fruiting occur early in the season before twig growth, which continues throughout the summer. In big sagebrush, seed formation is the last event in annual growth. One would expect that in shrubs, as in fruit trees (Finch, 1935), fruit formation would be accompanied by marked reductions in stored carbohydrates at whatever time it occurred. Because of these many factors, it is more difficult to generalize reserve-storage patterns in shrubs than in grasses. Percentage of nonstructural carbohydrates in the stems and roots of bitterbrush followed similar seasonal trends, both continuing downward until after seed formation. In contrast, inverse trends were observed in tops and roots of sagebrush (Fig. 4.3).

Effects of Forage Removal on Plant Carbohydrates

When actively growing grasses are clipped severely, there is an immediate reduction in nonstructural carbohydrates both in the remaining stubble and in the roots which, in the case of young, potted orchardgrass (*Dactylis glomerata*) plants, continued for 4 to 6 days (Davidson and Milthorpe, 1966a). However, this reduction did not take place in plants in which carbohydrate contents were under 2 percent. Sprague and Sullivan (1950) similarly reported that large reductions in nonstructural carbohydrates accompanied high concentrations at the time of cutting; smaller reductions were observed in plants having low initial concentrations. Reductions continued for 11 days after cutting in *Lolium perenne* (Sullivan and Sprague, 1943). Dramatic reductions were found in *Phleum pratense* cut for hay at early headings—from 40 down to 10 percent. A second cutting later in the season produced a smaller reduction (Fig. 4.4).

The intensity of clipping, expressed both in terms of stubble height and frequency, affects carbohydrate accumulation in the roots of previously ungrazed grasses. There was little difference in carbohydrate percentages between unclipped plants and those clipped at 3-week intervals to stubble height of 5 cm or above. Two years of clipping more closely at 2-week intervals resulted in reductions of 21, 24, and 26 percent in short grasses—blue grama (*Bouteloua gracilis*) and buffalo grass (*Buchloë dactyloides*)—western wheatgrass (*Agropyron smithii*) and big bluestem (*Andropogon gerardii*), respectively (Kinsinger and Hopkins, 1961).

The time of forage removal greatly influences the ability of the plant to recover afterward, and to achieve normal carbohydrate storage. Autumn carbohydrate concentrations in roots of crested wheatgrass were affected little when clipping took place in May; substantially lower

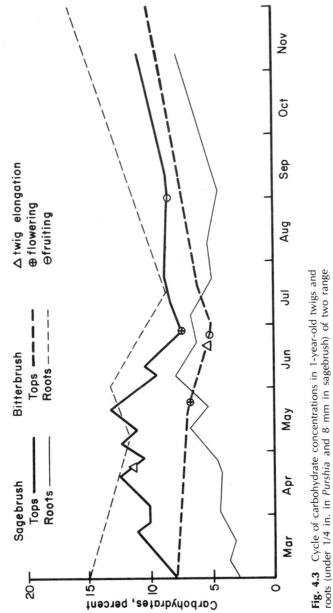

Fig. 4.3 Cycle of carbohydrate concentrations in 1-year-old twigs and roots (under 1/4 in. in *Purshia* and 8 mm in sagebrush) of two range shrubs, *Purshia tridentata* and *Artemisia tridentata*. Note reverse patterns of concentrations in roots and twigs in the two species. (*Data from Coyne and Cook, 1970; and McConnell and Garrison, 1966.*)

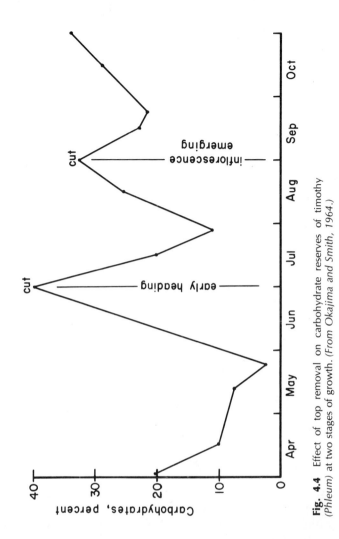

Fig. 4.4 Effect of top removal on carbohydrate reserves of timothy (*Phleum*) at two stages of growth. (*From Okajima and Smith, 1964.*)

root carbohydrates were found in plants clipped in June (Hyder and Sneva, 1963a).

McCarty and Price (1942) identified two critical periods in the growth of forage grasses to be (1) the period of active reproduction, from flower-stalk formation to and including seed ripening, and (2) the forepart of the normal carbohydrate-storage period. These may not be universally critical for all species judging from the results shown in Fig. 4.5.

Critical Carbohydrate-Reserve Levels   Although it is known that adequate food reserves must be present in plants to supply energy needs and to replace foliage removed by grazing, critical levels have not been established. Because of analytical obstacles, plant-food reserves are reported in concentrations (mg/gram or percent) in particular storage structures, rather than in total amounts even though the misnomer TAC (total available carbohydrates) often is applied to them. Comparatively low concentration in a large robust plant may better promote growth than high concentrations in a small plant. Ogden and Loomis (1972) found almost the same percentage of carbohydrates in roots of clipped *Agropyron intermedium* as in those unclipped, but less total carbohydrate in roots of clipped plants. When the carbohydrate reserves became as low as 1 percent, plants did not recover from etiolation. It is uncertain whether this provides a guide to minimum carbohydrate-reserve levels which plants can safely tolerate, for the role of carbohydrate reserves in replenishing lost foliage and the source of these reserves is not clearly established.

Davidson and Milthorpe (1966a; 1966b) attributed only a small part of the increased foliage production in the first few days following clipping of orchardgrass to reserve carbohydrates; the greater part they attributed to photosynthesis. Since carbohydrate percentages dropped both in the stubble and in the roots, except where carbohydrate concentrations were low, both stubble and root reserves are involved in foliage replacement. In a seeming contradiction to this, Sullivan and Sprague (1953) report the losses of stored carbohydrates to be considerably greater than the weight of foliar material that was produced in a 2-to-4 day period following cutting. They attributed the difference to respiration losses. Further, they found no relationship between the carbohydrate contents of the roots and stubble and the amount of aftermath subsequently produced. Thus, the carbohydrate content of plants, alone, does not well explain, at present, the observed effects of use on plant production. When the effects of moisture stress, temperatures, soil nutrients, and the physiological efficiencies of individual species are better under-

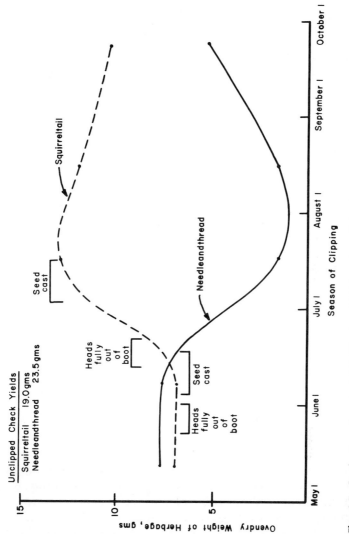

Fig. 4.5 Production of squirrel tail (*Sitanion histrix*) and needle-and-thread (*Stipa comata*) in the year following clipping at different times during the growing season. (*From Wright, 1967.*)

stood, carbohydrate-reserve data may become more useful in identifying grazing stress.

Physiology of the Plant in Relation to Grazing

Husbandmen long ago learned that animals perform inefficiently unless they are properly cared for and fed. This concept has, however, less commonly been extended to plant life. If appreciation of plants as growing organisms with biological limits had been widespread, much injury to rangelands might have been avoided. Plants are living organisms —fundamentally not unlike animals—converting stored food to energy to carry on their life functions by respiration. If too great demands are made upon them, premature death occurs. They require definite conditions for their proper development. Excessive grazing use or mechanical injury which prevents normal development, may result in decline of range vegetation.

Although there have been a great many studies devoted to the physiological processes of plants and the effects of grazing on these processes, there is much more to be known in order to define levels of use that do not seriously impair plant vigor. This is so because (1) species differ in their tolerance to use; (2) variations in environmental conditions, rainfall in particular, from year to year and from place to place affect tolerance levels; (3) simulated grazing by clipping does not well duplicate the effects of grazing; (4) morphological differences in plants make for differences in their vulnerability to any level of use; and (5) season of use differentially affects the plant's ability to respond.

MORPHOLOGY OF PLANTS AND GRAZING

Morphological as well as physiological characteristics determine the ability of plants to tolerate removal of foliage. Plants of different life forms have greater or lesser vulnerability to grazing. Even within life forms, specific differences in the number and position of perennating buds and growth sequences make plants differentially susceptible to injury from cropping. The location of meristematic tissues is important; greater impacts occur where active tissue can be readily removed. Shrubs can be eaten back into previous years' growth, but with herbs, only the removal of the current year's production is possible. However, under repeated use shrubs may become so compacted that animals can obtain forage from them only with difficulty, an effective means of protecting buds against removal. Plants with tough fibrous stems survive grazing better than do those with soft weak stems. Protective structures, such as thorns and spines limit use and promote survival.

Buds

In the established plant, growth following dormancy begins with the activation of buds. In grasses, and other herbaceous plants, these are found beneath the ground surface at the base of the stems and each, when activated by favorable ambient conditions, gives rise to a tiller. The particular combination of conditions required for activation differs among species. Cool-season grasses respond to warm days and cool nights and require vernalization; warm-season grasses respond to warm nights and do not require vernalization.

In shrubs, the buds are located at the tips of the branches and in the leaf axils of young stems. The latter do not commonly grow vegetatively so long as the apical bud is intact since growth is suppressed by hormones present in the apical bud, a process called *apical dominance*. Some grasses also have axillary buds which are released when the tops are removed. In cottontop (*Trichachne californica*), these axillary buds produce auxiliary stems in summer with or without top removal but are stimulated by it (Cable, 1971). It is the basal buds, however, that are most important in plant production and growth. Grasses exhibit great morphological differences among species with respect to the position and vulnerability of the terminal buds to grazing; in shrubs they are exposed to cropping at all times. These have serious consequences for the individual plant.

Growth Characteristics of Grasses

The beginning of growth in grasses is marked by the appearance of the outer (or lower) leaves, others developing successively inward. Three kinds of stems may be produced: (1) a vegetative stem without a culm, (2) a culmed stem which produces no inflorescence, and (3) a reproductive stem complete with inflorescence. Vegetative types may make up 25 to 90 percent of the stems present, depending upon the species and environmental conditions. Stems react quite differently to cropping due to two factors—position of the apical bud and duration of meristematic activity and foliar enlargement.

Individual leaves continue to enlarge until the ligule is formed (Sharman, 1947). If the leaf is cropped after ligule formation, no additional leaf growth is made. Growth continues if cropping takes place at earlier stages if the meristematic tissue at the base of the leaf blade and sheath remain intact. Vegetative stems can thus continue growth from the inner and later developing portions of the stems (Fig. 4.6).

Floral stems, for a short time after growth begins, behave like vegetative stems sending up shoots from the lowermost nodes. During this

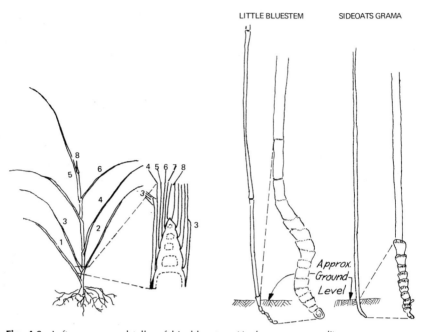

Fig. 4.6 Left, young seedstalks of big bluestem *(Andropogon gerardi)* with enlarged cross-section showing growing point with enclosing sheaths. Grazing at the seven-leaf stage does not remove the meristematic tissue and growth can continue. Right, seedstalks of little bluestem *(A. scoparius)* and sideoats grama *(Bouteloua curtipendula)* with leaves removed to show short basal nodes. Neither of these grasses elevate the growing point until well into the grazing season. *(From Rechenthin, 1956.)*

period they can be cropped without arresting growth. Individual grasses vary greatly with respect to the number of these short basal internodes, from 2 to 15, which remain belowground. Elongation of the stem takes place when upper internodes elongate although lower internodes may elongate slightly (Rechenthin, 1956). Once extension of the stem begins, the apical bud is elevated rapidly and becomes susceptible to removal (Branson, 1953). If the apical bud is removed, normal development of the stem is arrested (Cook and Stoddart, 1953). Further growth must come from axillary buds of the basal, unelongated parts of the stem or, in some grasses, from buds higher up on the stem. Cable (1971) found axillary buds in all but the internode immediately below the one supporting the inflorescence in cottontop, the uppermost having greater numbers than those lower down. In *Agropyron repens,* axillary buds produced vegetative stems only in the five lower leaf axils, and not in those above (Sharman, 1947).

These differences in growth and bud characteristics are of great

importance for the range manager. Grasses which have a high proportion of vegetative stems, those that delay the elevation of the apical bud,
and those that sprout freely from axillary buds are most resistant to
grazing and most productive under use. It is these characteristics that
influence differential responses among species such as the reduction of
midgrasses and the increase in low, turf-forming grasses on the short-
grass plains.

Grazing at some seasons may markedly affect subsequent tiller production. The magnitude of the effect varies with the stage of growth of
the plant; it varies among species as well, as shown by 3 years of clipping
on three prairie grasses. The number of tillers increased when clipping
took place at the flowering stage, was little affected in early vegetative
stages, and was reduced drastically by midseason clipping (Table 4.2).

McKendrick and Sharp (1970) compared tiller production of
crested wheatgrass that had been protected for 12 years and 1 year to
plants protected only in the year data were collected. There were 58
percent more tillers produced by the ungrazed plants than those that
had a prior history of grazing. Even 1 year protection resulted in 32
percent more tillers.

Growth in Forbs and Shrubs

In forbs and shrubs the meristematic tissue where growth occurs is at the
end of the stems and branches. Removal of the terminals halts growth,
and further growth can occur only through activation of lateral buds on
the branches below the point of cropping (most common with shrubs) or
from dormant buds at the base of the plant (most common with herbs).
In either case, growth lapses when cropping takes place, for a short time
if in the early growing season, or until the following year if cropping
takes place near the time of maturity. In shrubs, flower and seed formation are inhibited by heavy grazing, and production consists largely of
shoot growth (Fig. 4.7).

Other Factors Affecting Resistance to Grazing

Strength of grass stems also determines the severity of the effects of
grazing. Certain grasses, such as *Stipa lettermani,* have stout stems and
leaf bases and, when grazed, break at or near the point of cropping. By
contrast, *Festuca idahoensis* has weak stem bases which break at or below
the ground surface where regrowth must occur; hence it is more vulnerable to grazing. Consequently, Idaho fescue is a decreaser and needlegrass an increaser where they occur together. Hyder and Sneva (1963b)
observed similar vulnerability to grazing in *Agropyron inerme* in Oregon.

Table 4.2 Number of Tillers and Reproductive Culms of three Prairie-Grass Plants in 1963 in Percent of Number in 1961

Data from Vogel and Bjugstad (1968)

Time of clipping	Growth Stage	Little bluestem	Big bluestem	Indiangrass
Total tillers in May			(percent)	
May	Vegetative growth	94	111	95
June	Vegetative growth completed	55	96	63
July	Early reproductive	36	81	36
August	Flowering	80	75	73
October	Seed ripened	94	133	97
November	Plants dormant	112	154	142
	Unclipped	68	102	84
Reproductive culms at termination of vegetative growth				
May	Vegetative growth	40	60	43
June	Vegetative growth completed	0	50	20
July	Early reproductive	5	9	*
August	Flowering	33	27	15
October	Seed ripened	100	120	27
November	Plants dormant	80	64	91
	Unclipped	45	71	43

*Reproductive culms not identified in 1961.

Persistence of leaves on shrubs may also be a factor in tolerance to grazing. Big sagebrush, which retains its foliage during the winter, is particularly susceptible to heavy grazing, although factors other than leaf persistence may be involved also.

IMPORTANCE OF ROOTS TO PLANT SURVIVAL AND GROWTH

Roots, being underground, are not easily studied; yet on the range, where water is vital, there is no part of the plant more important. A knowledge of the anatomy and physiology of the root is elementary to the manager of the range, for its characteristics determine to a great extent the ability of a plant to thrive.

The underground parts of plants are layered much as are the aboveground parts, some plants rooting shallowly in the interstices between deeper-rooted plants and others possessing very deep taproots, which penetrate below the general level and form extensive networks there. Competition among plants is largely dependent upon rooting habits. Plants rooting in the same zone compete more than do those in different zones, though surface moisture may be intercepted by shallow-rooted plants to the exclusion of deeper-rooted plants. In the short-grass plains of the United States, the surface moisture, which comes largely during the growing season, is absorbed by the shallow-rooted grasses, which can easily dominate over deeply rooted forbs and shrubs. Conversely, in areas where heavy snow results in deep percolation, the surface-rooting grasses do not compete so effectively with the deep taproots of shrubs and forbs.

The root system is important in winter survival. Plants with deep roots are injured less than are shallow-rooted plants during cold weather, for they have access to moisture in the unfrozen subsoils and do not become desiccated (Weaver, 1930). Where precipitation falls mostly in the growing season, roots which spread widely at shallow depths are most efficient.

Size of Root System

The root is the medium whereby the plant makes contact with the soil, as the leaf is the medium whereby the plant makes contact with the air. Under drought conditions the demand for water in the leaves may ex-

Fig. 4.7 Cliffrose *(Cowania stansburiana)* with mass of flowers in the top above the browsing line and few flowers below. Browsed shrubs produce twig growth and few flowers; unbrowsed shrubs produce flowers and little twig growth. *(Photograph by C. H. Jensen.)*

ceed the capacity of the roots to provide it; then wilting and possibly death result. The rapidity of water loss and absorption is proportionate to the size of the transpiring and absorbing organs. Most drought-resistant plants have large and deep root systems and small aboveground parts, though there are many exceptions. Cacti, for example, depend not only upon ability to secure water quickly through their mat of roots near the surface, but also upon ability to hold water against the demands of the atmosphere.

To one unfamiliar with root characteristics, the magnitude of root systems comes as a surprise. Pavlychenko (1937) found that the total root length of a 3-year-old plant of crested wheatgrass was over 600 km, bromegrass was 106 km, and slender wheatgrass was 16 km.

Dittmer (1937) has shown that a single winter rye plant supported over 13 million distinct root members with a total surface of 237 sq m and a combined length of 623 km. A root-hair surface of 401 sq m was noted. The combined surface of roots and root hairs was found to be 130 times that of the aboveground parts.

Length of Root System

Measurements of root length or depth are seldom made because of the time and careful work required. Spence (1937) characterized fibrous-rooted annualgrasses as having few and short roots; cheatgrass (*Bromus tectorum*) with only seven major roots penetrated only about 30 cm. However, Harris (1967) found roots of this grass to a depth of 110 cm, a much deeper root penetration than reported by Spence. Short-lived perennials had more abundant but, nevertheless, short roots. *Poa secunda,* as an example of the latter type, produced many roots, but they penetrated only to 40 cm. Long-lived plants such as *Agropyron inerme* and *Carex geyerii,* conversely, produced more than 200 major roots and penetrated about 160 cm (Spence, 1937).

Root-distribution studies made of 43 representative prairie species showed that 14 percent absorbed almost entirely in the surface 60 cm, 21 percent between 60 and 150 cm, and 65 percent below 150 cm. Depths of 2½ or 3½ m are common, and maximums of over 6 m have been recorded (Shively and Weaver, 1939). Cannon (1911) reported a mesquite (*Prosopis*) root in southern Arizona that penetrated to 8 m and supported lateral roots 15 m in length. Not all desert plants have extensive roots, indeed, many of the succulents are unusually shallow-rooted. Certain species of cactus studied by Cannon did not root below 2 cm, though they had widely spreading root systems; and even the saguaro (*Carnegiea gigantea*) rooted to only 77 cm.

Longevity of Roots

Studies on length of life of grass roots (Weaver and Zink, 1946) indicate that relatively rapid replacement is usual; hence organic matter constantly is added to the soil. No roots of *Elymus canadensis* lived over 3 years, and in other species, 10 percent (*Andropogon scoparius*) to 45 percent (*Bouteloua gracilis*) survived 3 years. Kucera et al. (1967) estimated that the roots in tall-grass prairie were replaced at about 4 years. The significance of this for the range manager is that roots, except the large ones of shrubs, are not permanent fixtures, but must be replaced regularly. They require plant foods both for growth and respiration which must come from products of photosynthesis. Anything that affects photosynthesis thus affects the roots.

Effect of the Top Removal on Root Production and Growth

We have already seen that root carbohydrates are affected by top removal. A possibly more significant effect is that on root growth. By reducing the photosynthetic area through grazing, food manufacture is reduced, and there is a consequent reduction of the material available to the roots. The reaction of roots to clipping is both immediate and marked, as shown by the work of Parker and Sampson (1931) in which a single harvesting of foliage resulted in temporary cessation of root growth in grasses and was followed by immediate transfer of growth to the tops. So sensitive were the plants that a single harvesting at any time during a 120-day test resulted in a decrease in the root yield, the greatest decrease occurring during the periods of most rapid growth. Cessation of root growth was observed in orchardgrass immediately upon clipping of the tops (Davidson and Milthorpe, 1966a), although Crider (1955) reports the same species unaffected by the earliest of three clippings. Root growth of other species, however, was invariably halted by clipping for as long as 18 days in some species (Table 4.3). Growth stoppage was observed within 1 to 4 days after top removal.

Robertson (1933) noted reductions in the length of roots of grass seedlings following frequent clipping, to one-fifth of normal, an almost insurmountable handicap to a plant in an arid region. Other studies have shown that individual roots may be small and the production of rhizomes inhibited through heavy clippings (Biswell and Weaver, 1933). The height at which harvesting takes place influences root production (Table 4.4). More frequent clipping also reduces production (Table 4.5).

Black grama (*Bouteloua eriopoda*), blue grama, and galletagrass (*Hilaria jamesii*) excavated from properly grazed, overgrazed, and heav-

Table 4.3   Effect of Top Removal at Different Times in the Growing Season on Cessation of Root Growth and Root Production

Data from Crider (1955)

Species	Clipping		Root-growth stoppage		
	Date cut	Height cut, in.	Start, day	Dura- tion, days	Total days
Cool season					
Smooth brome	Apr. 28	2 1/2	2d	12	—
	May 20	2 3/4	1st	17	—
	July 7	3	1st	8	37
Tall fescue	Apr. 28	2 1/2	1st	13	—
	May 20	2 3/4	1st	12	—
	July 7	3	1st	7	32
Orchardgrass	Apr. 28	2 1/2	none	none	—
	May 20	2 3/4	3d	18	—
	July 7	3	4th	7	25
Warm season					
Florida paspalum	July 15	2	2d	11	—
	July 29	2	1st	6	—
	Aug. 4	2	1st	10	—
	Sept. 9	2	2d	18	45
King-Ranch bluestem	July 22	2	1st	8	—
	Aug. 12	2	2d	7	—
	Sept. 9	2	1st	18	33
Switchgrass	July 22	2	1st	18	—
	Aug. 30	2	1st	11	29
Blue grama	July 29	2	1st	17	—
	Aug. 23	2	2d	13	30
Bermuda grass	May 20	1	1st	9	—
	June 21	1	1st	6	—
	Sept. 28	1	2d	16	31

ily overgrazed ranges had root systems reaching depths of over 1.2 m, slightly over 0.6 m, and less than 0.3 m, respectively (Flory and Trussell, 1938). The short-root systems resulting from overgrazing supported only scattered clumps of grass, and there was not sufficient water-storage capacity within the absorbing zone of the restricted root system to maintain the grass through unprotected drought periods.

Comparable conditions have been observed in the case of *Agropyron spicatum;* the roots on overused ranges penetrated only 44 cm, whereas on protected ranges they penetrated over 65 cm (Hanson and Stoddart,

Table 4.4   Effect of Height of Clipping upon Weight of
Roots: Plants Clipped 16 Times at 5-Day Intervals

Data from Harrison (1931)

Height of cutting, in.	Root weights, gm		
	Red fescue	Blue-grass	Colonial bent
1/4	1.4	1.6	2.1
1 1/2	8.6	7.5	5.0
3	13.7	11.7	7.7

1940). Despite the effects on individual species, due to changes in plant composition, the root mass may not be reduced by heavy grazing even though there is a reduction in carbohydrate percentages in the roots (Table 4.6).

REPRODUCTION OF PLANTS

Since the life of the individual plant is not limitless, plants must be allowed to reproduce. Under normal conditions, native plants are lavish in their production of reproducing organs. Were all the seeds and specialized stems and roots of a single year to produce a new individual, there would be no space in which they could grow. Most plants reproduce by seed, but many range plants depend upon vegetative reproduction from specialized organs developing from the stem or the root. Vegetative reproduction is especially valuable to a plant when heavy grazing or other causes restrict seed formation. Among the reproducing

Table 4.5   Dry-Matter Weight Distribution of Roots of Russian
Wild Ryegrass following Four Different Clipping Frequencies

Data from Thaine (1954)

Number of clippings	0-6 in.	6-12 in.	12-18 in.	Total
	(Lbs/acre)			
1	1,885	783	668	3,336
2	1,734	724	462	2,920
3	1,528	688	513	2,729
5	995	425	349	1,769

Table 4.6   Weight and Water-Soluble Carbohydrate Content of
Belowground Plant Parts of Areas Grazed at Various Rates for 19
Years, Manyberries, Alberta

Data from Smoliak et al. (1972)

Measurement and depth	Grazing treatments			
	Ungrazed	Light	Moderate	Heavy
Belowground plant biomass (kg/ha)				
0-15 cm	14,955	16,163	18,946	24,038
15-30 cm	2,681	2,663	2,413	2,529
30-45 cm	2,120	2,495	1,877	2,299
45-60 cm	1,330	861	457	599
Total	21,086	22,182	23,693	29,465
Water-soluble carbohydrate content of roots (mg/g)				
0-15 cm	22.0	21.4	17.6	14.8

methods exhibited by range plants are seed, rhizome, stolon, tiller, layering, corm, and bulb.

Seeds

Seeding is the most usual method of reproduction among the higher plants and the sole method of many perennials and all annuals. Seed reproduction has some disadvantages. The seedling is dependent upon the rapid development of its own roots to supply moisture from the beginning of its active life. In drought years, the seed may fail to germinate or the young plant may be killed by drought before it can send its roots down to the moist subsoils. A seedling starting growth at the onset of the growing season must send its roots into the ground with sufficient speed to keep below the ever-deepening layer of dry soil which often reaches depths of 60 to 90 cm. The ability of the seedling to develop quickly a deep-root system determines its success, and often makes annuals like cheatgrass serious competitors to young perennial plants (Harris, 1967). Fortunately, root development in grass seedlings proceeds more rapidly than top growth (Whalley et al., 1966).

Seeds provide an efficient means of species survival over drought periods, as in the case of annual plants in the hot and dry summer climates. Annual plants spring to life when the rainy season arrives, rapidly mature seed, dry up, and die. The seed maintains its spark of life for many years. Seeds of native plants are known to remain viable for a period of 5 to 10 years or even longer; viability of seeds with hard seed coats after as long as 200 years has been observed (Weaver and Clements, 1938).

Rhizomes

A rhizome is an underground stem, as is evidenced by the presence of nodes and leaf scales. Rhizomes often spread many feet just under the surface of the soil, and young plants arise, sometimes from each node. Some vigorously rhizomatous plants may produce rhizomes in such abundance as to form a dense mat just below the soil surface and produce little foliage, a condition called *sod bound.*

The rhizome is valuable to range plants because it is below ground and, thus, seldom subjected to mechanical injury. Reproduction is possible, though physiologically restricted, under conditions of heavy overgrazing. Rhizomes, having a greater food reserve at their disposal than do seeds, are able to produce new plants from great depths such as occur when blowing soil is deposited upon them. Rhizome reproduction is reliable in dry years, for the already established root system of the mother plant provides ample moisture giving rhizomatous plants a competitive advantage over seedlings. The rhizome has a disadvantage in that it can form new plants only close to the parent plant.

Other Methods of Reproduction

Other methods of vegetative reproduction generally are inferior to rhizomes. The *stolon,* or runner, occurs in many desert-range plants. It is similar to the rhizome, except that it is an aboveground organ; in fact, some plants produce one or the other, depending upon soil conditions. Buffalo grass is an example of a plant which spreads by stolons (Fig. 4.8).

Layering is a process common among many shrubs. It is similar to the stoloniferous method, except that the normal stem takes root where the node touches the ground.

Tillering is not unlike reproduction by rhizome or stolon. A tiller is a basal branch originating from the lower nodes of the stem. These branches grow outward for short distances increasing the area of the plant. New plants are formed as the older material in the center dies leaving a ring of living tillers which may later break up into individual plants.

Corms and *bulbs* are not primarily means of reproduction, but are food-storage organs. However, from season to season smaller bulbs may form alongside the parent one which serve to multiply the plant if dis-

Fig. 4.8 Buffalo grass *(Buchloë dactyloides)* is strongly stoloniferous as shown by this photograph of a range badly depleted by drought, where buffalo grass is spreading rapidly. *(Photograph by J. E. Weaver.)*

turbance of some sort provides for dissemination. These organs are used by man in reproducing plants, such as the tulip and onion.

Bulbils are small aerial bulbs whose primary purpose is reproduction. The bulbil may grow in the axil of the leaf or in the flower head instead of seed, as in *Poa bulbosa*. The bulbil falls to the ground shortly after maturity and, much as a seed, it produces a new plant.

Root or *crown sprouts* are important means of regeneration among woody plants, especially after destruction, as by fire.

Effect of Grazing upon Reproduction of the Plant

Decrease in valuable forage plants on the range results not entirely from the death of established plants, although death is a significant factor (McKell et al., 1966), but also from a decrease in reproduction and consequently a smaller number of young plants available to replace those that are dying. The life span of a range plant varies from a few weeks in annuals to 50 years or more in shrubs and, possibly, in perennial grasses also. Perennials reproduce intermittently, each 5 or 10 years, when a sequence of ideal weather conditions results in free pollination of flowers and formation of seeds followed by a year favorable for seedling establishment. Often no seedlings of good forage plants will be in evidence because these conditions have not occurred. When it becomes evident that mature plants are not being replaced by young plants, maintenance of the range requires an immediate change in livestock management to allow normal reproduction.

A healthy range has a mixture of many age classes of plants. Old and senescent plants die and are being replaced constantly by new ones. The reproduction process must not be interrupted to the point where no new plants are available to replace those that are dying.

Studies of Utah deserts (Stewart et al., 1940) on heavily grazed areas showed that desirable climax woody plants were on the average much older than the less valuable invading species. Thus, 43 percent of the rabbitbrush (*Chrysothamnus*) was less than 10 years old, whereas none of the much preferred winterfat was less than 10 years old. The ratio of dead to living plants of rabbitbrush was less than 10 percent; whereas that of winterfat in heavily grazed pure types was 59 percent, as compared with 23 percent on adjacent moderately used ranges. About 90 percent of the Indian ricegrass (*Oryzopsis*) plants were dead on heavily grazed range, and only about 20 percent were dead on that moderately grazed.

Relation of Grazing to Seed Production   The influence of grazing upon seed production is twofold. The animals may graze the plant so heavily as to consume the seedstalks prior to the dropping of seed, or cropping may so disturb the physiology of the plant as to prohibit seed formation. Probably the latter is far more important than the former, despite the general opinion to the contrary. Many references to deferred grazing stress deferment until seed formation, as though that were the primary purpose of deferment. Actually, the production of seed probably is a minor benefit accruing from deferment, whereas the allowance of normal early-season food storage is a major benefit.

Very intense grazing would be necessary to consume all the seed produced by healthy plants. Because nature is so lavish with the number produced, the seed remaining after the usual grazing may produce all the young plants for which there is space. Some plants protect their seeds so as to make them inaccessible to ordinary grazing; on others they are easily accessible and highly preferred.

Since seed formation requires large quantities of concentrated food reserve, any depletion in the reserve of the plant interferes with normal seed formation. The work of Hanson and Stoddart (1940) showed that the seed of wheatgrass on improperly grazed range and that on correctly grazed range were of comparable vitality. Clipping at various intensities for 4 years produced no significant difference in the germination of filled seeds from clipped and unclipped plants.

A strikingly different picture is shown when the volume of seed production is considered. Lightly grazed stands of wheatgrass, compared with a heavily grazed stand, showed almost 17 times as many seed heads, which, in turn, produced over 50 times as many viable seeds (Table 4.7). In studies with *Oryzopsis miliacea,* Whalley et al. (1966) showed that moisture stress caused a reduction in seed numbers but not seed weight; whereas, low soil fertility reduced seed size and vigor.

In Oregon, decreased vigor of the plant induced by grazing not only reduced seed production and germination but also influenced the season of seed production (Sampson, 1914). Overgrazed ranges produced seedstalks 4 to 7 days later in the spring and also required a period of 6 to 10 days longer for completing seeding. Clipping experiments showed the production of forage and seed to be unaffected by fall grazing, however.

Seed production is especially important to annuals, since it is the only way they can reproduce. It has been shown that seed production in annualgrasses can be greatly reduced by clipping, especially late in the growth season. It is unlikely, though, that grazing can reduce seed production below the amount needed for production (Stechman and Laude, 1962). Grazing may influence species composition, however, because of its differential effect on seed production and other responses among various species.

Influence of Grazing upon Vegetative Reproduction

Many vegetative organs, especially rhizomes, bulbs, and corms, serve in food storage. These organs may be of small size or even entirely absent in instances where synthesis of food is curtailed. The reaction of Johnson grass to clipping (Table 4.8) indicates that frequent and early clippings are most detrimental to rhizome production. Also, close clipping is more detrimental than less close (Harrison, 1931). Heavy grazing reduced both the amount of the roots and rhizomes in *Andropogon scoparius*

Table 4.7 Seed Heads, Filled Florets, Percentage Germination, and Viable Seeds Produced on Grazed and Protected Stands of *Agropyron Inerme*

Data from Hanson and Stoddart (1940)

Condition	Number of heads per sq m	Filled Florets		Germination (filled florets)	Number of viable seeds per sq m
		Per sq m	Percentage		
Grazed	7.1	19.6	23.9	62.2	12.2
Protected	120.4	972.6	38.8	64.8	630.2

Table 4.8   Rhizome Yield of Johnson Grass (*Sorghum halepense*) Cut at Various Growth Stages

Data from Sturkie (1930)

Stage of cutting (plants were cut each time they reached designated stage)	Weight of rhizomes, gm				
	Cut continuously		Permitted to grow a crop of seed before cutting in 1927		Permitted to grow a year before cutting
	1927	1928	1927	1928	1928
1 ft high	130	10	658	11	45
2 ft high	262	51	608	75	110
Booting	321	139	604	74	383
Blooming	408	184	772	191	479
Late milk	589	495	811	345	799
Seed maturity	774	739	...	...	802
End of growing season	1,180	1,281	1,180	1,281	...

var. *littoralis,* although roots were more affected than rhizomes; and, in consequence, the ratio of rhizomes to roots increased (Bowns and Box, 1964).

Rhizomes of *Hilaria jamesii* in the Southwest were found to average 2 m in length when the plants were not impaired by heavy grazing; but when the plants were heavily grazed, rhizomes did not develop beyond a few centimeters in length, the grass being thus restricted to a bunchlike form much less efficient for erosion prevention (Flory and Trussell, 1938).

EFFECTS OF GRAZING ON RANGE PLANTS

The foregoing discussions have dealt with specific responses of plants to forage removal. Much of the data was developed from experiments involving potted plants and clipping, an almost necessary course for the objectives sought. It is important to keep in mind, however, that clipping, which removes all plant material to a given level, may not duplicate grazing where selected parts of the plant only are removed. Marked differences in the subsequent forage production of *Andropogon scoparius* occurred when leaves, stems, or leaves and stems were removed (Jameson and Huss, 1959). Leaf removal reduced yields much more than either of the other two treatments. Moreover, work done with cultured plants lacks an essential element to which plants in the field are subjected— competition. This can have great consequences for plants, as Mueggler (1972) points out.

The ability of a range plant to survive and produce forage under grazing cannot be explained on the basis of carbohydrate contents, morphology, root growth, or reproduction alone. All are interrelated and what adversely affects one affects another. For example, low carbohydrates and few vegetative buds may go together, and lowered forage production is not the result of one but of both factors. Forage production integrates all these separate factors together into a practical guide useful to the range manager when based on consecutive years of records.

Effects of Frequency and Intensity of Forage Removal on Grasses

With some exceptions, experience has shown that total forage yield from grasses on arid ranges decreases with increased frequency of clipping and closer harvesting in any one year. Usually, experiments are not designed to show clearly the effects attributable to these factors separately. In humid regions and with certain grasses which have a high percentage of vegetative shoots, greater production may be obtained by more than one harvest, but in range situations this is less frequently the case. Even more important is the cumulative effect of close and frequent forage removal over a span of years.

A single fall harvest produced much greater yields than did a series of cuttings spaced throughout the summer, but different types of grasses responded differently. Taller-growing species yielded more under a single harvest, but short grasses produced more under multiple cuttings (Table 4.9). Turner and Klipple (1952) obtained greater yields from blue grama by cutting twice during the summer than from a single fall cutting. In contrast to this, Thaine (1954) harvested more leaves and stems of *Elymus junceus* from five clippings than from a single one (Table 4.10). Total vegetative yield, due to the reduction in weight of the roots and crowns, was greatest when clipping took place only once, a factor of great importance to long-time yields.

Dramatic differences in yields resulted from mixed prairie cut once, twice, and three times over an 8-year period (Table 4.11). Some cumulative effects are indicated by these data. Yields from three cuttings averaged 62 percent of the single-harvest yields in the first 4 years; they were only 25 percent in the last 4. Peterson (1962) found that production from *Stipa comata* varied directly with the length of the period they had been protected and inversely with the intensity of the use that preceded

Table 4.9 Forage Yields in Wyoming under Frequent Harvesting Compared to Single Harvesting, in Pounds per Acre

Data from Lang and Barnes (1942)

Clipping dates	Midgrasses	Short grasses	Forbs
Frequently clipped:			
June 4	696.2	122.5	156.1
June 27	66.9	96.0	26.4
July 22	85.0	91.2	11.0
Aug. 23	21.9	50.4	15.1
Oct. 20	12.2	0.0	0.0
Total	882.2	360.1	208.6
End of growing season only:			
Oct. 20	1,564.1	273.7	69.6

protection. Regrowth following the first clipping, though smaller from previously grazed than from protected plants, more nearly equaled that produced in the first clipping, leading the author to the conclusion that the plants become conditioned to clipping.

Production in grasses is inversely related to the height to which they are cut as shown by clipping Bermuda grass (Fig. 4.9). The differences are both direct and marked on roots and tops.

Effect of Time of Forage Removal on Forage Production

At any given level of use, forage production is affected greatly by the time of forage removal, for plants are more vulnerable at some periods than others. If forage removal occurs in the early growth stages and while moisture is available, the healthy plant quickly replenishes lost foliage and there is little disruption of plant functions. The same level of clipping in midseason is much more critical, as illustrated by data for wheatgrass (Table 4.12). In general, defoliation early in the growing season is less detrimental than at a later time. This may be due to the increased level of physiological activity during early growth, or to inadequate soil moisture to support further growth later in the season. Grazing injury may be as much due to lack of soil moisture and opportunities for regrowth as to physiological factors (Blaisdell and Pechanec, 1949). Thus, in dry climates, time of cessation of grazing may be more important to plant welfare than time of beginning of grazing (Stoddart, 1946). Grazing which removes herbage just prior to the onset of the dry season prevents normal food storage, development of roots, and formation of

Table 4.10 Yield of Vegetative Material, Protein and Lignin Content in Leaves and Stems, and Available Carbohydrates of Russian Wild Ryegrass under Different Intensities of Clipping

Data from Thaine (1954)

| Number of clippings | Dry matter, lbs/acre | | | Protein in leaves and stems | | Lignin in leaves and stems, percent | Available carbohydrates in stubble and roots, lbs/acre |
	Leaves and stems	Crowns and stubble	Total vegetative material*	Percent	lbs/acre		
1	1,550	3,880	8,766	7.35	114	13.05	1,593
2	1,594	2,537	7,051	17.00	271	8.25	1,337
3	1,895	2,666	7,290	20.29	384	7.79	1,088
5	1,889	1,780	5,438	22.69	429	7.53	592

*Includes root material to 1 1/2-foot depth.

Table 4.11   Annual Dry-Matter Yields of Mixed Prairie for Each of Three Harvest Frequencies when cut at 1-in. Height in Each of 8 Years

Data from Lorenz and Rogler (1973)

	Harvest dates		
Year	August 1	June 1 and August 1	June 1, July 1, and August 1
		(lbs/acre)	
1958	1,496	1,027	901
1959	480	304	305*
1960	1,621	1,104	928
1961	416	322	286
1962	4,157	2,897	1,676
1963	1,356	1,081	896
1964	1,838	1,493	950
1965	2,131	1,654	1,089
Average[†]	1,687	1,235	879

*Cut only twice in 1959, June 1 and August 1.
[†]Least significant difference at 0.05, 139.

buds. Grazing or clipping after the food-storage cycle has been completed has the least effect. Combinations of drought and heavy grazing are particularly detrimental to plants.

Maximum yields of blue grama were greatest when cut in June or July, with both earlier and later harvesting giving lower total yields (Turner and Klipple, 1952). Remarkably different yields of three prairie grasses after 3 years of harvesting resulted from harvesting forage at different times during the summer. Yields are greater when plants were clipped late in the season than when unclipped (Table 4.13).

Effects of Grazing on Production from Shrubs and Forbs

Most shrubs and forbs, unlike grasses, are not well adapted to regenerate forage removed through grazing. It is well known that browsing or clipping shrubs encourages twig growth at the expense of flowers and fruits, but the effects only become apparent in following years. By thus keeping shrubs in a vegetative condition, increased forage production may result (Garrison, 1953). Willard and McKell (1973) reported a large increase in new shoots on two shrubs, *Symphoricarpos* and rabbitbrush, that were clipped at various times in the year over a period of 5 years.

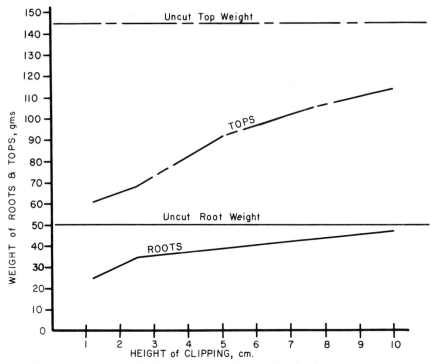

Fig. 4.9 Effect of repeated clipping to maintain a fixed stubble height on production of tops and roots of Bermuda grass *(Cynodon dactylon). (From Dittmer, 1973.)*

Certain shrubs can withstand heavy utilization year after year when cropping takes place during the winter. Conversely, heavy and repeated clipping during the growing season can cause rapid declines in subsequent forage yields and increase plant mortality (Lay, 1965). A single clipping which removed 80 percent of the foliage from two species of sagebrush reduced forage production the following year irrespective of when clipping had taken place, but the reduction in yields was greatest when clipping took place in July just as flower and twig growth was ending (Wright, 1970). Spring defoliation was even more damaging to big sagebrush (Cook and Stoddart, 1953).

Generally, forbs are less resistant to grazing than are grasses although many leguminous plants withstand repeated use. Many of the more productive forbs are weak-stemmed and are readily damaged through breakage and lodging by trampling. A single annual clipping (40 percent foliage removal) for 4 years reduced the production of *Mertensia leonardii* from one-third to two-thirds that of unclipped plants (Laycock and Conrad, 1969). Heavier foliage removal was even more damaging. For 70 percent removal, production was reduced to as little as 3 percent of that on unclipped plants. Flowering and fruiting stages

Table 4.12   Effect of Clipping upon Herbage Yield and Survival of *Agropyron spicatum* (yields are expressed in percent of original—(1943)—yield)

Data from Stoddart (1946)

	Percentage total annual yield was of original yield		Percentage of original plants alive, fall of 1945
	1944	1945	
Clipped at 1-in. height:			
Weekly, Apr. 15–May 7	41.34	20.49	100
Weekly, May 1–May 22	23.48	1.24	25
Weekly, May 15–June 7	40.95	4.25	50
Weekly, June 1–June 22	79.00	6.91	55
Biweekly, Apr. 15–May 15	35.03	11.41	75
Biweekly, May 15–June 15	45.13	2.99	20
Biweekly, Apr. 15–June 15	30.36	0.37	0
Biweekly, Sept. 15–Nov. 1	128.41	88.02	100
Clipped at 2-in. height:			
Weekly, Apr. 15–May 7	53.54	40.17	100
Weekly, May 1–May 22	38.80	9.93	90
Weekly, May 15–June 7	49.89	7.48	50
Weekly, June 1–June 22	70.30	8.53	75
Biweekly, Apr. 15–May 15	76.67	39.75	100
Biweekly, May 15–June 15	46.92	5.04	35
Biweekly, Apr. 15–June 15	37.20	3.35	45
Biweekly, Sept. 15–Nov. 1	109.18	73.21	100

Table 4.13   Yields of Three Prairie Grasses in 1963, in Percent of Yields in 1961 when Clipped at Different Phenological Stages

Data from Vogel and Bjugstad (1968)

Time of clipping	Growth stage	Little bluestem	Big bluestem	Indiangrass
May	Vegetative growth	70	72	65
June	Vegetative growth completed	53	63	56
July	Early reproductive	34*	52	34*
August	Flowering	76	59	56
October	Seed ripened	125*	159*	72
November	Plant dormant	152*	207*	132*
	Unclipped	77	97	81

*Significantly different from unclipped plant of the same species, 0.05 percent.

were especially vulnerable periods. Similar results were found in three other range forbs, although *Valeriana edulis* sustained 50 percent use for several years without marked effect, while *Geranium richardsonii* showed immediate reduction in yield after only a year of clipping (Julander, 1968).

Quality of Forage as Affected by Multiple Harvesting

More important than total herbage yield is the actual digestible nutrient yield from the range. Chemical composition and digestibility of herbage differ radically according to species, soil, season of the year, and other factors (see Chapter 7). The value of forage varies also with the frequency of harvesting. For example, forage removal may be followed by a regrowth of forage of high protein content and relatively low fiber (see Table 4.10). Even though herbage yield is less with frequent harvesting, total yield may be greater because of an increase in percentage of nutrients (Ellett and Carrier, 1915).

The above observations do not imply that heavy range use necessarily improves forage quality. Actually, the closer animals graze a range, the poorer quality of forage they receive, unless conditions are favorable for rapid regrowth. On winter sheep range, it was found that as the degree of utilization of shrub ranges during the nongrowing season increased, the content of the more desirable nutrients in the available forage decreased, and, in addition, the digestibility of these nutrients was decidedly lowered (Peiper et al., 1959). With heavier utilization, the animals were forced to consume the less nutritious portions of the shrubs, and as a result the available nutrients frequently were not adequate to meet the demands of the grazing animals.

Range Management Based on Physiology of the Plant

Any grazing, whether moderate or heavy and whether early or late, has a measurable influence upon the metabolism of a plant. With reduction of photosynthetic tissue comes reduction in carbohydrate and nitrogen reserves and decreased forage production. This is one of the most important effects of incorrect grazing. Critical to the range plant is the influence of grazing upon volume and depth of the root system. A reduction of food reserves slows the growth of the entire plant, and root growth is no exception. The acceptance of these facts is basic to range management. Provided that grazing is neither too frequent, too close, nor improperly timed, plants have great ability to survive use. Means by which the effects of use are ameliorated are treated in Chapter 9.

OTHER INFLUENCES OF GRAZING
UPON VEGETATION

Animals also influence vegetation indirectly. Grazing animals have an influence upon the soil, tending to compact it to surprising depths, especially during the spring or other moist season. Not only are compact soils poor absorbers of precipitation, but they restrict normal root development; the roots sometimes are only half their normal length. Compaction is greatest near the surface; thus the hard soil increases the difficulty of seedling emergence and establishment, and depresses vigor (Barton et al., 1966). Conversely, when the soil is not wet, animals are believed to be beneficial in loosening the soil surface and in covering seeds that have accumulated on the surface. In line with this theory, a herd of sheep often is passed over an area before and after broadcast seeding, to loosen the soil and cover the seeds.

The mechanical action of animals in loosening seed, in carrying burs, awned seeds, and the like, in their hair, in distribution of hard-coated seeds through the feces, and in loosening bulbs, corms, and bits of rhizome so that they may be transported elsewhere is probably of unsuspected importance. There are instances in which total protection of range from livestock has failed to result in the expected revival of the vegetation, presumably because of the lack of animal action in aiding reproduction.

It often is felt by stockmen that ungrazed plants are not so healthy and vigorous as lightly grazed plants, and research seems to indicate that properly grazed plants are as productive as, or more productive than, ungrazed ones (Nelson, 1934; Weaver and Rowland, 1952). Although this may seem an impossibility, there is some argument to support the theory. A grazed plant is like a pruned plant, and every nurseryman or orchardman knows the value of pruning to the woody plant. Upon grazing, the plant will branch and become more bushy, and the forage will be less rank and fibrous. Grasses tend to spread and form a more dense sod when grazed, provided that sufficient soil moisture is available. Farmers sometimes graze fall wheat in the belief that they obtain a greater tillering and hence more grain stalks. Increased tillering has been demonstrated in seedlings of *Phalaris* by clipping (Laude et al., 1968).

In arid regions, it is possible that grazing induces better moisture relations since, with the removal of herbage, the transpiring surface is reduced and plants may be better able to withstand drought. Where great quantities of plant material accumulate, production may be lowered. Ungrazed Nebraska prairie has been shown to produce less herbage than comparable grazed prairie (Weaver and Rowland, 1952). Fifteen years without grazing or fire resulted in deep mulch which promoted infiltration and reduced evaporation. Lower soil temperature under the mulch delayed spring growth 3 weeks and herbage yields were from 26 to 57 percent less than those of the grazed prairie.

Often grazing animals are credited with beneficial effects by manuring the range. Considering the entire ecosystem, this can hardly be true. The grazing animal, in reality, removes rather than adds fertility unless there is no animal harvest. Ozanne and Howes (1971) found that the phosphorus requirements were greater for grazed than for ungrazed pastures, due to disruption of normal phosphorus redistribution in the plant. It is true that animals, through digestion of organic material, make the elements more immediately available to plants, but they cannot add any material that would not, through normal decay, ultimately become available. It is possible that movement of nitrogen through its cycle may be expedited, the reduction of plant materials in the digestive tract hastening decay. Still, the mummification of animal feces in extremely dry climates makes this suspect as a generality. Under dry conditions fecal material may remain on the soil surface for several years. Moreover, nutrients may be altered so as to be less available; Bromfield and Jones (1970) concluded that phosphorus was converted to resistant organic forms in the dung by the digestive tract.

Animals do redistribute organic matter over the range. This is especially notable near water holes and places where animals collect for protection in inclement weather. In areas where livestock are penned at night and the feces burned or hauled to cultivated fields, as in much of Africa, there is a net loss of nutrients to the range ecosystem.

Graber (1931) found that frequent harvesting of plants made them more susceptible to winter killing and to insect injury, possibly because of reduced levels of stored carbohydrates.

BIBLIOGRAPHY

Arizona Interagency Range Committee (1972): Proper use and management of grazing land, U.S. Government Printing Office, Region 8 (multil.).

Barton, Howard, Wayne B. McCully, Howard M. Taylor, and James E. Box, Jr. (1966): Influence of soil compaction on emergence and first-year growth of seeded grasses, *Jour. Range Mgt.* **19:**118–121.

Biswell, H. H., and J. E. Weaver (1933): Effect of frequent clipping on the development of roots and tops of grasses in prairie sod, *Ecology* **14:**368–390.

Blaisdell, James P., and Joseph F. Pechanec (1949): Effects of herbage removal at various dates on vigor of bluebunch wheatgrass and arrowleaf balsamroot, *Ecology* **30**:298–305.

Bowns, James E., Jr., and Thadis W. Box (1964): The influence of grazing on the roots and rhizomes of seacoast bluestem, *Jour. Range Mgt.* **17**:36–39.

Branson, Farrel A. (1953): Two new factors affecting resistance of grasses to grazing, *Jour. Range Mgt.* **6**:165–171.

Bromfield, S. M., and O. L. Jones (1970): The effect of sheep on the recycling of phosphorus in hayed-off pastures, *Austral. Jour. Agr. Res.* **21**:699–711.

Cable, Dwight R. (1971): Growth and development of Arizona cottontop (*Trichachne california* [Benth, Chase]), *Bot. Gazette* **132**:119–145.

Cannon, W. A. (1911): The root habits of desert plants, *Carnegie Inst. Pub.* **131**:1–96.

Cook, C. W., and L. A. Stoddart (1953): Some growth responses of crested wheatgrass following herbage removal, *Jour. Range Mgt.* **6**:267–270.

Coyne, Patrick I., and C. Wayne Cook (1970): Seasonal carbohydrate reserve cycles in eight desert range species, *Jour. Range Mgt.* **23**:438–444.

Crider, Franklin J. (1955): Root-growth stoppage resulting from defoliation of grass, *U.S. Dept. of Agr. Tech. Bull.* **1102**.

Davidson, J. L., and F. L. Milthorpe (1966a): Leaf growth in *Dactylis glomerata* following defoliation, *Ann. Bot., New Series* **30**:173–184.

————, and ———— (1966b): The effect of defoliation on the carbon balance in *Dactylis glomerata*, *Ann. Bot., New Series* **30**:185–198.

Dina, Stephen J., and Lionel G. Klikoff (1973): Effect of plant moisture stress on carbohydrate and nitrogen content of big sagebrush, *Jour. Range Mgt.* **26**:207–209.

Dittmer, H. J. (1937): A quantitative study of the roots and root hairs of a winter rye plant (*Secale cereale*), *Am. Jour. Botany* **24**:417–420.

———— (1973): Clipping effects on Bermuda grass biomass, *Ecology* **54**:217–219.

Donart, Gary B. (1969): Carbohydrate reserves of six mountain range plants as related to growth, *Jour. Range Mgt.* **22**:411–415.

Dyksterhuis, E. J. (1945): Axillary cleistogenes in *Stipa leucotricha* and their role in nature, *Ecology* **26**:195–199.

Ellett, W. B., and L. Carrier (1915): The effect of frequent clipping on total yield and composition of grasses, *Jour. Am. Soc. Agron.* **7**:85–87.

Finch, Alton H. (1935): Physiology of apple varieties, *Plant Physiology* **10**:49–72.

Flory, E. L., and D. F. Trussell (1938): Preliminary notes on important vegetative species of Region 8, *U.S. Dept. Agr. Regional Bull.* **23,** Soil Conservation Service 2 (mimeo.).

Garrison, George A. (1953): Effects of clipping on some range shrubs, *Jour. Range Mgt.* **6**:309–317.

Graber, L. F. (1931): Food reserves in relation to other factors limiting the growth of grasses, *Plant Physiology* **6**:43–71.

Hanson, W. R., and L. A. Stoddart (1940): Effects of grazing upon bunch wheatgrass, *Jour. Am. Soc. Agron.* **32**:278–289.

Harris, Grant A. (1967): Some competitive relationships between *Agropyron spicatum* and *Bromus tectorum, Ecol. Monogr.* **37**:89–111.

Harrison, C. M. (1931): Effect of cutting and fertilizer applications on grass development, *Plant Physiology* **6**:669-684.

Hyder, D.N., and Forrest A. Sneva (1959): Growth and carbohydrate trends in crested wheatgrass, *Jour. Range Mgt.* **12**:271–276.

———, and ——— (1963a): Morphological and physical factors affecting the grazing management of crested wheatgrass, *Crop Sci.* **3**:267–271.

———, and ——— (1963b): Studies of six grasses seeded on sagebrush bunchgrass range: Yield, palatability, carbohydrate accumulation, and developmental morphology, *Ore. Agr. Expt. Sta. Tech. Bull.* **71**.

Jameson, Donald A. (1963): Responses of individual plants to harvesting, *Bot. Rev.* **29**:532–594.

———, and Donald L. Huss (1959): The effect of clipping leaves and stems on number of tillers, herbage weights, root weights, and food reserves of little bluestem, *Jour. Range Mgt.* **12**:122–126.

Jensen, Charles H., Arthur D. Smith, and George W. Scotter (1972): Guidelines for grazing sheep on rangelands used by big game in winter, *Jour. Range Mgt.* **25**:346–352.

Julander, Odell (1968): Effect of clipping on herbage and flower stalk production of three summer forbs, *Jour. Range Mgt.* **21**:74–79.

Kinsinger, Floyd E., and Harold H. Hopkins (1961): Carbohydrate content of underground parts of grasses as affected by clipping, *Jour. Range Mgt.* **14**:9–12.

Kramer, Paul J., and Theodore T. Kozlowski (1960): *Physiology of Trees,* McGraw-Hill, New York.

Kucera, C. L., Roger C. Dahlman, and Melvin R. Koelling (1967): Total net productivity and turnover on an energy basis for tallgrass prairie, *Ecology* **48**:536–541.

Lang, Robert, and O. K. Barnes (1942): Range forage production in relation to time and frequency of harvesting, *Wyo. Agr. Expt. Sta. Bull.* **253**.

Laude, Horton M., Guillermo Riveros, Alfred H. Murphy, and Robert E. Fox (1968): Tillering at the reproductive stage in Hardinggrass, *Jour. Range Mgt.* **21**:148–151.

Lay, Daniel W. (1965): Effects of periodic clipping on yield of some common browse species, *Jour. Range Mgt.* **18**:181–184.

Laycock, William A., and Paul W. Conrad (1969): How time and intensity of clipping affect tall bluebell, *Jour. Range Mgt.* **22**:299–303.

Lorenz, Russell J., and George A. Rogler (1973): Interaction of fertility level with harvest date and frequency on productiveness of mixed prairie, *Jour. Range Mgt.* **26**:50–54.

McCarty, E. C., and Raymond Price (1942): Growth and carbohydrate content of important mountain forage plants in central Utah as affected by clipping and grazing, *U.S. Dept. Agr. Tech. Bull.* **818**.

McConnell, B. R., and G. A. Garrison (1966): Seasonal variations of available carbohydrates in bitterbrush, *Jour. Wildlife Mgt.* **30**:168–172.

McKell, Cyrus M., R. Derwyn Whalley, and Victor Brown (1966): Yield, survival, and carbohydrate reserve of Hardinggrass in relation to herbage removal, *Jour. Range Mgt.* **19:**86–89.

McKendrick, Jay D., and Lee A. Sharp (1970): Relationship of organic reserves to herbage production in crested wheatgrass, *Jour. Range Mgt.* **23:**434–438.

Miller, E. C. (1938): *Plant Physiology,* 2d ed., McGraw-Hill, New York.

Mueggler, W. F. (1972): Influence of competition on the response of bluebunch wheatgrass to clipping, *Jour. Range Mgt.* **25:**88–95.

Nelson, E. W. (1934): The influence of precipitation and grazing upon black grama grass range, *U.S. Dept. Agr. Tech. Bull.* **409.**

Ogden, Phil R., and Walter E. Loomis (1972): Carbohydrate reserves on intermediate wheatgrass after clipping and etiolation treatments, *Jour. Range Mgt.* **25:**29–32.

Ojima, Kunihiko, and Takeshi Isawa (1968): The variation of carbohydrates in various species of grasses and legumes, *Can. Journ. Bot.* **46:**1507–1511.

Okajima, Hidea, and Dale Smith (1964): Available carbohydrate fractions in the stem bases and seed of timothy, smooth bromegrass, and several other northern grasses, *Crop Sci.* **4:**317–320.

Owensby, C. E., and Kling L. Anderson (1969): Effect of clipping date on loamy upland bluestem range, *Jour. Range Mgt.* **22:**351–354.

Ozanne, P. G., and K. M. W. Howes (1971): The effects of grazing on the phosphorus requirement of an annual pasture, *Austral. Jour. Agr. Res.* **22:**81–92.

Parker, K. W., and A. W. Sampson (1931): Growth and yield of certain gramineae as influenced by reduction of photosynthetic tissue, *Hilgardia* **5:**361–381.

Pavlychenko, T. K. (1937): The soil-block washing method in quantitative root study, *Can. Jour. Res.* **15:**33–57.

Peiper, Rex C., C. Wayne Cook, and Lorin E. Harris (1959): Effect of intensity of grazing upon nutritive content of the diet, *Jour. Animal Sci.* **3:**1031–1037.

Peterson, Roald A. (1962): Factors affecting resistance to heavy grazing in needle-and-thread grass, *Jour. Range Mgt.* **15:**183–189.

Rechenthin, C. A. (1956): Elementary morphology of grass growth and how it affects utilization, *Jour. Range Mgt.* **9:**167–170.

Robertson, J. H. (1933): Effect of frequent clipping on the development of certain grass seedlings, *Plant Physiology* **8:** 425–447.

———, D. L. Neal, K. R. McAdams, and P. T. Tueller (1970): Changes in crested wheatgrass ranges under different grazing treatments, *Jour. Range Mgt.* **23:**27–34.

Sampson, A. W. (1914): Natural revegetation of rangelands based upon growth requirements and life history of the vegetation, *Jour. Agr. Res.* **3:**93–148.

Sharman, B. C. (1947): The biology and developmental morphology of the shoot apex in gramineae, *New Phyt.* **46:**20–34.

Shively, S. B., and J. E. Weaver (1939): Amount of underground plant materials in different grassland climates, *Neb. Conserv. Bull.* **21.**

Smith, C. C. (1940): The effect of overgrazing and erosion upon the biota of the mixed-grass prairie of Oklahoma, *Ecology* **21**:381–397.

Smith, Dale (1972): Carbohydrate reserves of grasses, in V. B. Younger and C. M. McKell (eds.), *The Biology and Utilization of Grasses,* Academic Press, New York, pp. 318–333.

Smoliak, S., J. F. Dormaar, and A. Johnston (1972): Long-term grazing effects on *Stipa-Bouteloua* prairie soils, *Jour. Range Mgt.* **25**:246-250.

Spence, L. E. (1937): Root studies of important range plants of the Boise River watershed, *Jour. Forestry* **35**:747–754.

Sprague, V. G., and J. T. Sullivan (1950): Reserve carbohydrates in orchard grass clipped periodically, *Plant Physiology* **18**:656–670.

Stechman, John V., and Horton M. Laude (1962): Reproductive potential of four annual range grasses as influenced by season of clipping or grazing, *Jour. Range Mgt.* **15**:98–103.

Stewart, G., W. P. Cottam, and S. S. Hutchings (1940): Influence of unrestricted grazing on northern salt desert plant associations in western Utah, *Jour. Agr. Res.* **60**:289-316.

Stoddart, L. A. (1946): Some physical and chemical responses of *Agropyron spicatum* to herbage removal at various seasons, *Utah Agr. Expt. Sta. Bull.* **324.**

Sturkie, D. C. (1930): The influence of various topcutting treatments on rootstocks of Johnson grass (*Sorghum halepense*), *Jour. Am. Soc. Agron.* **22**:82–93.

Sullivan, J. T., and V. G. Sprague (1943): Composition of the roots and stubble of perennial ryegrass following partial defoliation, *Plant Physiology* **18**:656–670.

———, and ——— (1953): Reserve carbohydrates in orchard grass cut for hay, *Plant Physiology* **28**:92–102.

Taylor, W. P., C. T. Vorhies, and P. B. Lister (1935): The relation of jack rabbits to grazing in southern Arizona, *Jour. Forestry* **33**:490–498.

Thaine, R. (1954): The effect of clipping frequency on the productivity and root development of Russian wild rye-grass in the field, *Can. Jour. Agri. Sci.* **34**:299–304.

Turner, G. T., and G. E. Klipple (1952): Growth characteristics of blue grama in northeastern Colorado, *Jour. Range Mgt.* **5**:22–28.

Vogel, W. G., and A. J. Bjugstad (1968): Effects of clipping on yield and tillering in little bluestem, big bluestem, and Indiangrass, *Jour. Range Mgt.* **21**:136–140.

Weaver, J. E. (1930): Underground plant development in its relation to grazing, *Ecology* **11**:543–557.

——— (1963): The wonderful prairie sod, *Jour. Range Mgt.* **16**:165–171.

———, and F. E. Clements (1938): *Plant Ecology,* 2d ed., McGraw-Hill, New York.

———, and N. W. Rowland (1952): Effects of excessive natural mulch on development, yield, and structure of native grassland, *Bot. Gaz.* **114**:1–19.

———, and Ellen Zink (1946): Length of life of roots of ten species of perennial range and pasture grasses, *Plant Physiology* **21**:201–217.

Wells, T. C. E. (1970): A comparison of the effects of sheep grazing and mechanical cutting on the structure and botanical composition of chalk grassland, in E. Duffey and A. S. Watt (eds.), *The Scientific Management of Animal and Plant Communities,* The 11th Symposium of the British Ecological Society, University of East Anglia, Norwich, Blackwell Scientific Pub., Oxford, pp. 497–515.

West, Neil, Russell T. Moore, K. A. Valentine, Lamont Law, Phil Ogden, Fred Pinckney, Paul Tueller, Joseph Robertson, and Alan Beetle (1972): Galleta: Taxonomy, ecology and management of *Hilaria jamesii* on western rangelands, *Utah Agr. Expt. Sta. Bull.* **487.**

Whalley, R. Derwyn B., Cyrus M. McKell, and Lisle R. Green (1966): Seedling vigor and the early nonphotosynthetic stage of seedling growth in grasses, *Crop Sci.* **6:**147–150.

White, Larry M. (1973): Carbohydrate reserves of grasses: A review, *Jour. Range Mgt.* **26:**13–18.

Willard, E. Earl, and Cyrus M. McKell (1973): Simulated grazing management systems in relation to shrub growth responses, *Jour. Range Mgt.* **26:**171–174.

Wright, Henry A. (1967): Contrasting responses of squirreltail and needle and thread to herbage removal, *Jour. Range Mgt.* **20:**398–400.

———— (1970): Responses of big sagebrush and three-tip sagebrush to season of clipping, *Jour. Range Mgt.* **23:**20–22.

Younger, V. B., and C. M. McKell (1972): *The Biology and Utilization of Grasses,* Academic Press, New York.

Chapter 5

Ecology in Relationship to Grazing

Ecology is that phase of biology which deals with the mutual relationships among organisms and between the organisms and their environment. Since range plants and range animals are biological organisms, their interrelationships are ecological in nature. Range management is applied ecology, since it consists of manipulating the environment in which both plant and animal live in such a way as to provide each, as far as practicable, with its most favorable habitat.

The grazing animal is a part of the plant's environment and the plant a part of the animal's. So long as the two live together, the welfare of each is dependent upon the other. This concept is fundamental in range management. Never can the forage and the animal be considered separately. Each of these must be looked upon as part of a great and intricately related biological complex, or *ecosystem.*

This complex system includes both living organisms and the nonliving environment. The parts of this complex, called *habitat factors,* can be classified as climatic, edaphic, biotic, physiographic, pyric, and an-

thropeic (Fig. 5.1). All are interrelated and a change in any one changes
the relationship of all other factors in the system.

Nature tends to balance each member of this complex one with
another, a phenomenon sometimes referred to as the *balance of nature* or
web of life. This principle of ecology assumes that any organism must
reach a point of equilibrium between related factors, such as its food
supply, its predators, its diseases, and its physical needs, including temperature, moisture, and protecting cover favorable to such life processes
as reproduction. In ecological terms, this equilibrium is termed *climax.*
This does not imply a static relationship—far from it! Nature is dynamic.

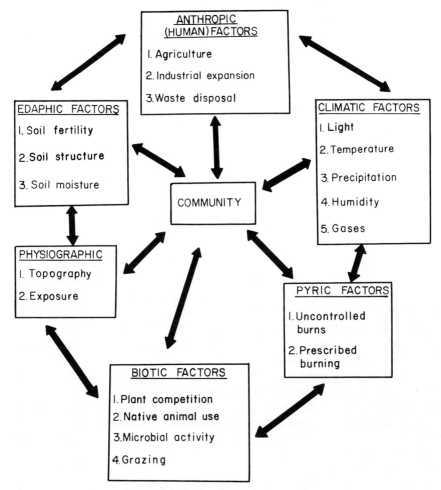

Fig. 5.1 Diagramatic scheme of the habitat factors in an ecosystem.

Constant fluctuation is the rule. Wet cycles or cold cycles induce widespread cycles of disease, population change, reduced food supply, etc. The important point is that such changes are the product of a changed environment. Interrelationship is inevitable. Whenever we change through management any one factor of this complex habitat, we must expect change elsewhere.

Range livestock, although not a natural factor of the habitat, become, nonetheless, a part of the ecosystem. Stockmen and range managers must realize that this change, this introduction of a new element into nature's balance, must cause widespread changes elsewhere in the habitat. No one knows all these complex interrelationships, but the range manager must be aware that the relationships exist. Ecology is the foundation of intelligent range management.

ECOLOGICAL CONCEPTS USEFUL IN RANGE MANAGEMENT

Ecology is a relatively new science, having developed from concepts of plant and animal geographers around the turn of the twentieth century. It has developed from a largely descriptive science for classifying communities to a highly quantitative science dealing with the functions and dynamics of ecosystems.

Ecology traditionally is divided into two phases: *autecology* and *synecology*. Autecology deals with the response of an organism as an individual to its environment. It is related closely to plant physiology. Chapter 4 dealt in large measure with autecology. The present chapter deals largely with synecology, the response of plants and animals *as a group* to their environment. This is akin to the science of sociology among human beings—how they react as a group. Indeed, synecology sometimes is called *plant sociology*.

Just as the individual plant grows, matures, and reproduces, so also does the community develop. The occurrence of a certain plant in a certain place, or the grouping of plants, usually does not come about by chance location of propagules. Rather, it is the direct result of a long series of developments controlled by climate in which plants have interacted with disintegrated rock materials to form soil. Climate and soil are the environmental factors which determine the kind of vegetation that can develop. Lewis (1959) identifies three controlling factors of an ecosystem: climate, geological materials, and organisms available to seed the area.

The Ecosystem Concept

The term *ecosystem* first was used in 1935 (Tansley, 1935) to describe the entire complex of organisms and their environment, but the idea of an ecological complex is much older and is found in the writings of many early ecologists. The system can be divided many ways. One division that is useful to range managers is to separate the living or *biotic* portion from the nonliving or *abiotic* portion (Fig. 5.2). Soil, composed of mineral elements, humus, and living organisms, acts as a figurative bridge between the living and nonliving portions.

The abiotic factors form the setting or environment in which the biotic factors operate. With the exception of fire, the manager has little control over them. They enter into his decisions in determining the suitability of a site for various uses, but they are not easily manipulated.

The biotic factors can more easily be controlled. Their manipulation is the basic tool used in determining productivity and usefulness to man. Four basic functions are performed by organisms in the biotic portion of an ecosystem.

Producer organisms are plants that capture the energy from the sun. They are the only major agent for converting the sun's energy into food for animals. The number of livestock or wildlife that a particular range can support depends directly upon the plants' ability to synthesize food by fixing light energy through photosynthesis.

Consumer organisms are animals that eat, rearrange, and distribute the energy captured by plants. *Primary consumers* are herbivores that live directly off plants such as some insects, livestock, and large ungulates. *Secondary consumers* are those animals that eat herbivores, i.e., carnivores. They are represented in range ecosystems by such animals as coyotes, lions, jackals, and vultures.

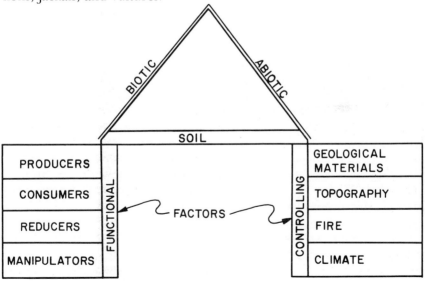

Fig. 5.2 Relationship between the biotic and abiotic factors in an ecosystem.

Reducers decompose and rearrange organic bodies of both producers and consumers. These include bacteria and fungi, and they perform vital functions in cycling minerals and organic matter, thus maintaining soil fertility.

Manipulators are organisms which deliberately rearrange the factors of the ecosystem for their own benefit. Man is the master manipulator. As a thinking, planning animal he brings his knowledge, experience, and technology to bear on the ecosystem. By his activities he may impinge on the system at almost any level, and in doing so, he may augment or destroy natural processes and functions. His ultimate success as manager may not be known for centuries.

Functioning of the Ecosystem

Atoms of hydrogen, carbon, oxygen, nitrogen, and other elements are cycled continuously in the range ecosystem among plants, animals, microorganisms, soil, and the atmosphere. In a stable system the amount of each element remains fairly constant, but the amount within each division of the ecosystem may vary. Losses occur when volatile elements such as nitrogen are released into the atmosphere, as by fire. Streamflow may remove water-soluble elements. Eventually, these are replaced from reservoirs, such as the atmosphere or parent rock. The rapidity of movement of nutrients varies with the climate, principally with temperature and precipitation. Oftentimes most of the particular nutrient may be tied up at one place (Table 5.1).

Energy, on the other hand, does not cycle, but flows through the system following the first law of thermodynamics. Energy transference within the system is never 100 percent efficient, following the second law of thermodynamics. Energy lost in each transfer is not destroyed, but is changed biologically, through respiration, to heat energy.

Energy enters the system as sunlight. Plants capture some of this energy and convert it to plant materials. From there it flows by a number of paths until it is released from the ecosystem. For example, plant materials may fall to the ground and be reduced by microorganisms, or they may be eaten by animals. In turn, the energy stored in the animals may be released by microorganisms when the animal dies, removed from the ecosystem when animals are harvested by man, or eaten by other animals. The paths through which energy may be transferred are innumerable. Each level of energy storage in the ecosystem is called a *trophic level* (Fig 5.3).

The amount of stored energy in the *biomass* is always greater in the producer level than in consumer levels, and each consumer level becomes smaller as one proceeds up the pyramid. Where forest products

Table 5.1 Nitrogen Relationships of Some Rangeland Vegetation Types

Data from Rodin and Bazilevich (1967) *

	Steppe (U.S.S.R.)			Cold deserts (U.S.S.R.)		Hot deserts (Syria)		Savanna
	Meadow	Medium	Dry	Shrub	Ephemeral-half-shrub	Ephemeral-half-shrub	Lichen-half-shrub	Dry (India)
Biomas (g/m)	27.4	23.6	10.3	6.1	14.2	7.9	1.10	23.8
Annual uptake (g/m)	16.1	12.2	4.5	1.8	10.8	2.6	0.65	8.1
Annual litter fall (g/m)	16.1	12.2	4.5	1.8	10.8	2.6	0.60	8.0
N retained in growth (g/m)	0	0	0	0	0	0	0.05	0.1
Retention relative to biomass (percent)	0	0	0	0	0	0	4.55	0.4

*From English translation, Scripta Technica limited, by permission of Oliver and Boyd, Edinburgh and London.

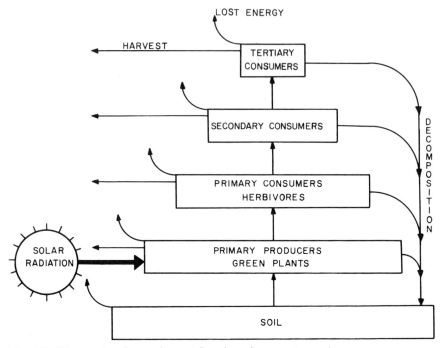

Fig. 5.3 Diagramatic scheme of energy flow through successive trophic levels of an ecosystem.

or prairie hay are harvested from a range, a greater energy efficiency is attained than by harvesting forage with grazing animals, since the product is removed at a lower trophic level. Trapping of fur bearers, especially if they are carnivores, is a very inefficient use of the sun's energy.

Complete data on energy transfer within an entire ecosystem do not exist, although partial analyses have been made which demonstrate the essential features of an ecosystem. The prime characteristic of the ecosystem is its inefficiency, and wastage at each level is high. Plants capture only a small fraction of the sun's energy, as little as 1.21 percent in a tall grass prairie (Kucera et al., 1967). Efficiency of energy transfer between other trophic levels may be even less, as data for small mammals show (Table 5.2). (See also Fig. 5.4.)

Similar data are not now available for livestock on the major range ecosystems. Major ecosystem analysis and function studies being conducted by the International Biological Program offer promise in collecting and synthesizing information useful to range managers in the improvement of the capture, transfer, and harvest of energy from the ecosystem. This remains the major challenge to today's range manager.

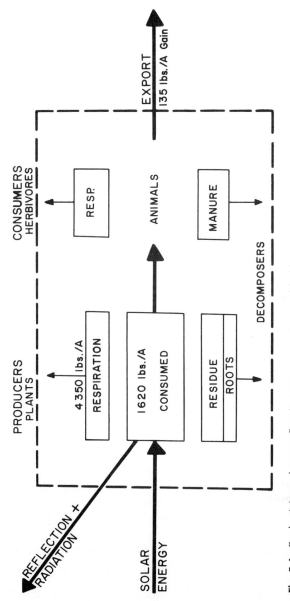

Fig. 5.4 Productivity and energy flow in a range ecosystem modified by introduction of rose clover and application of sulfur fertilizer. *(From Williams, 1966.)*

Table 5.2   Energy Flow in Calories per Hectare per Year in a Terrestrial
Community

Data from Golley (1960)

Trophic level	Flux-incident solar radiation, gross photo-synthetic production, or consumption (P)	Respiration (R)	Ratio R/P	Efficiency ratio between adjacent trophic levels
Sunlight	47.1×10^8			
Vegetation	58.3×10^6	8.76×10^6	0.150	0.012
Herbivore (*Microtus*)	250×10^3	170×10^3	0.680	0.004
Carnivore (*Mustela*)	5824	5434	0.933	0.023

It must be remembered, however, that efficiency is not the sole control-ling factor. The most efficient course would be for man to become a vegetarian, a condition not all would accept. Furthermore, any major alteration in functioning of the ecosystem would result in its destruction.

Energy flow in a rangeland ecosystem is complex. At equilibrium, or climax, the rate of turnover and cycling of matter and nutrients is at a maximum. Energy flow through the system is rapid; yet all of the energy fixed by the primary producers is dissipated in the maintenance cost of the extensive and diverse mass of consumers and reducers (Lewis, 1969). As Odum (1957) points out, there is no net output from an ecosystem in a natural state of equilibrium.

Communities in climax conditions are more stable than those of lower stages. The maximum diversity of the high stages of succession offers maximum resiliency and freedom from degradation. Thus it may be desirable to manage a critical watershed for climax conditions even though forage production may not be maximized.

Products from rangelands come from controlled unbalancing or harvesting from ecosystems. Solar energy, wind energy, and precipita-tion enter the system as inputs. These inputs may be stored in soil, plants, animals, or organic litter. Topsoil and nutrients may be blown from the system or harvested and removed for human use in another system (Fig. 5.5). The range manager's job is to minimize energy and nutrient drain on the ecosystem and maximize system health. One of the major ways to insure ecosystem health is through manipulating the pro-ducer level of the ecosystem. The range manager, unlike the agronomist who works against natural processes to establish a new vegetation, works with the natural process of succession.

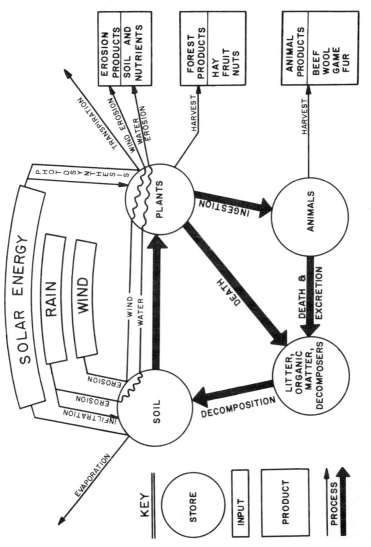

Fig. 5.5 Relations of inputs, stores, and products from a rangeland ecosystem. (*Adapted from Perry, 1970.*)

SUCCESSION

Succession is the orderly process of community change. It is the process whereby one association of species replaces another. Such a succession usually is gradual and involves a series of changes which follow a more or less regular course. The word *succession* comes from *succeed*. In other words, one type of vegetation succeeds or follows a former type into the area. Succession results from a change in habitat and invasion of new species. Plants are always seeking balance with their environment. Change of environment or habitat results in change of the plant cover adapted to the area. Change in habitat *(reaction)* sometimes results from action of plants upon soil and microclimate. Thus the plants themselves may initiate the change which ultimately will result in succession and their own destruction.

Understanding of succession is basic to range management. The range manager works with the plant habitat to direct plant succession toward his desired objective. He needs to see and understand successions in early stages as the succession is the measure of the effectiveness of his management. He must recognize it early if he is to avoid costly loss of time. Jameson (1970) questions the utility of the concept of succession, feeling that it has lost much of its usefulness and suggests that "trophic-dynamic ecology" offers promise of being a more useful tool of range managers. This remains to be seen. However useful the systems approach proves to be, especially for the researcher and theoretician, it is, perhaps, more likely to supplement rather than supplant the concept of succession for the practicing range manager.

Succession may be either natural or induced. *Natural succession* takes place until climax conditions are reached. It results from soil changes in the process of soil succession. Also, both before and after climax is reached, advancement or recession of the process may result from fluctuation in the habitat. For example, extensive decline in vegetation during unusual drought periods is a perfectly natural phenomenon.

Distinguishing between natural and induced succession is important. *Induced succession* results generally from the action of man, and hence is not a condition imposed by nature. As such, it can be modified by man more readily than can natural succession. Abnormal vegetation cover may remain for many years, especially in instances where soil erosion follows destruction of the climax plants and induces subclimax soil. Such a condition easily may be confused with a soil-plant complex which has never reached climax, unless careful study is made. The presence of a climax soil definitely will indicate the former existence of a climax-plant cover, for the two can never develop separately.

Natural plant succession, on arid lands, moves slowly and the stages are only relative, the end of one and the beginning of a new one being ill defined. The trend is toward a more exacting plant group, the individuals tending to become more specific as to their requirements.

Large plants have a positive advantage in their ability to shade out competing species of lower stature, but they have an inherent disadvantage in that a large surface area means large transpiration loss. Succession on dry land changes the habitat from the xeric to the more mesic condition; hence, the vegetation changes are from small and drought-resistant species to larger and less drought-resistant species. Other factors influencing plant competition, such as root spread, reproduction capacity, and shade tolerance, modify the trend, but generally succession follows a more or less similar pattern.

Succession involves a change in species composition and also a change in plant abundance. As soil develops and its moisture-holding capacity increases, greater plant density results. Soil, like vegetation, goes through successional stages. These stages develop concomitantly and closely parallel each other.

Soil Development

Soil undergoes a series of developmental stages as climate and organisms act upon the original rock or *parent material*. Ultimately a *climax soil* or one that is in a state of relative stability or balance with weathering and plant action is formed. On a climax soil, erosion is at a minimum; horizonal development has progressed as far as possible under existing climate; and downward movement of soluble materials by leaching is in balance with the upward movement by water and plants. Organic deposition is at a balance with decay. A biological balance is attained among the many minute plants and animals which inhabit the soil. Such a balance comes about only after many years of weathering. The nature of the product is determined primarily by climate but also is influenced by physiographic position, parent material, age, etc.

Jenny (1941) has suggested the formula

$$\text{Soil} = \text{function of } cl, o, r, p, t$$
$$\text{where } cl = \text{climate}$$
$$o = \text{organisms}$$
$$r = \text{relief or topography and exposure}$$
$$p = \text{parent material}$$
$$t = \text{time of soil formation}$$

Soil is a function or product of these factors, and a change in any one will induce a change in soil development.

Since soil and vegetation are so intricately related, it becomes obvious that these five soil-forming factors are the same ones which determine vegetation (Major, 1951) and the following formula is suggested:

$$V = \text{function of } cl, \, p, \, r, \, o, \, t$$

where V = any property of vegetation that can be expressed in quantitative terms. Thus, in a sense, vegetation is not determined by soil, and soil
is not determined by vegetation; vegetation and soil develop concomitantly (Major, 1951). At any one stage the quantitative aspects of vegetation are determined by soil; and if the soil is defined within areas of
uniform climate and relief, then vegetation also is defined.

Following this reasoning, it is obvious that, as the soil develops and
approaches climax, so also does the vegetation. With each phase of soil
development, a specific plant development is found—though the flora
differs according to the climate under which it develops. If undisturbed,
a climax soil will support a climax vegetation, both of which are in a
condition of approximate stability, at balance with the climate. A climax
vegetation is undisturbed by man or man's activities. It is a natural vegetation which has completed its development to a condition of relative
balance. It changes or fluctuates; but it is no longer following a trend
toward a fundamentally different condition. As the climate fluctuates, so
does the vegetation and, to a lesser extent, the soil. Nature is ever variable, and annual or cyclic waves, in the main controlled by weather, are
always present. A static condition is never known in nature. Climax
vegetation is in a state of dynamic equilibrium (Phillips, 1935).

Temperature and, especially, precipitation determine more than
any other factors the speed with which soil and vegetation develop toward the climax. Because of low precipitation, in most rangelands
movement toward a climax is slow. Steep topography has not been conducive to stable soils; hence, much of the mountainous Western United
States has not yet reached the climax condition. This fact is one of
extreme importance to the range ecologist, and he should not be misled
into believing that all subclimax conditions are due to retrogression from
a once superior condition rather than a normal lack of development.
Only careful studies of relict plants, root remnants, protected areas, and
historical records will show whether a higher development of soil and
vegetation once existed. Such an analysis is not easy, but it is basic to
determining whether the plant community is in a stage of primary or
secondary succession.

Stages of Primary Succession

The term *primary succession* generally is applied to natural plant succession on previously unvegetated areas leading to a climax. *Secondary succession* refers to a succession, usually induced, on land which previously has been occupied by more highly developed vegetation destroyed by some unusual circumstances, such as fire. Often, soil already is developed beyond presently existing vegetation.

Because of the great variety of habitats upon which succession can begin, there is also a vast number of possible vegetation combinations (Oosting, 1948).

Large areas of rangeland originated from deposits of wind- or water-moved soil. Under such conditions, soil formation and modification were not extensive. Other rangelands originated from dry rock surfaces where a typical xeric succession (Phillips, 1935), involving thousands of years, was necessary for formation of even a thin layer of soil. Under such conditions, few species are able to survive the aridity of an almost soilless environment. Lichens and xeric mosses which are able to absorb and hold precipitation against desiccation may be the sole occupants until soil depth is sufficient to store water for less specialized plants. These early invaders form carbonic acid from carbon dioxide and water which, together with mechanical forces of the rhizoids, tend to erode away the rock surface. These rock particles, together with organic and inorganic dust particles from the atmosphere, are physically retained by the plant thallus to form shallow layers of what ultimately is to become soil.

Larger and more advanced plants are able to invade as soil and moisture conditions become more favorable through the process of reaction. The soil development thus progresses at an ever-increasing rate. Plant roots penetrate minute cracks with such force as to break rocks. The increase of soil depth and the greater shade resulting from the taller growing plants creates an increasingly mesic habitat. Generally, longer-lived plants and those of greater stature gain in prominence. A dense forest may be the final product. Evaporation and extremes of temperature are decreased, humidity is increased, and drought periods are shortened.

Early plant succession progresses at a rate controlled by the rate of soil succession, but climate determines the end point of succession, and ultimately arrests the succession at a point which we call climax. This, in arid regions, is a point where available precipitation is used effectively, and invasion of species requiring more favorable moisture relationships will never be possible under present climate because of deficient precipitation. Weathering and plant action previously produced soils that could support a successively higher type of vegetation; now, although there will be variation from year to year, there no longer will be active trend and development toward another type of vegetation requiring more favorable growing conditions.

Competition as a Factor in Succession

An understanding of plant competition is important to understanding succession. In general, the best-adapted plant can compete best, because it can make most efficient and full use of the resources offered by the environment. Trees compete best because they are tall and can shade out smaller species. But the very size of the tree prevents its growth in dry areas because of its tremendous transpiring area. Usually the largest plants which can thrive under existent soil moisture will dominate. These will be perennial so they can guard their ground throughout the year, instead of just temporarily. They may be slow, but they will be efficient in occupying new areas by seed or by rhizome. The flora which they create generally will be the densest possible on the area forming a *closed* community in which soil and moisture resources are fully used, and there is no room for additional plants. This does not mean the forage is dense. Dry areas may be fully occupied and still display bare surface soil (Fig. 5.6) which will be filled underground by roots. Bare ground should not be considered an indicator of unoccupied area or an ecologically open community.

INTERACTIONS BETWEEN ANIMALS AND THE PLANT COMMUNITY

Disuse is not normal for vegetation. Plant communities evolved under the grazing pressures of large herbivores. The type of plants and their density largely reflected the grazing habits and food preferences of the animals present.

Plant communities change in an orderly way when grazed by a particular kind of animal. Those plants most preferred by the animals are the first to show signs of grazing stress. They lose vigor, little annual growth is produced, and reproduction is almost absent. As grazing is continued these palatable plants, or *decreasers,* die.

With the death of the more palatable plants, the less palatable members of the plant community *(increasers)* increase in abundance, resulting in a change in community composition. As the community progresses toward the less palatable plants, the grazing animals must change their

Fig. 5.6 Arid rangelands support sparse stands of shrubs as shown in this photograph of creosotebush *(Larrea divaricata)*. Bare ground does not mean there is room for other plants; the widely spreading root systems effectively prevent invasion of other plants.

food habits, move to new areas, or die. In turn, populations of new animals may develop that prefer the dominant species of the altered community. In that case, the increaser plants may become decreasers under the grazing pressures of the new herbivores.

Thus, plant communities have constantly changed through geologic time. At any particular time, the flora available to constitute the vegetation is a product of the climate, soil, and organisms available. The composition of the vegetation, however, is determined largely by grazing pressures from major herbivores.

With the advent of modern man, plant communities changed radically. Man brought new species of both plants and animals. He reduced the population of native fauna. Plant populations changed under the reduced native animal impact and the increased grazing pressure of domestic animals. Those plants that were most palatable to domestic livestock declined in vigor and abundance and became minor elements in the community. Less palatable plants increased. Entire plant communities changed their composition and brush and woody plants increased on many rangelands. In many instances, changes in the plant community induced by cattle grazing improved the habitat for deer and other wildlife (Leopold, 1939).

As habitats changed, some graziers changed from cattle to sheep or goats to better utilize the vegetation. This shift from cattle to sheep and goats is especially noticeable in Southwest Asia and Africa. In Texas, not only did sheep and goat numbers increase as grasslands deteriorated, but white-tailed deer *(Odocoileus virginianus)* numbers increased as well (Box, 1968).

Range vegetation, then, is being influenced constantly by the kind and amount of animals present. Indeed, the native rangelands of the world are the result of different types of grazing pressures. New Zealand, for instance, evolved in the absence of a large herbivore. The Australian rangelands developed under the grazing of large marsupials. In North America, pronghorn antelope *(Antilocapra americana)*, bison *(Bison bison)*, and usually either white-tailed or mule deer *(O. hemionus)* kept constant pressure on developing vegetation.

Grazing by Native Herbivores

In Africa where more than 20 species of large herbivores graze the same range, niche segregation among different species is highly developed. Talbot and Talbot (1963) reported that in Kenya there are more than 70 species of grasses available for wildebeests *(Connochaetes)*; however, 5 plant species comprised from two-thirds to three-fourths of the diet and 10 species made up 90 percent of the diet. Talbot (1962) reported that the diets of animals grazing the East African rangelands were complementary and nonduplicating. Most of the large grazing species ate different plants, but where they did eat the same plants, they ate different portions of the same species. For instance, wildebeests and topi *(Damaliscus korrigum)* eat the same species, but wildebeests eat the green leaves and topi eat the stalks and dry materials. When the preferred forage species of the highly selective African ungulates are utilized, the animals move to new areas rather than eat forage not normally eaten.

In contrast, white-tailed deer in South Texas may eat as many as 160 different species of plants within a 3-month period (Chamrad and Box, 1968), and as many as 58 species in a single month (Drawe and Box, 1968). As the supply of one plant becomes limiting, white-tailed deer readily shift to other plants.

Apparently white-tailed deer have evolved in an area with little or no competition for forage. Their habits allow them to pick and choose species and portions of species throughout the year and change diets with phenological changes in the vegetation. Conversely, African species have evolved a highly selective diet and instead of accepting other plant species, or portions of plants, they migrate. It remains to be seen what will happen when an African species is introduced into a new habitat, is confronted with new plants, and is unable to move with rains.

Grazing by Introduced Herbivores

When grazing animals are placed in range environments where they had not been before, the effects of grazing are intensified. An example of extreme habitat destruction from introduced animals was described by Howard (1964). Exotic animals were introduced in New Zealand where there had been no history of large animal grazing and the native plants had not been subjected to defoliation. The plant species found in New Zealand had evolved in the absence of browsing animals, and they had not developed resistance to use. Native plants died due to browsing and were replaced by exotic plants. In some of the heavily overgrazed areas of New Zealand, a new and stable equilibrium of animals, soils, and vegetation has been reached, more browse-resistant or less palatable plants having replaced the original vegetation. The introduction of domestic livestock in America had similar effects, not because the vegetation had not been conditioned to use but because domestic livestock had quite different foraging habits, and in some instances, were present in far greater numbers.

PLANT RETROGRESSION

Any of a great number of actions may disturb the climax plant cover and bring about *retrogression* leading away from the climax. Retrogression may be caused by drought, fire, or grazing. If this action is temporary, a succession leading back to climax follows.

Causes of Retrogression

By far the most important of the factors bringing about retrogression on range is improper grazing. The retrogression of a plant cover under grazing does not follow in the reverse order to the succession that gave rise to it, because the retrogression is usually of vegetation and not of soil. Since the climax soil is damaged less easily, it is more permanent than the vegetation, and its retrogression lags far behind. The stages of grazing retrogression in vegetation, then, are determined not by climate or soil, but by the newly introduced biotic factor, usually livestock.

Unfortunately, with continued weakening of the soil-protecting vegetation by grazing, soil deterioration also occurs. Johnston et al. (1971) documented soil changes in a prairie soil heavily grazed for 17 years and found reductions in organic matter and phosphorus, less soil moisture, and higher soil temperatures. They concluded that the soil changes observed made for a drier microclimate.

Water or wind may move away the developed surface soil to the point that exposed subsoil is incapable of again supporting climax plants. Succession to the climax, therefore, must again await development of a new soil mantle. Especially in arid areas, soil formation is a very slow process involving hundreds, or even thousands, of years. This fact should be kept constantly in the mind of the range manager, for no amount of management will return the range to full productivity if the soil cannot support the desired plants. Soil retrogression caused by erosion and trampling may progress so far that vegetation may be held in a subclimax stage, even though grazing has ceased entirely. Absence of a rapid secondary succession following good management often confuses the range manager, since vegetation cannot respond to improved grazing conditions as he expects.

Retrogression of vegetation under grazing may follow a multitude of courses, dependent upon vegetation and type of grazing. Grazing during a restricted season may harm only certain species, whereas others may be benefited because of reduced competition. If a short grazing season results in use of a certain species during a critical growth stage, that species may disappear. Another plant, fully as palatable, may thrive or even increase its numbers, because grazing does not occur in its critical period. Balsamroot *(Balsamorhiza)* may disappear from spring range under such conditions because its flower heads are highly preferred by livestock.

Similarly, because of forage-preference differences among kinds of livestock, grasses may increase on a sheep or deer range at the expense of forbs and brush; conversely, on cattle ranges grass may disappear. As Klapp (1964, p. 316) observed: "Simply stated the grazing animal eats what it likes and disdains what it does not like. That means protection for the disdained plant insofar as it is not hurt by trampling."

Too intensive grazing is marked by a disappearance of the preferred plants or of those physiologically less resistant to grazing. Retrogression, thus, involves plant competition. The removal of climax plants by abuse beyond their endurance leaves space for other plants. Less preferred or more resistant plants may survive and replace the removed plants. These species sometimes are referred to as *increasers*, because they increase under heavy grazing. Continued grazing will cause an influx of species, often annual, which are not a part of the climax. These are called *invaders*.

Stages in Vegetation Retrogression Induced by Grazing

Some stages of retrogression following improper grazing are recognized easily and are characteristic of most retrogressions. A complete understanding of these is essential to the range manager.

Physiological Disturbance of Climax Plants The most preferred climax plants, under stress of heavy grazing, lose vigor, as evidenced by reduction in annual growth; reduction or complete absence of reproduction activity; and, in woody plants, abnormality of growth induced by removal of the growing tip and excess stimulation of lateral buds.

Composition Changes of the Climax Cover   Continuance of physical disturbance of the preferred plants results in their death. Death and disappearance may result from starvation following reduced photosynthesis, competition from other plants less weakened by grazing, natural old age accompanied by lack of reproduction, or drought made more serious by a weakened root system. Composition change on the range usually is gradual, marked first by a decrease in (1) the most preferred plants, and (2) the plants physiologically and anatomically most susceptible to grazing injury (decreasers). Accompanying the decrease in numbers is a decrease in competition, which results in an increase of less preferred or more resistant individuals (increasers). Animals change their diet, because of increasing shortage of desirable species, to those less preferred. Thus, succession continues, with better climax plants progressively becoming fewer.

Invasion of New Species   Following, or simultaneous with, these composition changes comes the invasion of new species, which may or may not have been present in the primary succession but which were not constituents of the climax cover. The first invaders are mobile annuals; but, later the invasion of herbaceous or woody perennials of low grazing value often takes place (Fig. 5.7). The annual invaders may be highly preferred by stock for a short season, but they often are adapted to thrive despite grazing. Most invading perennials are not highly prefer-

Fig. 5.7 Burroweed *(Aplopappus tenuisectus)*, a low-value shrub which has increased on desert grasslands in the Southwest with overgrazing.

red by stock, and many are valueless. The first three stages in retrogression are marked more by decreased quality than by decreased quantity.

Disappearance of Climax Plants Climax plants ultimately may disappear. They leave first from the most accessible and, hence, most grazed areas, and soon are evident only under the protection of a stout shrub or thorny cactus. Later, even these disappear, often leaving nothing but annual invaders (Fig. 5.8).

Decreased Density of Invaders Continued heavy grazing forces stock to consume the invading species which suffer as did the climax species. The most preferred and most susceptible species are removed first, and the less valuable temporarily increase in numbers. As grazing continues, these may bear the brunt of the grazing, and their numbers will decrease. If these are not followed by new invaders such as shrubs of low palatability, the land approaches a barren state, with soil deteriorating rapidly (Fig. 5.9).

Secondary Succession Following Retrogression

The secondary succession following improved grazing conditions usually differs from primary plant succession since good soil conditions may

Fig. 5.8 The relatively bare ground which follows excessively heavy grazing is invaded by undesirable annual and biennial plants such as *Bromus tectorum* and *Grindelia squarrosa*, as shown here.

Fig. 5.9 Bare ground is the ultimate stage in retrogression as shown by this photograph taken in the vicinity of an old Roman cistern at El Quasr. (*Courtesy F. A. O.*)

remain. Frequently, however, soil retrogression follows plant retrogression, because of erosion and trampling. In such cases, secondary succession may be almost as slow as primary succession, and follow in very similar steps toward the climax.

When soil has not deteriorated along with vegetation, succession of vegetation, upon removal of grazing stress, may be very rapid, especially in areas of high precipitation. It is especially rapid if climax plants remain to seed the area, but slower when all climax plants have been removed. The speed, in the latter case, is dependent upon the mobility of the propagules of climax plants, and it may involve many decades. Indeed some areas appear to be so completely and effectively dominated by exotic invaders that it seems doubtful whether climax plants ever can reoccupy the area under any economic form of use. It is doubtful, for example, that the Mediterranean annualgrass ranges should be grazed with an objective of returning them to perennial cover. The range manager must assess each such area to determine practical feasibility of his management objective. Invaders may be highly productive and perhaps ideally suited to certain purposes. Annual grasses, for example, may be excellent spring-lambing range but less valuable summer range.

Practical management may maintain climax cover but, often, a vegetation cover lower than climax proves most practical. It is impossible to obtain the best use of a range without some disturbance, and the rancher cannot always have climax vegetation as his goal. There is error in a too liberal viewpoint, however, and the many tragic results of inadequate attention to maintenance of good forage far outweigh the short-lived profit from excess numbers of stock. When maintenance of maximum soil protection is wanted, it may be desirable to manage for near-climax conditions and capitalize on the diversity of higher stages. Changes in plant composition may not result in reduced plant cover, however; cover may actually increase (Table 5.3).

Forage Value of Invading Species

The preference which an animal displays for a plant is not an accurate index to its value for grazing. Animals can be forced to eat almost any plant, and some of the less-liked species are as nutritious as are the preferred ones and animals sometimes do as well on them. A slight decrease in palatability of the plant cover after excess grazing may, in itself, be no indication of reduced value. Usually, however, invasion of less-preferred species is accompanied by marked reduction in grazing capacity, independent of volume yield. This, probably, is attributable to the fact that animals eat less of feed which they do not like, rather than to nutritional difference. Many invaders actually are highly palatable and valuable forage (see Fig. 2.7). Notable examples are bur clover (*Medicago hispida*) and alfilaria (*Erodium cicutarium*). Nutritional studies have failed to show consistent differences between climax and invading

Table 5.3   Basal Area of Vegetation under Different Grazing Treatments from 1950 to 1969

Data from Smoliak et al (1972)

Species	Study area in 1950	Study areas in 1969 by grazing treatments			
		Ungrazed	Light	Moderate	Heavy
		(Percent)			
Blue grama	2.0	0.3	2.4	2.8	3.6
Needle-and-thread	0.6	2.2	1.3	1.2	0.5
Western wheatgrass	1.4	1.3	1.7	1.2	0.7
Other grasses	1.3	1.4	1.6	1.5	1.1
Low sedge	0.9	0.4	0.5	0.8	1.1
Forbs and shrubs	0.5	0.4	0.7	0.9	0.4
Little clubmoss	7.1	9.6	18.7	22.6	26.0
All vegetation	13.8	15.6	26.9	31.0	33.4

species, except that invading ephemerals are likely to become dry earlier and to deteriorate more upon drying than are long-lived perennials.

In some cases, particularly in the leached soil of monsoonal tropics such as northern Australia, the climax vegetation is low in nutritive value during the dry season. The introduction of weedy annuals from other continents such as Townsville lucerne (*Stilosanthus sp.*) results in improved forage quality during the dry season.

Unfortunately, most rangeland climates are hazardous for establishment of young plants and, since this process is a frequent necessity for short-lived species, their dependability is greatly reduced. Annual plants depend upon a favorable period each spring during which they can germinate and send their roots to the moist subsoils. Such a period is far from a surety over most ranges, and, hence, failures are common. Fluctuation in forage volume from year to year is much greater on ranges high in annual plants. Grazing capacities and animal gains from cheatgrass range in southern Idaho may be three times greater one year than another (Murray and Klemmedson, 1968).

Annuals are, likewise, most variable in their season of growth. Perennials, having deep roots already established, are less dependent upon current precipitation and more upon temperature, which is less variable, for their start of growth. Annuals are dependent upon precipitation for initiation of growth and may reach their period of productivity at vastly different dates from one year to another.

Annuals are short-lived and are best grazed during their green period, often only 6 to 8 weeks, which may not fit well with the management scheme.

Most poisonous plants are low in palatability, hence increase upon heavy grazing is inevitable. It is believed that losses from many poisonous plants have greatly increased with depletion of the ranges. Retrogression following misuse is the greatest single factor contributing to poisoning. Gradual invasion of low-quality species and decline of good forage, indeed even serious decrease in total production, may escape notice of the range manager. This decrease in forage ultimately forces animals to eat plants which normally remain untouched.

Many losses attributed to other factors are indirectly a result of forage deficiency. The losses of livestock from increased disease accompanying malnutrition are high. Losses from predators and parasites are known to be large in weakened flocks. Drought and severe winters result in more damage if animals are in poor condition. Malnutrition causes physiological disturbances, decreased gains, decreased calf and lamb crops, and decreased quantity and quality of wool. Healthy animals and maximum production cannot be expected on sick ranges.

MANAGEMENT OF RANGE ECOSYSTEMS

The range manager today does far more than attempt to maximize animal yield or even forage production. He must be aware of the many interlinking functions in the ecosystem of which his range is a part, and he usually attempts to optimize the combined output of several goods and services. He may use many of the quantitative decision-making tools available to foresters, wildlife managers, and other natural resource specialists (Van Dyne, 1969; Watts, 1968)—only his goal will be different. In any case a detailed analysis of the structure and function of the particular range ecosystem will provide a basis for developing a productive system.

A productive system of products from rangelands implies that outputs are expected from the system. As with a checking account at a bank, if outputs are withdrawn without "deposits" in the form of inputs, the system will deteriorate. In intensive-cropping agriculture, the system is sustained through inputs of commercial fertilizer, irrigation, gasoline for the tractor, etc. The addition of such nutrients and energy is not practical on most rangelands. Therefore, the range manager depends on plants to capture energy and manipulates animal use of the plants to minimize drain on the system.

In systems where native plants make up most of the biomass, as on the Nkomazi ecosystem of East Africa (Harris, 1972), nutrients may be redistributed and recycled through the native fauna, but nutrients remain within the ecosystem. When the game animals are removed through hunting or game cropping, those animals represent a net loss of nutrients from the ecosystem. The harvesting of forage with domestic animals that are almost all removed from the ecosystem is an even more serious nutrient drain. Because this drain is not as rapid or dramatic as from those that are cropped, the loss has not received adequate attention, but is no less real. This fact is being recognized and research projects are being designed to supply information to range managers on this phase of ecosystem dynamics (Blaisdell et al., 1970).

Control of the ecosystem by the range manager is practiced through manipulating both the biotic and abiotic factors. Of the abiotic factors, fire is the most easily manipulated to gain range improvement. Burning as a range-management tool is widely used from Africa (Phillips, 1972) to grasslands of the Southwestern United States (Wright, 1973).

The biotic factors most often are manipulated to gain the objective of the range manager. Range seeding and brush control alter the populations and composition of producer organisms. Stocking rates are set for livestock, combinations of species of grazing animals are selected, and supplemental feeding is practiced to control the primary consumer level. Predators may or may not be controlled, thus influencing the tertiary consumer trophic level. Decisions about all levels are made by man, the master manipulator, who must live within his social, political, and economic framework. Most of the actions listed will receive detailed treatment in later chapters. But, the range manager must always keep in mind that an action on one part of the ecosystem, e.g., seeding a spring range for sheep, affects the entire system. Thus, he must be prepared to bear the consquences of his action.

BIBLIOGRAPHY

Blaisdell, James P., Vinson L. Duvall, Robert W. Harris, R. Duane Lloyd, Richard J. McConnen, and Elbert H. Reid (1970): Range ecosystem research—the challenge of change, *U.S. Dept. Agr. Inform. Bull.* **346.**

Box, Thadis W. (1968): Introduced animals and their implications in range vegetation management, *Proceedings of Symposium on Introduction of Exotic Animals,* Texas A & M University Press, College Station, Tex., pp. 17–20.

Chamrad, Albert D., and Thadis W. Box (1968): Winter and spring food habits of white tailed deer in South Texas, *Jour. Range Mgt.* **21:** 158–163.

Drawe, D. Lynn, and Thadis W. Box (1968): Forage ratings for deer and cattle on the Welder Wildlife Refuge, *Jour. Range Mgt.* **21:** 225–228.

Golley, Frank B. (1960): Energy dynamics of a food chain in an old field ecosystem, *Ecological Monogr.* **30:**187–206.

Harris, Lawrence D. (1972): An ecological description of a semi-arid east African ecosystem, *Range Science Department Science Series,* No. 11, Fort Collins, Colo.

Howard, W. E. (1964): Introduced browsing animals and habitat stability in New Zealand, *Jour. Wildlife Mgt.* **28:** 421–429.

Jameson, Donald A. (1970): Land management policy and development of ecological concepts, *Jour. Range Mgt.* **23:**316–321.

Jenny, Hans (1941): *Factors in Soil Formation,* McGraw-Hill, New York.

Johnston, A., J. F. Dormaar, and S. Smoliak (1971): Long-term grazing effects on fescue-grassland soils, *Jour. Range Mgt.* **24:**185–188.

Klapp, Ernst (1964): Features of a grassland theory, *Jour. Range Mgt.* **17:**309–322.

Kucera, C. L., Roger C. Dahlman, and Melvin R. Koelling (1967): Total net productivity and turnover on an energy basis for tall grass prairies, *Ecology* **48:**536–541.

Leopold, A. (1939): A biotic view of the land, *Jour. Forestry* **37:**729–730.

Lewis, James K. (1959): The ecosystem concept in range management, *Amer. Soc. Range Mgt. Proc.* **12:**23–25.

———— (1969): Range management viewed in the ecosystem framework, in George M. Van Dyne (ed.), *The Ecosystem Concept in Resource Management,* Academic Press, New York, pp. 97–188.

Major, Jack (1951): A functional approach to plant ecology, *Ecology* **32:**392–412.

Murray, R. B., and J. O. Klemmedson (1968): Cheatgrass range in southern Idaho; seasonal cattle gains and grazing capacities, *Jour. Range Mgt.* **21:**308–313.

Odum, Howard T. (1957): Trophic structure and productivity of Silver Springs, Florida, *Ecological Monogr.* **27:**55–112.

Oosting, Henry J. (1948): *The Study of Plant Communities,* W. H. Freeman and Co., San Francisco.

Perry, Rayden A. (1970): Arid zone newsletter, *Commonwealth Scientific and Industrial Research Organization,* Canberra.

Phillips, John (1935): Succession, development, the climax, and the complex organism: an analysis of concepts, Part II, Development and climax, *Jour. Ecology* **23:**210–246.

———— (1972): Fire in Africa—a brief re-survey, *Tall Timbers Fire Ecol. Conf. Proc.* **11:**1–9.

Rodin L. E., and N. I. Bazilevich (1967): *Production and Mineral Cycling in Terrestrial Vegetation,* G. E. Fogg (trans.), Oliver and Boyd, London.

Smoliak, S., J. F. Dormaar, and A. Johnston (1972): Long-term grazing effects on *Stipa-Bouteloua* prairie soils, *Jour. Range Mgt.* **25:**246–250.

Talbot, L. M. (1962): Food preferences of some East African ungulates, *East African Agr. For. Jour.* **27:** 131–138.

Talbot, L. M., and M. H. Talbot (1963): The wildebeest in Western Masailand, East Africa, *Wildlife Monographs* No. 12.

Tansley, Arthur D. (1935): The use and abuse of vegetational concepts and terms, *Ecology* **16:**284–307.

Van Dyne, George M. (ed.) (1969): *The Ecosystem Concept in Natural Resource Management,* Academic Press, New York.

Watts, Kenneth E. F. (1968): *Ecology and Resource Management,* McGraw-Hill, New York.

Williams, A. William (1966): Range improvement as related to net productivity, energy flow, and foliage configuration, *Jour. Range Mgt.* **19:**29–34.

Wright, Henry A. (1973): Fire as a tool to manage tobosa grasslands, *Tall Timbers Fire Ecol. Conf. Proc.* **12:**153–168.

Chapter 6

Range Inventory and Evaluation

The management of rangelands depends upon a knowledge of the physical and biological characteristics of the land. Unless these are known, prescribing kinds and levels of use appropriate to a specific area is impossible. Particularly is this so with respect to livestock grazing, for the effects of too large numbers of animals or improperly timed grazing can be devastating to the often fragile environments of the range. Effective range-management practices require careful inventories of resource characteristics. The concepts and techniques developed or adopted by range managers to do this are the subjects of this chapter.

PURPOSES OF RANGE INVENTORIES

Range inventories are made for a number of reasons. They may be extensive, or general; they may be intensive, or detailed. They may include a wide spectrum of data ranging from surveys to determine the livestock productivity of an entire nation to a survey of a ranch or a range allotment. Each individual and agency tailors basic methods to fit the particular circumstances at hand. The following cover the general purposes for which inventories are made. They are seldom mutually exclusive.

Ecological classifications deal with the physical and environmental factors. Precipitation, topography, soils, and broadly defined vegetation communities are the usual components. Commonly, they are undertaken to provide a framework for other more intensive inventories. The ecological survey techniques developed for Australia (Christian, 1959) now are used by many developing countries where detailed range analyses are impractical. As more information is needed, more detailed methods can be used.

Range-forage inventories most commonly are for the purposes of determining the grazing capacity of rangeland either for domestic livestock or wild herbivores. Special consideration is given to plant species and types which are preferred as forage, though other physical and biological data bearing on range productivity and management procedures may be included.

Utilization surveys are ways of assessing the current grazing pressure exerted by foraging animals as a means of determining appropriateness of current stocking levels or management systems. Utilization in excess of physiological requirements of the vegetation requires adjustments in livestock numbers. Unequal utilization throughout a grazing unit suggests the need for improvement in management practices.

Condition and trend analyses are at present the most important range evaluations. The data compiled from them enable the range manager to judge the adequacy of stocking and management practices. Based on successional and community-dynamics concepts, they are designed to assess whether range sites within the range ecosystem are at, or depart from, accepted standards and capabilities based on their potential for production.

Multiple-use surveys are whole-spectrum analyses done to determine the entire biological and physical resource base with the objective of integrating all capabilities and uses of the range into a comprehensive and coherent plan.

Rangeland appraisals are for the purpose of determining economic productivity of a range area. The physical inputs come from information developed in inventories described above. These are related to other factors which, though outside the range area, are intimately associated with it. For example, in the case of range-livestock production the facilities of the base ranch must be considered; in the case of public ranges, other uses and resources must enter into the evaluation. In the latter case, important factors are size and proximity of neighboring communities, and their dependency on the range for nonforage resources.

METHODS USED TO INVENTORY RANGELAND

Ecological surveys provide useful basic information to other kinds of range inventories, though they are not strictly essential to them. Commonly, ecological information is gathered coincident with other inventory efforts. They are often broadbrush treatments in connection with land-resource surveys, as in Australia; they may involve personnel of several disciplines; and they are largely descriptive (Perry, 1960; 1961). These broad surveys often form the first step in range-management planning.

Vegetation Inventories

A vegetation inventory provides data on the absolute or relative abundance of plant species by vegetation types. Data may be estimates or they may be quantified by numeration, by ground cover, by volume, or by weight. Normally the data are derived from sample plots positioned throughout the area sampled. Vegetative types most often form the basis of the sampling unit.

Kinds of Measurements Used in Plant Inventories The simplest inventory is a *species list* in which consideration is given not to relative amounts, but only to presence of a plant. *Number lists* are made by counting the number of individual plants of each species occurring in sample plots. When referenced to a unit area, number lists yield a measure of plant occurrence called *density*, as 20 plants per square meter. Often inventories are based on the surface area occupied by each species, called *cover*, stated in percentage of area. In older range literature, cover was called *density* which should not be confused with its present usage. Cover may be measured at crown level (Fig. 6.1) or, especially in the case of grasses, at ground level.

None of the above measures supplies information on amounts of plant material. Plant species vary greatly in size. A low-growing plant may cover a large area; an erect plant may cover little surface area. Either density or cover data, however, provide useful information on plant changes over a time span and record the disappearance of plants from the species mix or the appearance of plants not previously present.

Volumetric measurements of vegetation are seldom made in range analyses; rather, weights are used to record the amounts of plant material produced (*biomass*). This can be done most precisely by *clipping* sample plots, but this is time-consuming and costly, and, where the vegetation is heterogeneous, requires a large number of sample plots. Thus,

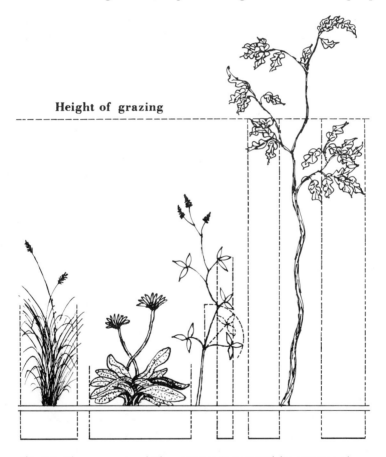

Height of grazing

Fig. 6.1 The manner in which vegetation is projected for purposes of cover determination. If other than forage cover is being estimated, the foliage in the top of the shrubs would also be included.

except in research, and not always then, weights most often are determined by estimation. By repeated estimation followed by clipping and weighing, many individuals can become quite proficient in making reasonably accurate weight estimates. Others are never quite able to make estimates either in weight or cover with any degree of reliability (Smith, 1944).

Often clipping and estimating are combined, a procedure called *double sampling*. Part of the plots are clipped and weighed, and estimates are made on all of them. The clipped weights are compared to the estimated values to correct, usually by regression analyses, the production estimates for the unclipped plots. Students wishing more detailed information on these and other methods of vegetation analysis should refer to the National Academy of Sciences—National Research Council (1962).

Sample Plots   Sample plots vary in size depending primarily on the kind of vegetation studied. Tree and shrub stands require larger plots than herbaceous vegetation. The most effective sampling of an area can be obtained by the use of numerous small plots, rather than fewer and larger plots, but the plot chosen must be large enough to encompass individual plants of the larger species present. Spacing of individual plants and the number and distribution of species are important in determining plot size. The plot size required increases both as distance between plants and the number of species increase.

Plots commonly used in range analyses are 1 sq m, a milacre (6.6 ft sq), 9.6 sq ft, or 96 sq ft (8.36 sq m). The last three of these were adopted to facilitate conversion of data to an acre basis. Forage yield in grams within a 96-sq-ft plot is equivalent numerically to yield in pounds per acre (Frischknecht and Plummer, 1949). Where a percentage figure is desired a plot 100 sq ft is most efficient. With wider use of the metric system, plot sizes based on English measurements will occur only in older literature.

Shape of Plots   Plots may be round, square, or rectangular. Sometimes rectangular plots are elongated greatly in length and narrowed in width (belt transect), sometimes to a mere line (line transect). Transects are especially useful to sample across ecotones where one vegetation type intergrades into another. Permanent plots are commonly square or rectangular since marking the corners of the plots with stakes ensures more accurate placement of a plot marker at subsequent visits. A stake marking the center of a round plot accomplishes the same thing, except there is greater chance for loss of a single stake than four if the plot is to remain for a long time.

Irrespective of shape, permanent plots commonly are referred to as *quadrats* especially when the position and area of each plant are mapped. Round plots are used more frequently for temporary than for permanent plots. As in the case of a line transect, a circular plot may be reduced to such a size that no significant area is represented—e.g., a 19 cm diameter loop (Parker, 1951) or a point (Levy and Madden, 1933). Neither the loop nor the point method gives volumetric data, only frequency occurrence, unless accompanied by other inventory data. They do provide data on plant composition, most nearly in terms of cover.

Sampling Vegetation. Whatever the size of plot and the area sampled, effort must be made to obtain a representative sample. Though carefully selected plots may fulfill this criterion, statistical theory requires that plots be randomly located, since all measures of statistical reliability are based on chance occurrence. As a practical matter, a purely random arrangement of plots makes plot location much more time-consuming than a mechanical arrangement in which plots are spaced regularly at fixed intervals. Consequently, elements of the two methods often are combined. This is done by randomly selecting lines through the area under study and randomly locating plots along these. Thus, plot location is simplified, and chance is permitted to operate so that statistical analyses are possible.

Stratification of Vegetation to Facilitate Sampling Since the number of plots required to obtain an adequate sample depends on the heterogeneity of vegetation, this number can be minimized by reducing the heterogeneity of the population being sampled. This can be done by dividing the area under study into subareas on the basis of differences in the vegetation. Plant communities differ from place to place due to edaphic factors—slope, exposure, and soils. Recognizing these differences beforehand and making these subareas the sampling unit reduce the variability among individual plots which, in turn, enables one to attain a given degree of reliability with few plots (a smaller sample). The outlines of these sampling areas should be drawn on a base map, preferably aerial photos, before collection of vegetation data is begun.

Vegetation Typing As with plot size and shape, there is no precise pattern which must be followed in delineating vegetation types. Controlling factors are the intensity and the specific purposes of the study. Thus minimal areas may be in hundreds of hectares (or acres) or only a few meters. Type distinctions may be based only on the dominant species which give characteristic aspect to the area, or they may be based on secondary species which are discernible only on close inspection. Within types of the same composition or aspect, differences in cover or density may be recognized which may provide further bases for type differentiation.

There are, however, more or less standard types which have been found useful in range analyses of public lands in the United States (Table 6.1). Though these range types were based on different ecological criteria (Humphrey, 1949), they served a useful purpose. Using a four-letter symbol for plant species in place of one of three letters, a type symbol might be 4-Artr Stco Agin. This describes an area dominated by *Artemisia tridentata* with *Stipa comata* and *Agropyron inerme* as principal understory species. Wnen the specific objective of the range inventory is grazing capacity, acreages and forage production figures can be entered underneath each type designation. Thus:

4-Artr Atco
640

210

Table 6.1 Standard Vegetation Types Adopted for Use in Range Surveys

Type number	Name	Standard map color
1	Grassland	Yellow
2w	Wet meadow	Orange
2d	Dry meadow	Orange
3	Forb	Lake red
4	Sagebrush	Brown
5	Browse	Olive green
6	Coniferous timber	Light green
7t	Heavy timber	Blue green
7	Other	Blue green
8	Barren	Uncolored
9	Piñon-juniper	Verdant green
10	Deciduous trees	Pink
11	Creosotebush	Bottle green
12	Mesquite	Yellow earth
13	Saltbush	Slate grey
14	Greasewood	Royal purple
15	Winterfat	Light tan
16	Desert shrub	Dark tan
17	Half-shrub	Wisteria
18	Annuals	Red terra cotta

indicates an area dominated by big sagebrush and *Atriplex confertifolia* of 640 acres with the equivalent of 210 acres occupied by usable plant forage (forage acres). A completed range-reconnaissance map without coloration might appear as in Fig. 6.2.

Whatever the purpose and intensity of a range-resource inventory, a map makes the most suitable and widely used means of summarizing and depicting the essential facts pertaining to the area studied.

Photography and Remote Sensing

Photographs are a useful aid to range analyses because they provide visual evidence which is difficult to convey by data alone. They are especially useful in reconstructing changes in vegetation over a period of years. Close-range photography has not, however, proved helpful as an analytical device because of the distortion resulting from varying proximity to the camera of vegetation occupying different height strata.

Aerial photographs taken at greater heights and which cover considerable area are more useful, their quality and usefulness varying with the type of imagery used. Most useful are low-level aerial photographs (1,500 to 2,000 m) taken for the specific purpose of making vegetative-resource surveys. Careful analysis of stereo pairs of such photographs can provide estimates of many of the vegetative parameters on which range decisions are made. They can be valuable aids to mapping vegetation and even to indicating utilization in the California-annualgrass type. Differences in vegetation density or kind of vegetation appear as differences of tone and texture on the print (Fig. 6.3). Dudzinski and Arnold (1967) used aerial photographs to study the grazing patterns and behavior of sheep in Australia.

High-elevation photographs taken from supersonic aircraft or satellite coverage provide for synoptic or broad-scale evaluation of drainage patterns, soil groups, and vegetation types, but they lack sufficient details for intensive management. As experience is gained, and technology and interpretive skills improve, they will become more useful. Whatever kinds of photographic aids are used, they provide only a preliminary evaluation; specific details must be derived from on-the-ground sampling and conventional inventory techniques, although as experience is gained in interpretation more information can be derived from the photographs themselves if one is acquainted with the ecology of the area.

Color photography and thermal sensing have opened new but as yet largely unexplored horizons for range analysis. Color photography provides much sharper contrasts than are possible with black-and-white photographs, especially when filters are used. Thermal sensing employs invisible wave lengths, infrared rays (Avery, 1968). Because of wide

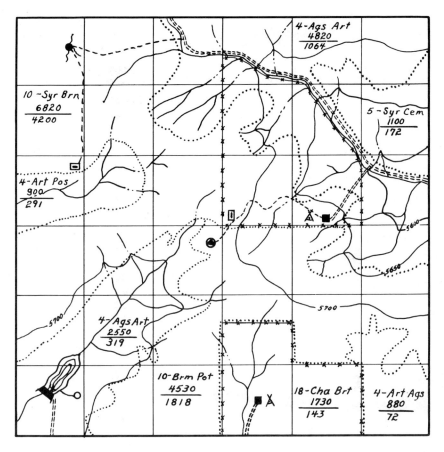

LEGEND

■	BUILDING	=====POOR MOTOR ROAD	
▭	SALT TROUGH	-----TRAIL	
◉	PERMANENT LOOKOUT	-*-*-*-FENCE	
Å	WINDMILL	PERMANENT STREAM	
	SPRING DEVELOPMENTS	INTERMITTENT STREAM	
	PERMANENT	........TYPE LINE	
	SEASONAL	-5700-CONTOUR	
Q	NATURAL SPRING		

4-ART AGS	TYPE NUMBER & MAJOR SPECIES
890	SURFACE ACRES
52	ANIMAL UNIT MONTHS

Fig. 6.2 A range-survey map with acreages and carrying capacity compiled by vegetative types.

Fig. 6.3 A section of an aerial photograph showing dotted vegetative-type lines. Letter-number symbols are index numbers which refer to inventory data for each type.

differences in the degree to which vegetation of different kinds reflect infrared rays, sharp contrasts result, although the colors shown are false and one must learn to interpret these differences by inspections on the ground or through experience.

The field of remote sensing of natural events is growing rapidly. New developments in this area offer promise of exciting new tools, especially for research, that will make range surveys more accurate and easier to accomplish.

DETERMINATION OF GRAZING CAPACITY

A major problem for the range manager is determination of grazing capacity of the land. Many range administrators begin their careers without the background of experience which some stockmen possess and through which they have developed a facility for judging rangeland potentialities. However, some ranchers through lack of observation have failed to acquaint themselves with facts essential to good management. These problems have led to development of procedures for inventorying the range-forage resource to determine proper and safe levels of use.

Concept of Grazing Capacity

Weather in range areas is subject to great annual variability, and annual forage production tends to vary even more, being correlated especially with seasonal distribution of precipitation. The production may be 2 to 3 times as much one year as another on perennial forage; on annual vegetation it may be 10 times as much one year as another. Such variation necessitates ultraconservative use in productive years or flexibility in number of livestock from year to year.

Great variation of stock numbers permitted on federal lands is infeasible, however. Federal range managers often must grant term permits for specified livestock numbers. Specific numbers are demanded by the livestock growers to maintain stability, and only through occasional and minor adjustment in the duration of the grazing period in a given year can grazing be adjusted fully to annual production of forage. The problem of the administrator, then, is one of designating a specific number of animals which can graze on a unit of land year after year without injury to the land and which he designates as the grazing capacity of that land (Stoddart, 1952).

Grazing capacity, then, has come to be regarded as the maximum animal numbers which can graze each year on a given area of range, for a specific number of days, without inducing a downward trend in forage production, forage quality, or soil. Perhaps it should be pointed out that this is, in many ways, a wasteful concept of grazing capacity, because, if forage is to be adequate in years of low production, then there must be excessive amounts of ungrazed forage in years of high production. If economically full use is made of forage in good years, overgrazing must result in poor years. We cannot be sure that overuse in one year can be compensated by underuse in another year; hence, stocking that would result in correct use in the average year may not suffice. The practice generally is to stock so as to prevent overutilization of the available forage in all but the most extreme drought.

Estimating Grazing Capacity

Early efforts to determine grazing capacities focused on examination of the vegetation. Data on plant occurrence of each species were combined with forage-value ratings to arrive at an index of forage productivity.

Plant abundance was measured in terms of cover (then called density). Two procedures were used to make this "density" inventory. In one, the cover of all vegetation first was estimated without recourse to plots; then the proportion each species made up of all vegetation was estimated. Both were expressed in percent. Later plot sampling was introduced to obtain greater precision and sophistication. A series of 100-sq-ft plots were located in the area being sampled, and the area within the plot occupied by each species was estimated in square feet. Since the plots were 100 sq ft in size, the square-foot value was also a percentage value. Summing the mean values for each species from all the plots yielded the density figure for the entire area. The grazing-capacity rating in either procedure was arrived at by summing the products of the cover ratings and the forage-value ratings for all species. The forage-value rating or *proper-use factor* (originally called palatability factor) was in percent and was assumed to represent the part of the plant's annual production eaten by animals under proper use.

This summation yields the *forage-acre factor* (FAF) which, since it is derived from two percentages, is always less than unity. Thus, it indicates to what degree the area under study approaches a theoretical standard acre completely covered by vegetation all of which is acceptable to and usable as forage by animals. When the FAF is multiplied by the area of the range unit to which it applies, the number of such theoretical *forage acres* is determined. Fig. 6.4 shows the calculations made when plots are used as the basis of the vegetation estimates.

This procedure falls short of being a precise guide to grazing capacity because (1) cover is not a good index to production, low-growing spreading plants being overvalued as compared to tall erect ones (Table 6.2); (2) there are great variations in individual estimates on the same plots (Smith, 1944); (3) forage-rating indexes (proper-use factors) vary greatly from place to place and are often overly optimistic; and (4) forage acres thus determined are not of equal value in support of grazing animals among different range types and areas even when numerically equal.

The problems arising from these factors were surmounted by establishing a *forage-acre allowance* which was determined by actual performance. This required an area where livestock numbers were known and where a range survey had been conducted. If, for example, 100 cows had spent 4 months on a range having 120 forage acres, the forage-acre allowance for that area, and presumably all other areas similar to it, would be 0.30 forage acres per cow month ($120 \div 100 \times 4 = 0.30$). Other areas and forage types might have quite dissimilar allowances depending largely on the productiveness of vegetation per unit area of ground cover.

RANGE SURVEY WRITE-UP SHEET

Project ..Utah Co., Utah..   Transect No. .....D - 24

Date ..July 17, 1948..   Location ..6 N 5E 24

Examiner ..John Doe..   T. R. Sect. , or Aerial Photo No.

Type ..4 Atr Aco..   Timber ..Douglas fir.. poor.

Forage Density ..1437..   (Composition) (Condition)

Total Density ..1431..   ..none.. (Reproduction) (Density) (Age)

F.A.F. ..0.270.. For ..Sheep..   Slight   Porcupine

(C, S, C.U.)   (Injury)   (Cause)

Utilization cuts: Slope ..0.. % Timber ..0.. % Rocks ..0.. % Erosion ..15.. %

Lack of water ..10.. % Unstable soils ..0.. % Total cut ..35.. %

PLOT NUMBER															Total Sq.ft.	Ave. Dens.	Times Pal.
Artemisia tridentata	8	0	6½	5	4	3½	6	2½	0	1	3	48½	.0482	.0048
Atriplex confertifolia	2½	1	0	5	6½	2	3	3½	1	2	24½	.0243	.0049
Eurotia lanata	0	0	1½	0	0	0	0	0	0	0	1½	.0012	.0007
Agropyron spicatum	0	¼	3½	0	0	¼	½	2	0	0	6½	.0065	.0039
Oryzopsis hymenoides	0	0	0	0	0	½	0	0	3½	2	6	.006	.0048
Achillea lanulosa	6	0	1½	0	0	0	0	<¼	2	1	3½	.0185	.0061
Salsola pestifer	5½	6	1¼	2	0	7½	2	4	3½	2	44	.044	.0110
												1431	.0070

Fig. 6.4 Data sheet and compilation used in plot surveys, showing the method of computing the forage-acre factor.

Although the method was not a precise means for determining grazing capacities, it did provide an inventory of vegetation present, data on plant cover, and observations on soil erosion and other factors pertinent to proper range use. Despite its shortcomings, it was useful as a means of allocating portions of a large range unit among several claimants when individual range allotments were made. It proved particularly

Table 6.2 Dry-Weight Production, Grams per Estimated Square Foot, of Several Important Forage Plants; Cover Obtained by Averaging Estimates of 27 Individuals

Plant species	Yield per sq ft, gm
Aspen (Populus tremuloides)	61.0
Snowberry (Symphoricarpos sp.)	49.6
Geranium (Geranium sp.)	113.3
Sweet clover (Melilotus alba)	208.8
Great Basin ryegrass (Elymus cinereus)	248.1
Bluegrass (Poa pratensis)	16.4

useful during the years when the public domain in the United States was being brought under administrative control by the Bureau of Land Management for the reason that, even though it had inaccuracies, the inaccuracies applied more or less evenly over all parts of a particular range. One can, for example, divide a stack of hay quite evenly between two or more individuals without knowing how many tons of hay are in the stack.

Weight Methods

With the passage of time, attempts were made to compensate for the fact that cover did not give a precise measure of yield. One proposal was to determine a mean yield per plant for each species and combine this with the proper-use factor into what was called a *volume-palatability* rating for each species (Standing, 1933). This approach was little used. It required a great deal of harvesting and weighing of plants to establish mean yields, and site and annual variations prevented establishing a mean figure of general applicability.

Another solution was to estimate weight of forage produced (Pechanec and Pickford, 1937). The weight estimate for each species was then multiplied by the *proper-use factor* for that particular species to determine the amount of harvestable forage each species contributed. Summing these products for all species gave forage production for the area. Because of the time required for individuals to reach proficiency, weight estimations are most used by researchers.

For administrative purposes, clipping of forage often is resorted to on federal ranges to provide an index to forage production, although seldom is it done at an intensity to provide precise estimates of forage yield. Usually, clipping is confined to key areas (page 212) to provide factual data to support more general information developed in extensive surveys. The U.S. Soil Conservation Service employed clipped plots to develop grazing-capacity guides. Once forage-production standards are determined for each range site, clipping may be used to establish the condition of other ranges.

Production data can be used as guides to grazing capacity by equating them to the estimated forage consumption for the particular kind and class of animal being considered. The U.S. Forest Service allows 24 pounds (1 kg) per cow and 4.1 pounds (0.2 kg) per ewe, dry weight (U.S. Department of Agriculture, 1969).

Regardless of the technique used, all methods thus far developed based on vegetation analyses yield only an estimate of grazing capacity. True grazing capacity can be determined only by stocking with an estimated number of animals and watching the range trend (page 198). If range condition improves through a climatic cycle, then the estimated grazing capacity is too low and should be adjusted upward. If the range condition declines during wet and dry years, the stocking rate is too heavy and should be reduced.

RANGE-CONDITION ANALYSES

The term *range condition* to the range manager has a special meaning and relates current condition of the range to the potential of which the particular area is capable. It should not be confused with immediate availability of succulent forage, which may be a reflection of general weather conditions for that season or year. If rains have been frequent and temperatures favorable, ranges are good under this usage irrespective of what plants make up this crop. The range manager attempts to look beyond the immediate greenness of the herbage and to discover whether the plants which characteristically should grow in a particular situation are present and in good vigor. He notes the quantity of each species present as a basis for determining the degree to which the productivity of the range has been impaired. Range condition in this sense is best described as the state of health of the range.

Range-condition analysis, though little used during the early years of the development of the science of range management, has become the basis for adjustment of stocking figures and revision of management plans. The basic concept was not a new one, for it had been recognized early by F. E. Clements. Sampson (1919) proposed that the story of range use could be read in the plant cover that was found upon the range for, as he pointed out, continued heavy use results in retrogression of the plant cover. Although the general idea was accepted widely by range ecologists, a considerable period elapsed before the principles enunciated were applied generally and the stages in range deterioration were identified for other areas of different composition than the one used by Sampson.

Range-condition classification is based upon the theory that vegetation is the product of environment—a cause-and-effect relationship. The ecological concepts of *plant succession* and *climax* were the foundations for the procedures developed. Environmental factors of most direct importance to range vegetation can be classified as climatic, edaphic (soil), and biotic (chiefly grazing by animals). Fire also must be considered as an important habitat factor locally. Because of these factors, much of the world's rangeland may never have reached climax status, but other parts have reached the stable climax only to retrogress because of a change in environment, such as increased grazing or fire. Thus deviation from climax may involve past use, but time must also be considered. Time is necessary for vegetation to reach climax initially; and time is necessary for the return to climax following disturbance. Sometimes the great time involved makes climax an impractical objective of management. Occasionally a high-producing subclimax may be a satisfactory range, and economic factors make it unwise to manage so as to return to the theoretically possible climax. The management objective should be clearly outlined, but only after careful study of environment and current and potential productivity of the site. Determining climax or normal conditions is difficult and sometimes almost impossible. Protected areas, conservatively grazed range, ungrazed fence corners or cemeteries, and other such indicator areas must be sought out and studied carefully, because often even these have been altered (Fig. 6.5).

Factors Used in Range Condition Analyses

Several schemes have been developed for judging range conditions which, though they had some common characteristics, varied in details,

Fig. 6.5 Sagebrush *(Artemisia tridentata)* becoming established in an almost pure grass stand in a cemetery at Mona, Utah, in spite of competition. Repeated visits to this cemetery confirm that periodically sagebrush are chopped out by man. Great care must be exercised in selecting areas for determining climax conditions.

complexity, and reference to climax conditions. Range condition is described by means of *condition classes,* variously four or five, which roughly correspond with stages in secondary succession, i.e., in departure from climax conditions (Fig. 6.6). Ranges in an early weed stage are considered poor; those with climax vegetation are excellent. Ranges in intermediate stages of succession are considered fair or good. The kind and amount of vegetation and soil conditions in each condition class differ for each region and vegetative type.

Some early range-condition methods included vegetative cover measurements as range condition guides, but this proved unproductive. Often, percent of cover increases as the condition of the range declines due to the replacement of tall, erect species with low-growing, spreading species (Hanson, 1951) (Table 6.3).

Observation of the behavior of individual species led to categorizing them with respect to their response to grazing and their presence in the climax. If they were not present in native vegetation, they are called *invaders.* Those normally present in the climax are classed as (1) *increasers,* those which increase under heavy use, and (2) *decreasers,* those that diminish under heavy use. Generally, increasers are the less palata-

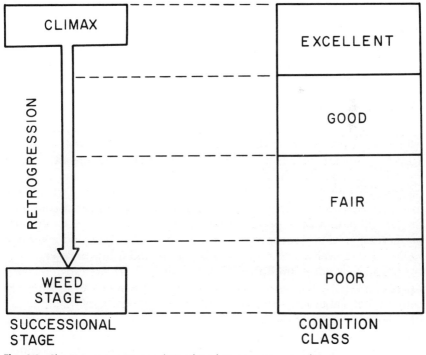

Fig. 6.6 Showing approximate relationships between range condition and degree of retrogression from climax conditions.

Table 6.3   Basal Area in Percent under Different Grazing Use from 1950 to 1969

Data from Smoliak et al. (1972)

Species	Study area in 1950	Study areas in 1969 by grazing treatments			
		Ungrazed	Light	Moderate	Heavy
Blue grama	2.0	0.3	2.4	2.8	3.6
Needle-and-thread	0.6	2.2	1.3	1.2	0.5
Western wheatgrass	1.4	1.3	1.7	1.2	0.7
Other grasses	1.3	1.4	1.6	1.5	1.1
Low sedge	0.9	0.4	0.5	0.8	1.1
Forbs and shrubs	0.5	0.4	0.7	0.9	0.4
Little clubmoss	7.1	9.6	18.7	22.6	26.0
	13.8	15.6	26.9	13.0	33.4

ble plants, and decreasers the more palatable ones, although resistance to grazing is also a factor in the response of plants to use. This response to grazing is not fixed for each species, but varies with its location and the plants with which it is intermixed. Brown and Schuster (1969) consider blue grama (*Bouteloua gracilis*) a decreaser on the Southern plains while Smoliak et al. (1972) found that it increased on the Northern plains. Galleta (*Hilaria jamesii*) generally is regarded as an increaser in Utah and Arizona; in New Mexico it is an increaser, decreaser, or invader depending upon the location and site (West, 1972).

Judging Range Conditions from Plant Composition

By comparing the expected percentage of the climax composition contributed by each species to the actual composition, it is possible to derive a numerical value which indicates the degree of departure of a range from climax and thus its condition rating. This index includes all decreasing climax species, that portion of the increasing climax species not in excess of climax percentages, and no invaders. Excellent-condition ranges will have a large percentage of decreaser plants, a lesser amount of increasers, and practically no invaders (Fig. 6.7). Good-condition ranges have fewer decreasers than increasers and a few invaders.

As an example of how plant-composition data are used to determine range conditions, assume a range in Texas has a plant cover as shown in column 3 of Table 6.4. Since side oats grama is a decreaser, any remaining plants are part of the original climax and would be entered in column 4 as climax. Perennial threeawn is an increaser and the present 10 percent cover is an increase of 5 percent over that expected in climax cover. Therefore, only 5 percent is entered in column 4, since this por-

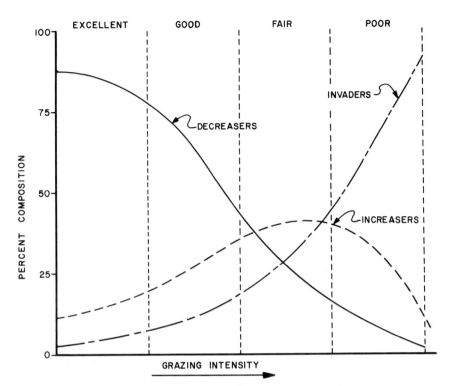

Fig. 6.7 Showing the relationship of intensity of grazing to range condition and the relative proportion of decreasers, increasers, and invaders. *(Data from Sims and Dwyer, 1965.)*

Table 6.4 Calculating Range Condition from Composition of the Present Cover in Percent Near San Angelo, Texas

Data from Dyksterhuis (1949)

Species or group	Contribution to climax composition*	Composition of present cover	Climax portion
		(Percent)	
Sideoats grama	100	10	10
Perennial threeawns	5	10	5
Texas grama	5	5	5
Forb increasers	10	5	5
Woody increasers	5	20	5
Hairy triodia	0	15	0
Annuals	0	35	0
		100	30

*Maximum percentage to be expected under undisturbed conditions.

tion is not considered abnormal. Annuals are not a part of the climax; hence, none of the present 35 percent annual cover is considered. Since the total of column 4 is 30 percent, the range can be said to support only 30 percent of the normal climax cover. Ranges can be put into condition classes using this index; thus an index of from 0 to 25 percent of climax indicates poor condition, 25 to 50 percent fair condition, 50 to 75 percent good condition, and 75 to 100 percent excellent condition (Dyksterhuis, 1949). Other schemes employ different condition classes and numerical values, but basically the method is widely applied in this manner.

The relationship between condition class and grazing capacity is determined from local stocking experience, experimentation, and by determinations of differences in plant production under various situations. Such data are tabulated in a technician's guide to average annual precipitation as shown in Table 6.5. Local site conditions, such as soil depth and physiographic location, may further increase or decrease these grazing estimates. Each guide has a range of safe stocking rates for each site and condition. Recommendations for stocking usually are made from the more conservative estimates for the particular range condition and are adjusted upward if the range improves.

Quantification of Range-Condition Ratings

Because there are no strictly objective means of assessing the several factors entering into condition analyses and because there are several factors to be considered, numerical ratings have been assigned to each factor to provide a means of aggregating the separate value judgments into a single, overall numerical figure. This then is compared to a standard which establishes the condition rating for the area.

Table 6.5 Relationship between Range-Condition Class and Grazing-Capacity Estimate Near San Angelo, Texas

Data from Dyksterhuis (1949)

Average annual precipitation, in.	Climax portion, percent			
	100	75	50	25
	Animal-unit months per acre			
14–18	0.6	0.45	0.3	0.15
19–24	0.8	0.60	0.4	0.20
25–29	1.0	0.75	0.5	0.25
30–34	1.2	0.90	0.6	0.30

The U.S. Forest Service developed a four-factor system for quantifying range condition based on two vegetation factors, *vegetal composition* and *plant production,* and two soil factors, *ground cover* (both living vegetation and litter) and *soil erosion,* which illustrates how integration of several factors can be achieved (U.S. Department of Agriculture, 1969).

Vegetal composition ratings are arrived at by consideration of the desirability of the species going to make up the stand. There are three plant classifications: *desirable, intermediate,* and *least desirable.* Plant lists are prepared to fit different areas, forest regions, or individual forests. Depending upon the relative proportion and dominance of each of these classes of plants, range-condition rating points are assigned according to the following scale:

1. Desirable species dominant — 49–60 points
2. Desirable and intermediate species mixed — 37–48 points
3. Intermediate species dominant — 25–36 points
4. Intermediate and/or least desirable dominant — 13–24 points
5. Least desirable species dominant — 0–12 points

These ratings are combined with a similar point scale based on the nearness of approach to a standard forage-production scale for the particular site to obtain the overall vegetation rating. Other descriptive guides to soil conditions provide rating points for the two soil factors considered which, when summed, provide the soil rating.

The maximum number of points for each of the four factors are as follows: 60 for vegetal composition; 40 for plant production; 50 for ground cover; and 50 for erosion index. The first two are combined into a vegetation rating and the latter two into a soil rating. If all factors were at their best, vegetation could receive 100 points and soil 100 points. Point totals for either vegetation or soil are converted to condition ratings based on the following scale:

Excellent	81–100 points
Good	61–80 points
Fair	41–60 points
Poor	21–40 points
Very poor	Less than 20 points

It is possible to have two different condition ratings, e.g., a good rating for vegetation and a poor rating for soil. These are indicated by a fraction such as 70/30 in which the numerator indicates the vegetation rating, and the denominator gives the soil rating. The lesser value of the two governs in arriving at an overall condition rating when vegetation and soil ratings differ.

Influence of Range Site upon Range Condition

Within any broadly defined area dominated by a particular kind of vegetation (shrub, grassland, etc.) or even by a single species, there exist great differences in productive capacities. These are occasioned by climatic and topographical differences from place to place and further by soil characteristics. These areas of differing potentialities are called *range sites* and are defined on the basis of their potential productivity. They are comparable to site indexes used by foresters and recognize that better growth can be expected in a good than a poor site. For example, in some areas sagebrush *(Artemisia tridentata)* will produce thick stands which may reach heights exceeding that of a man on horseback, or it may grow less than a meter in height. In the settlement of the Western United States, settlers judged what land was potentially suitable for clearing and cropping, where sagebrush occurred, on the basis of its appearance. Thus, within any vegetative type, different standards must be developed for forage production and plant composition (Humphrey, 1945). A good site in poor condition still may be more productive than a poor site in its best possible condition. These differences prevent the use of any one cover-class or species-composition guide to range condition and also prevent adoption of a universal scheme for range-condition determination (Fig. 6.8).

One cannot successfully manage rangelands without consideration of these differences in site quality. Range management is based on normal undisturbed development of vegetation. Unless these range sites are recognized, an area may be regarded as deteriorated when in fact it may be a poor site with limited potential. Conversely, failure to recognize a below-standard condition on a top site results in acceptance of production which is below maximum. Range-condition analyses are useful only to the extent that site variations are recognized.

The identification of range sites is a laborious and painstaking enterprise. Since soil forms a more permanent record of site differences than does vegetation, the identification of range sites is accompanied by soil surveys. It is assumed that within a particular vegetation type the soil development will reveal those areas where conditions for growth are similar, even though the vegetation may not at present be the same. Thus, by comparing the vegetative composition and production on one area to that found on another area showing the same site characteristics and known to be in top condition, it is possible to judge range condition.

To the present, no universally accepted basis for vegetation analysis

Fig. 6.8 Above is a protected area in which *Artemisia tridentata* and *Agropyron spicatum* grow in near climax association. The lower left shows a severely grazed area nearby with the grasses gone, leaving only the sagebrush. If the sagebrush is killed as in the lower right, low-value invaders occupy the area.

has been developed for the determination of range condition. Although the principles are widely accepted, experience has not yet supplied adequate information about the behavior of individual species in all the vegetation types and sites to make range-condition ratings completely objective and accurate. In the meantime, range-condition schemes are undergoing constant revision. Launchbaugh (1969), for example, found conventionally used schemes inadequate in clay upland sites in Kansas and proposed percentage yields of major grass dominants as being more useful guides to range conditions.

Climax as a Criterion of Range Condition

Among other objections to condition classification is that it presupposes knowledge of climax or normal conditions. It is not always possible to determine the climax with reasonable assurance. There is also the question of whether climax is the most productive or desirable condition. Perhaps it is an uneconomical goal.

Another problem inherent in range-condition analyses, in addition to those so far discussed, is whether under a livestock-grazing situation a climax cover can be maintained. Climax is the result of many forces operating in an ecosystem and ultimately reaching an equilibrium. Climate, soils, native animals, and fire have all contributed to determining what this climax condition is for any area. This is illustrated by Fig. 6.9 in which the "pull" from one set of factors is counterbalanced by that of another set. This equilibrium may be upset by changes in the force exerted of any of the factors involved. Drought may weaken the force of climate causing a shift in the position of climax; animal numbers may rise or fall, thus changing the pressure exerted by them and causing the equilibrium point to shift. If these perturbations are of short duration, the original equilibrium again is restored.

Enter man and his domestic animals. Native animals are removed or greatly curtailed in species and numbers. They are replaced by domestic animals with differing forage preferences, and, almost always, in far greater numbers than native herbivores. Fire prevention may alter the frequency of fires. Consequently a new set of forces or forces of different magnitude now move toward a new equilibrium. Obviously, it cannot be expected that the original climax condition can be maintained under a now-altered ecosystem. Theoretically, the new ecosystem will reach a point of equilibrium and stability, but this may not be acceptable. In this case, periodic manipulation must take place to correct the system toward the original climax, or to a condition which man will accept.

There is much evidence to suggest that plant invasions, particularly by shrubs, on rangelands in the Western United States are explained by the restructuring of forces operating in ecosystems, not simply by "overgrazing." Perhaps any appreciable level of grazing with exotic animals

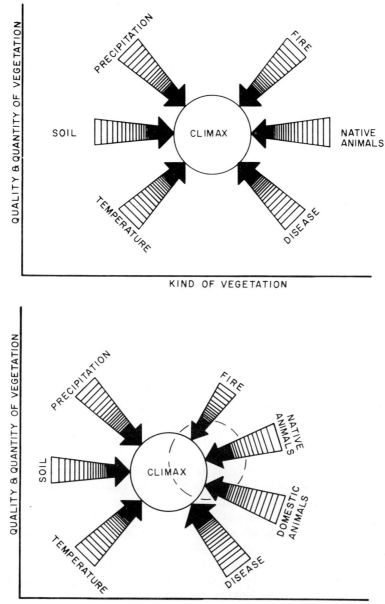

Fig. 6.9 Above, climax equilibrium in the absence of man in which biotic and fire factors come to balance with climatic and edaphic (soil) factors and produce a relatively stable vegetation. Below is shown the effect of man and livestock in which the forces to the right are amplified causing a shift to the left and, presumably, a new point of equilibrium. If this equilibrium is not an acceptable one, periodic manipulation of the vegetation may be required.

when accompanied by reduction in fires would be marked by vegetative changes of great magnitude. Both sagebrush in the Great Basin and juniper wherever it occurs invade in response to this reordering of the climax determinants; and perhaps no range where they occur, however carefully managed, can be expected to remain free of them. Periodic manipulation by one means or another may thus be necessary to maintain an acceptable subclimax condition.

Range-Condition Trend

Condition ratings, even though they are accurate, are of little usefulness without knowing the trend in condition. A range in poor condition which is still deteriorating requires different treatment from a poor range which is in the process of improving. Trend can be determined only by careful analyses of the range. Among the most important factors for determining trend are vigor and reproduction of both desirable and less desirable plant species, and the amount of litter.

Determining trend is highly important. Generally, livestock reductions and wide-scale changes in management are unnecessary if condition trend is upward, although, by improved management, *rate* of range improvement may be increased. Poor range condition does not mean that current management practices are wrong. Only trend of condition will reflect correctness of current grazing practice.

Judging range trend is even more hazardous than judging range conditions, because there are few objective means for assessing trend. Commonly soil and vegetative factors are listed on a score card to ensure consideration of a uniform set of factors. Soil factors include among others presence of litter, evidence of soil trampling, and presence and condition of gullies. Plant factors include such things as plant vigor, seedling establishment, degree of present utilization, and evidence of past utilization, especially on browse plants. Obviously, each of these factors may be judged positively or negatively. Gullies may be new and raw indicating deterioration, or they may be filling in indicating improvement. If seedlings of better species are present, conditions are favorable. Seedlings of poor or invader species are an indication of deterioration.

Admittedly condition and trend analyses are gross measures. Although rooted in sound ecological concepts, the tools have not heretofore existed for complete documentation of the multitude of responses in complex ecosystems. It was possible only to observe changes of considerable magnitude occurring among the more obvious components of the system. Hopefully, in the future the availability of sophisticated tools such as computers will make possible the inclusion of a much broader spectrum of inputs into the data. The result may be that changes in the ecosystem can be detected much earlier and even eventually that they may be predicted beforehand. If computer technology does no more than direct us to questions that should be researched and answered, the contribution will be enormous.

RANGE UTILIZATION

Utilization, as used by rangemen, refers to the percentage of the annual production of forage that has been removed by animals throughout a grazing period or grazing season. If the permissible level of use for range plants in terms of forage removal is known, *percent utilization* at once reveals whether forage remains for use. Thus utilization surveys are made, usually toward or at the end of a grazing season, to judge the appropriateness of existing levels of stocking and management practices.

Unfortunately, utilization has been used to express two things: (1) the use of the range unit as a whole, and (2) the use made of plant species individually. Except in general terms, the former usage is confusing and should not be used, although it occurs in older range literature. Under the second usage, utilization is defined as *the degree to which animals have consumed the total current production, expressed in percent.* In the case of herbs, percent use is based upon weight of the entire plant and measures what portion has been consumed *without regard to what is correct utilization* (proper use). In the case of browse, only current year's growth of twigs and leaves is considered.

Utilization Determination Based on Heights

First attempts at determining utilization were merely ocular judgments. Estimates were based on heights of ungrazed and grazed plants. A plant normally 25 cm high grazed until it averaged only 5 cm in height was 80 percent utilized. When the need for allowing range plants to produce seed was recognized, utilization standards were set to permit a certain percentage of the seedstalks to remain ungrazed or propagate new plants. This had the effect of reducing utilization, for example, leaving 20 percent of the seedstalks on *Agropyron smithii* resulted in only 52 percent utilization (Crafts, 1938c). Similarly blue grama reached 40 percent utilization when 25 percent of the seedstalks were left (Crafts, 1938b).

There is a fundamental error in judging utilization directly from the percentage reduction in height. The distribution of weight with height in plants is not linear. When the weights of forage produced in segments of equal length are determined and plotted, the resulting diagram is roughly conical (Fig. 6.10); that is, there is much greater weight in the bottom inch of height than in the top inch. By cutting harvested plants into short lengths and weighing each segment, the weight distribution can be determined (Lommasson and Jensen, 1938). Crafts (1938a) prepared charts which showed for a number of species the relationship of height reduction to utilization (Fig. 6.11). With these it is possible to extrapolate utilization from measurement of stubble heights and height of ungrazed plants. Because of differences in growth habit among species, individual charts must be prepared for every species before this method can be used.

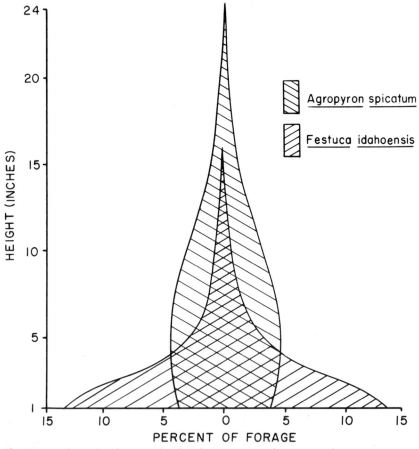

Fig. 6.10 Relationship between height of two species of grasses and volume of plant material aboveground. *Festuca* would be little utilized when reduced by two-thirds. The precise shape of these varies somewhat with the year and height of the plants. *(Adapted from Heady, 1950.)*

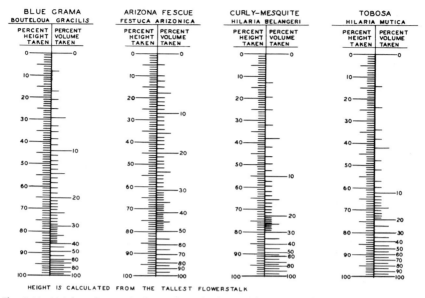

HEIGHT IS CALCULATED FROM THE TALLEST FLOWERSTALK

Fig. 6.11 Height-volume calculation charts for four Southwestern grasses showing the relationship between height removed and volume utilized. *(Data from Crafts, 1958a.)*

Determining Utilization from Weights

Weight provides another means of determining utilization. Weights may be arrived at by estimation or by clipping and weighing. Either requires an ungrazed, fully developed stand of vegetation from which the appearance and yield of the vegetation can be inferred or measured. If growth is complete at the start of the grazing period, as on winter ranges, pregrazing measurements are possible. When these are compared to measurements taken after grazing is completed, the level of utilization is established. Where grazing is going forward during the growing season, however, before-and-after grazing examinations are not useful. In this case it is customary to protect sample areas from grazing to provide an estimate of forage production for that year. Enclosures, fenced areas which exclude livestock, or cages placed over sample plots serve this purpose. Utilization determined in this way is probably biased upward, because ungrazed plants may produce more forage than grazed plants (see Chapter 4).

Grazed Stem and Grazed Twig Counts

The inadequacies of ocular-utilization estimates and the difficulties inherent in weight determinations led to proposals for counting the grazed and ungrazed stems in grasses (Stoddart, 1935) and twigs of browse plants to derive a utilization estimate. This method has greater objectivity than ocular methods, and is more rapid than weight determinations. However, the relationship between stems or twigs eaten and percent use is not linear; 100 percent of the twigs are used well before utilization is 100 percent. Stickney (1966) reported that when 100 percent of the twigs on *Prunus virginiana* had been browsed, utilization was under 50 percent. Thus, grazed stem or leaf counts have disadvantages: (1) charts showing utilization plotted over percent of stems grazed must be prepared, and (2) accuracy is possible only at the intermediate levels of use. There is no way to extrapolate utilization at the higher use levels, and it is here that the need for information is most critical to proper range management.

Grazed Plant Method for Determining Utilization

Percent of plants grazed has also been used to judge utilization (Table 6.6). Theoretically, the same relationship exists with plants as with grazed stems or twigs, but there is conflicting evidence. Holscher and Woolfolk (1953) in Montana, and Roach (1950) in Arizona reported a straight-line relationship between utilization and percent of plants grazed. In constrast to this, Hurd and Kissinger (1953) show a curvilinear relationship. This disparity is probably due to the range of utilization values considered. One can get a satisfactory straight-line fit within intermediate levels of use, but neither low nor high levels of utilization can be determined accurately. Gierisch (1967) proposed plotting the percent of plants grazed more than 30 percent which, although it gave better agreement at low levels of use, did not register use above 50 percent on *Festuca thurberi*. It has also been shown that separate regression curves must be prepared for sheep and cattle, and that cattle show different relationships as between spring and fall grazing (Springfield, 1961). These disparities arise from the differences in intensity of

Table 6.6 Guide to Proper Utilization of Northern Great Plains Ranges in Summer Showing Number of the Plants to Be Grazed, in Percent

Data from Holscher and Woolfolk (1953)

Species	Upland subtype	Hills subtype	Bottom subtype
Wheatgrass	55	55	75
Blue grama	40	45	65
Needlegrass	55	60	—
Buffalo grass	55	50	75

use by cattle and sheep. At a given level of use, cattle use fewer individual plants heavily; sheep use more plants but less heavily.

Terms and Concepts Related to Utilization

Raw utilization data have little utility in range management. They depend for their usefulness on (1) knowing what level of use plants can withstand, and (2) knowing how animals select among plant species occurring together in a mix. Since neither of these is constant in space or time it is not possible to set universal limits to acceptable levels of use. This leads to a consideration of the meaning and use of *palatability, preference, preference index,* and *use factor* which, in the literature, are used quite differently and often inaccurately.

Palatability and Preference   *Palatability* refers to the attractiveness of a plant to animals as forage. What factors make plants desired are not clearly known and different kinds of animals are differentially attracted by a particular species. Palatability once had a meaning for rangemen equivalent to that now reserved for proper-use factor and was expressed in percentage use, but it should not be so used.

Preference refers to the selection of plants by animals. Often the term is used incorrectly when palatability is meant. In response to some plant characteristics as yet not well known, animals clearly prefer some species over others. Preference may vary throughout the season or from year to year. Whether one speaks of palatability or preference, the measure of either is the percent utilization observed at a particular time or place; 70 percent utilization of a species commonly is taken to mean both the palatability of that plant and the preference animals show for it vis-á-vis other plants. This is incorrect for reasons discussed later in connection with the use factor.

Preference Index   Because abundance of a plant influences percentage utilization, it has been thought that animal preference could be determined from utilization data by making a correction for plant volume. Under this concept, a plant which is utilized to 50 percent and only represents 10 percent of the forage would have a high preference index ($50 \div 10 = 5$); a plant utilized 50 percent that made up 25 percent of the vegetation would have a much lower preference index ($50 \div 25 = 2$). This concept is of questionable validity or utility. A good case can be made for the idea that the amount of forage an animal will consume from a given species, when there is a choice and the supply available is not limiting, is a better measure of food preference than the index arrived at by the method shown above. It would be possible, by the preference-index approach, to show that the American public prefers lamb to beef when the usual interpretation is that beef is the preferred meat, since the per-capita consumption of beef is over 100 lbs, and that of lamb is but a fraction of it. Were it not for the fact that some species in the forage mix are in limited amounts, percent of diet would accurately represent relative preferences for plants. In cafeteria trials, or on a previously ungrazed range, where availability is not a limiting factor to consumption even though species may be present in differing amounts, relative consumption is an accurate index to plant preference (Smith, 1953). Paulsen (1969) used forage-value ratings that were derived from the product of utilization and production for each species. These are essentially dietary indexes.

Factors Affecting Choice of Plants

It is not clearly known why certain plants are selected over others. It is clear that no universal plant factor is responsible, for some plants are readily eaten by one kind of animal and not by another, while other plants may be readily eaten by more than one kind of animal. *Geranium fremontii*, for example, is eaten readily by mule deer in northern Utah and is used only lightly by domestic livestock. The reverse is true for *Senecio serra* which is relished by sheep, eaten moderately by cattle, and not at all by deer. *Aster chilensis* is eaten readily by sheep, cattle, and deer (Smith, 1958). There may even be differences among breeds of livestock; Herbel and Nelson (1966) report some differences in species selection by Hereford and Santa Gertrudis cows.

The following reasons have been advanced for plant selectivity by animals: (1) nutrient content, especially protein; (2) taste, such as salty, bitter, sour, sweet, etc.; (3) moisture content; (4) mineral content; (5) essential oils; (6) fiber or lignin content; and (7) texture. The evidence supporting any of these factors is conflicting. Crampton (1959) observed that the most readily eaten forages among a selected group of roughages were high both in protein and lignin. Efforts to relate selectivity for certain tussock grasses in New Zealand to chemical makeup led to the conclusion that no chemical factor could be said to explain why sheep selected certain plants (O'Conner, 1971).

Part of the problem may be the fact that these factors are interrelated. High-protein contents occur in plants in the early growing season when moisture contents are also high and fiber and lignin contents are low. Thus it is not certain whether an animal selects for moisture or protein or against fiber.

It has been shown that animal species vary in the ability to discriminate among certain taste factors, although the ranking among kinds of animal varied with the threshold chosen, i.e., lowest concentrations discriminated or point at which feeds were rejected (Goatcher and Church, (1970). Among four kinds of animals the following rankings were given, with respect to when rejection was observed as molar concentrations of the substances used to produce the taste were increased:

Salty	cattle > sheep > normal goats > pigmy goats
Sour	cattle > sheep > normal goats > pigmy goats
Bitter	sheep and cattle > normal and pigmy goats
Sweet	no rejection thresholds established

Cattle were least tolerant and pigmy goats most tolerant for salty, sour, and bitter tastes. Black-tailed deer were placed between cattle and goats in sensitivity (Crawford and Church, 1971). Arnold and Hill (1972) have reviewed the findings with respect to chemical factors in plants in relation to animal selection and conclude that it is not possible to relate chemical composition of a plant to the preference animals show for it.

Nutritional Wisdom   A frequently held view is that some sort of feedback mechanism operates to guide animals to the most nutritious plants. Since some animals reject plants which are equally nutritious, measured chemically, to others which are taken, there is reason to doubt that nutritional selection takes place on the range. Some evidence can be found both for and against this theory but Arnold and Hill (1972) concluded that no "really relevant data" exist to support it.

The Proper-Use Factor

The proper-use factor is the percentage use that is made of a forage species under proper management. It is expressed as the percentage that is consumed of the current year's forage production of a particular species which is within reach of stock. A proper-use factor of 50 percent indicates that a plant will have half its total available annual production of vegetation removed by livestock at the end of the grazing season. Numerically, utilization and use factors are identical on properly used range. The use occurring on overgrazed ranges is not, correctly, the proper-use factor of a species.

The physiological ability of a plant to withstand grazing is considered only with respect to the major species in determining proper-use factors. Those plants which are abundant and palatable must not be utilized beyond their limit of tolerance, but it may not be feasible to thus limit use on the more highly preferred and less abundant species. Sometimes it may not be possible to graze any individual species to its physiological-tolerance level because of such factors as steep slopes and fragile soils. The proper-use factor, then, is dependent upon the ability of the range as a whole to withstand grazing, and not necessarily upon the ability of particular species to withstand cropping. It is not a constant but is dependent upon many modifying conditions.

When an animal is placed on previously ungrazed range, he may skim off the most preferred forage, which may make up the majority of his diet. As the season progresses and the range approaches full use, the preferred species will constitute increasingly less of the diet and less desirable plants will make up increasingly more of the diet. The proper-use factor is the cumulative utilization of a species when full (correct) use has been made of the range; it is a product of preference and quantity available.

Variability in the Proper-Use Factor   Because of the many variables that influence the actual use made of a species on a specific range, there is no universal proper-use factor which applies to a given species. Yet, it is most important in range management to have in mind an objective in utilization. Correct intensity of use is likely more important than any other item in good range management.

The use factor may vary according to (1) associated species, (2) kind of stock, (3) season, (4) year, (5) past grazing use, (6) undefined local conditions, and (7) familiarity with plant. Most of the variation in the proper-use factor is attributable to variation in animal preference under different conditions.

1.   A given plant species varies in its degree of use by livestock according to the plants with which it is associated, the use varying both with the quality and quantity of other species present. Sagebrush will be used heavily when it occurs in small quantities associated with other species; yet, in dense and almost pure stands, only a small percentage will be utilized. Relatively unpalatable plants such as *Distichlis* and *Hilaria mutica* can be used heavily when they occur in pure stands by fencing the type or otherwise holding animals on the area, thus limiting animal choice.

A plant of medium palatability growing in association with a highly preferred species in about equal proportion should be grazed lightly to prevent damage of the better-liked species. This same species of medium palatability, when growing in association with a species of very low palatability, however, could be used more heavily without damage to the associated species. Thus its proper-use factor in these two situations would be quite different.

2. Proper-use factors vary according to kind of animal. Sheep and goats can be and are raised on grass range, but they do not, on mixed vegetation, use as large a proportion of grasses as do cattle and horses. Horses are the most selective of the domestic grazing animals, the major part of their diet being grass. Goats are probably least selective, for, although they eat a larger proportion of forbs than do other classes, they also consume grasses and shrubs with about equal avidity. The goat is famed for his consumption of browse, a feat at which he is undeniably adept, although browse may not be forage of first choice (Malechek and Leinweber, 1972).

Wild animals likewise show distinct differences in preference; thus deer eat more shrubs than do elk, and elk eat more grass than do the deer (see Chapter 11). Among grass-eating wild ungulates in Africa, selectivity for specific grass species differs among different animals (Field, 1972).

3. The season during which they are grazed profoundly influences the proper-use factor of most plants. Some plants are highly preferred in early season and much less so in late season, especially after maturity. Most annuals fall in this class. Other plants are more preferred in late season, as *Sambucus microbotrys* which is grazed during late summer only. Some plants, such as *Balsamorhiza sagittata* and *Salsola*, are preferred early, then become dry and too harsh for stock consumption; but, when softened by winter snows, again are consumed in large quantities.

Seasonal differences in proper-use factors may be due to differences other than animal preference and availability. Much heavier utilization can be made of a species without injury in fall and winter than in spring and summer. At these times growth has been completed and forage removal is less injurious than during the active growth and food-storage phases; this is especially true with respect to herbaceous species, but it is true also for browse. It is possible to utilize browse species much more heavily on winter game ranges, 70 to 80 percent (Smith and Urness, 1962), than is permissible in summer, usually judged as 50 to 60 percent.

4. Climatical variations may alter the proper-use factor for a particular species. In good years, when moisture is abundant, rank growth tends to decrease the attractiveness of some plants to grazing animals. An example is Russian thistle which is consumed in large quantities when it remains small and, hence, does not develop abundant prickles. During extreme drought, plants that are fleshy are more attractive because of their high moisture content. Other plants are improved by excess precipitation; short plants may gain sufficient height to become valuable forage, whereas in dry years they are too small for animals to consume more than a small percentage of their foliage. Filaree *(Erodium)* is an example. Rainy periods may change animal preferences; Bowns (1971) reports that sheep turn from forbs, their most preferred forage, to browse when forage is wet.

Even the dry mature forage from herbaceous species on salt desert winter ranges show great variations in palatability from year to year, though outwardly there is no detectable difference in it. Presumably, weather factors related to cessation of growth and curing are responsible for this phenomenon.

5. Past grazing use influences animal preferences. Animals tend to return to a previously grazed plant because its regrowth is more tender and leafy, since dry stems and leaves are absent. Such repeated grazing results in *spot overgrazing* where the same patches are grazed repeatedly and others are unused. Individual plants, so-called wolf plants, may remain ungrazed year after year (Fig. 6.12).

6. Some plants vary in use factor within the species for no apparent reason, sometimes despite the fact that the individuals grow in close proximity. Both deer and sheep have been noted to graze closely one sagebrush and to ignore completely another growing alongside. Possibly the salts or minerals taken up by a plant from certain soils increase or

Fig. 6.12 Seeded range, tall wheatgrass to left and crested wheatgrass to right. Note unused plants (wolf plants) of each species where companion plants of the same species are grazed repeatedly. The same plants are untouched year after year.

Fig. 6.13 These two branches of *Juniperus osteosperma* were placed before hungry, penned deer. One was untouched, the other fully consumed. The trees from which they were removed on the range were similarly treated.

decrease its use by their influence upon taste. For example, it has been suggested that phosphorus-fertilized plants develop a high sugar content and that sugar content is the most reliable index to palatability of the forage (Plice, 1952).

Deer have been observed to show this marked preference for individual plants of juniper, leaving one tree untouched while consuming all available forage from another. When branches from grazed and ungrazed trees were cut and offered to deer in pens, the animals completely consumed branches from preferred trees (Fig. 6.13) but ignored branches from ungrazed trees of the same species (Smith, 1950). When forage samples from these differentially eaten trees were analyzed for volatile-oil content, small but significant differences were found. *Juniperus scopulorum* averaged 1.84 and 2.27 percent oil in the grazed and ungrazed trees, respectively, and *J. osteosperma* averaged 2.13 and 2.60 percent.

7. Just as individuals of a species vary locally, so do they vary from region to region. *Artemisia frigida* is regarded as an indicator of overgrazing in the northern plains, as fair forage in the central plains, but as good forage in the Southwest. Many cases are known where animals shipped from one region to another at first refuse to eat plants that are highly attractive to local animals.

Value of Proper-Use Factors   There are other reasons why the proper-use factor may not always be an accurate index to the grazing value of a plant. Since the proper-use factor indicates only the percentage of the total accessible current growth of plant material that is consumed, it does not give any indication of the actual amount of vegetation eaten or of the quality of the material in terms of food value. It is thus possible for a plant to have a high rating on the basis of the percentage eaten and yet be unimportant in the total diet and of inferior nutritive value.

The value of the proper-use factor is that it provides some guide to severity of range use. If utilization of important forage species exceeds a proper level of use as indicated by their use factors, the range manager is made aware of the need for adjusting the numbers or the handling of animals; he is alerted to possible deterioration of the range before it takes place and can act accordingly. Proper-use factors are of less utility under certain grazing systems where rest or deferment are expected to compensate for too heavy use.

APPLYING UTILIZATION DATA IN MANAGEMENT

Because of the complexity of range-utilization data in all situations except that of a monoculture, there is need for simplifying procedures for practical application. This is accomplished by basing judgments on indicator species or areas.

Indicator Species in Utilization Determination

Unless he resorts to a few indicator plants, the range owner or administrator will be confronted by a maze of plants, each species being utilized to a different degree, depending upon its availability and upon animal preferences. It is very difficult for even the experienced person to arrive at a correct estimate of the percent utilization that has been made, especially on ranges of heterogeneous vegetation. For this reason, it is usually the practice of range managers not to include all the forage plants in the area in the calculation of utilization but, rather, to select a few important plants. These plants are known as *utilization-indicator species* or *key species,* and they should be the most important forage species on the range. Since they usually furnish the bulk of the forage, it is obvious that when they are used to capacity, the range in entirety must be considered as correctly used. On most ranges, correct grazing for the two to four most important forage plants means correct grazing for the entire range. Highly preferred species that occur in only small quantities will be sacrificed (i.e., allowed to be overused), in order to obtain fuller use of the more abundant species. These so-called ice-cream plants cannot be used as utilization-indicator plants, despite their high palatability if, because of their scarcity, they are not important forage species (Fig. 6.14).

The ecological status of a plant must be known before a decision can be made concerning the advisability of sacrificing it. A climax plant or former dominant species that, because it is highly preferred by livestock, is rare on an overused range may still remain the best forage plant for the area and, in such a case, definitely should not be sacrificed in order to make better use of less desirable plants. It may be desirable to take extreme measures even to reducing grazing in order to bring the range back to its former cover. Temporary stock reductions can almost always be justified if they lead, ultimately, to an increased range capacity. There are instances, however, in which the climax plants are not the most desirable or in which mere stock reduction cannot restore the climax plants within an economically feasible time. In such cases, the remnant climax plants cannot be justified as key species. Only a careful ecological analysis will determine whether the greatest yield can be obtained by protecting a certain preferred species or by sacrificing it.

Forsling and Storm (1929) working on southern Utah browse range found that despite the fact that *Quercus gambelii* made up 30 percent of the cover—far more than any other species—it should not be fully

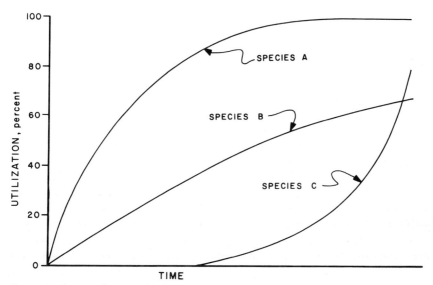

Fig. 6.14 Showing theoretical utilization of three plant species. Neither species A nor C is useful as a key species for there is no unique relationship between range use and grazing pressure. Species B would be the one selected as the key species. *(From Smith, 1965.)*

grazed. Continued heavy use of the oak would result in permanent removal of all the more desirable shrubs as well as the grasses. Further, cattle made unsatisfactory gains when the oak brush was used heavily, they lost weight during the latter part of the grazing season, and calf crops declined. Hence, the authors concluded that "the best summer use that may be made of browse range is to graze it only to the degree that will not injure the more palatable browse and the grasses and weeds of approximately equal palatability, even though this will entail but light use of the less palatable species (p.27)."

Indicator Areas in Utilization Determination

Analogous to the indicator-plant theory of determining utilization is the *indicator area,* or *key area,* theory. Just as one may select a single species or a specific group of species as a key to correct use, so may one select a representative area within which to study utilization. The principle behind this theory is that no range of appreciable size can be uniformly utilized. Heavy use is inevitable around water holes, salt grounds, driveways, level valley floors, and more accessible ridge tops. Likewise, lighter use or possibly even nonuse prevails at great distances from water and salt and on very steep hillsides. These inequalities are emphasized when the range is grazed by free-roaming cattle, but even herded sheep graze the average range unevenly.

Because of this lack of uniformity in grazing, certain areas may be *sacrifice areas* and be overused, just as the rare and highly preferred species may be overused. Other areas may be underused, for their proper use would cause serious overuse and range damage on more accessible areas. The intermediate areas, then, become the indicator areas because they are to be properly used.

Proper determination of the indicator area is extremely important in range management. One must be liberal or the grazing will be unreasonably light. But one must be conservative or excess use may be the cause of damaging erosion and permanent decreases in forage yield. Certainly the rule of thumb of the conservationist should be: "Keep the sacrifice area at the practicable minimum." The first obligation of a public-grazing administrator is to protect the resources, and this should also be the goal of a private landowner if he is to protect his land investment and maintain a stable business.

The Key-Plant and Key-Area Principles Applied

In order to obtain a clearer concept of the indicator plant and indicator-area principles, let us assume a hypothetical area upon which are three plants, *Agropyron, Osmorhiza,* and *Rudbeckia.* Further, let us assume that the range includes a water hole, an adjacent level bench, and a distant steep hillside. After the range had been fully utilized, we might expect the utilization shown in Table 6.7 (assuming 60 percent to be the correct use to maintain the plant).

Despite the fact that there are utilization percentages from zero to 100, confusing when looked at as a group, all that the observer need consider is the 60 percent use of *Agropyron* on the benchland. When this plant on this area is used properly, the range is used properly. Under continued use, one would expect that *Agropyron* and *Osmorhiza* would disappear from the area around the water hole, *Osmorhiza* would disappear on the benchland, and all three would survive on the hillside. This hillside could not be fully used if the *Agropyron* on the benchland is to be maintained. Underuse of forage on the hillside should, however, give the maximum forage yield in the long run and the minimum of damage to plant cover consistent with a reasonable use. Full utilization of steep slopes should not be expected, and the forage growing on such slopes should be discounted or, on extreme slopes, disregarded when carrying capacity is estimated.

Safe or Proper Forage Use

One of the questions ever present before the range manager and stockman is: "How much can the range be grazed without permanent damage to the good forage plants?" It is very important that this desirable use be known in order that a minimum of forage will be wasted and yet no damage will be done.

The objectives of proper use are clear: preservation of the range values, protection to the soil, and production of the maximum animal product consonant with these objectives. The latter cannot be achieved merely by observing animal condition, for good animals may be produced under a condition in which the range is deteriorating. Proper use may not result in climax vegetation. Soil protection, rather than vegetation, may be the criterion. Careful observation of the range as a whole is

Table 6.7

Plant	Around water hole	Adjacent bench	Distant hillside
		(Percent)	
Agropyron	100	60	30
Osmorhiza	100	90	60
Rudbeckia	60	10	0

needed. This was well stated by Crafts and Wall (1938):

> Proper use is the utilization of the range forage that is attained when all the land services are given due consideration. The interrelated values of soil, grazing, watershed, timber, recreation, wildlife, and other land services all enter into the determination of proper use. Proper use implies more than resistance to grazing, which is the degree of use a plant can stand year in and year out and still maintain its vigor and forage productivity.

How much grazing individual plants can withstand is not known (see chapter 4), but certainly there are great differences among species in their grazing tolerance. Percent utilization alone does not give an exact picture of the stress to which the plant is subjected. Among the factors affecting plant response to a given grazing use are:

1. *Current growing conditions.* In drought years plants can stand less use than during favorable years.

2. *Habitat or environment.* On fertile, moist soils regrowth is abundant and herbage soon is replaced. On dry, eroded hillsides, plants are less able to withstand grazing shock.

3. *Season of grazing.* During seasons of depleted food reserves, plants can be injured by grazing intensity that would not injure them during late fall or other less vulnerable seasons.

4. *Duration of grazing.* What appears to be full use at the end of a summer grazing season may result from continued grazing throughout the spring and summer; from heavy spring use followed by rest and regrowth, followed by heavy regrazing; from complete rest during spring and fall use just prior to the close of the summer grazing season; or from innumerable other conditions. None of these factors is considered in the usual expression of utilization. Continued research will be necessary to derive conclusive proper-use standards for all situations.

5. *Amount of forage remaining.* What is removed from vegetation may be of less importance physiologically than what remains. Regrowth following cropping is dependent upon the presence of buds and stored foods. These are more dependent upon the actual amount of plant materials that remain than upon what has been eaten. For example, a shrub with 5-cm twigs and browsed to 60 percent will have only 2 cm of twig left; a shrub with 15-cm twigs browsed to 75 percent will have 11 cm left. Both with respect to the number of buds and food reserves in the twig, the more heavily used plant is in better condition for regrowth than the more lightly used one.

6. *Parts of plant eaten.* Preference for certain portions of a plant over others is apparent in livestock, especially sheep. This becomes important in measuring physiological responses of plants and in interpreting correct grazing use. Studies on northern Utah summer ranges showed that sheep preferred leaves over stems (Table 6.8).

Utilization of leaves of grasses and forbs averaged 29 and 70 percent, respectively, but utilization of the entire plant was only 17 percent for grasses and 26 for forbs. In some species, only about 25 percent of the entire plant was utilized when 70 to 80 percent of the foliage was removed. Since the photosynthetic tissue in most forage plants is almost entirely in the leaves, basing degree of utilization on the percent removal of the leaves might better indicate the severity of grazing use, in those cases where leaves primarily constitute the forage supply. Since, however, this is true for only some plants, using leaves alone as a basis for calculating utilization would only result in greater confusion. We need to remember that light percentage utilization for some species may be virtual defoliation.

Table 6.8   Production and Utilization of the Parts of Plants for Some of the More Important Species on Summer Sheep Ranges of Northern Utah*

Data from Cook and Stoddart (1953)

Species	Percent of total production			Percent of utilization			
	Stems	Leaves	Heads	Stems	Leaves	Heads	Total
Agropyron subsecundum	63	19	18	20	39	25	25
Elymus glaucus	75	19	6	6	20	20	9
Average grass	69	19	12	13	29	22	17
Agastache urticifolia	57	33	10	6	60	36	26
Senecio serra	55	42	3	1	70	54	31
Valeriana occidentalis	80	17	3	10	80	2	22
Average forbs	64	31	5	6	70	31	26
Amelanchier alnifolia	20	80	...	26	62	...	55
Symphoricarpos vacciniodes	24	76	...	7	29	...	24
Average browse	22	78	...	16	45	...	39

*Data based upon current year's growth only.

BIBLIOGRAPHY

Aldon, Earl F., and George Garcia (1971): Stocking rangelands on the Rio Puerco in New Mexico, *Jour. Range Mgt.* **24**:344–345.

Arnold, G. W., and J. L. Hill (1972): Chemical factors affecting selection of food plants by ruminants, in J. B. Harborne (ed.), *Phytochemical Ecology,* Academic Press, New York, pp. 72–101.

Avery, T. Eugene (1968): *Interpretation of Aerial Photographs,* 2d ed., Burgess Publishing Co., Minneapolis, Minn.

Bowns, James E. (1971): Sheep behavior under unherded conditions on mountain summer ranges, *Jour. Range Mgt.* **24**:105–109.

Brown, Jimmy W., and Joseph L. Schuster (1969): Effects of grazing on a hardland site in the southern High Plains, *Jour. Range Mgt.* **22**:418–423.

Christian, C. S. (1959): The eco-complex and its importance for agricultural assessment, in W. Junk (ed.), *Biogeography and Ecology in Australia,* Uitgeverij, Netherlands.

Cook, C. W., and L. A. Stoddart (1953): The quandary of utilization and preference, *Jour. Range Mgt.* **6**:329–335.

Crafts, E. C. (1938a): Height-volume distribution in range grasses, *Jour. Forestry* **36**:1182–1185.

—— (1938b): Tentative range utilization standards, *Southwestern Forest Range Experiment Station Research Note 32.*

—— (1938c): Tentative range utilization standards, *Southwestern Forest Range Experiment Station Research Note 44.*

——, and L. A. Wall (1938): Tentative range utilization standards—fundamental concepts, *Southwestern Forest Range Experiment Station Research Note 25.*

Crampton, E. W. (1959): Interrelations between digestible nutrient and energy content, voluntary dry matter intake, and the overall feeding value of forages, in Howard B. Sprague (ed.), *Grasslands,* American Association for the Advancement of Science Washington, pp. 205–224.

Crawford, James C., and D. C. Church (1971): Responses of black-tailed deer to various chemical taste stimuli, *Jour. Wildlife Mgt.* **35:** 210–215.

Dudzinski, M. L., and G. W. Arnold (1967): Aerial photography and statistical analysis for studying behavior patterns of grazing animals, *Jour. Range Mgt.* **20**:77–83.

Dyksterhuis, E. J. (1949): Condition and management of range land based on quantitative ecology, *Jour. Range Mgt.* **2**:104–115.

Field, C. R. (1972): The food habits of wild ungulates in Uganda by analyses of stomach contents, *E. African Wildlife Jour.* **10**:17–42.

Forsling, C. L., and E. V. Storm (1929): The utilization of browse forage as summer range for cattle in southwestern Utah, *U.S. Dept. Agr. Circ.* **62.**

Frischknecht, Neil C., and A. Perry Plummer (1949): A simplified technique for determining herbage production on range and pasture land, *Agron. Jour.* **41**:63–65.

Gierisch, Ralph K. (1967): An adaptation of the grazed plant method for estimating utilization of Thurber fescue, *Jour. Range Mgt.* **20:**108–111.

Goatcher, W. D., and D. C. Church (1970): Taste responses in ruminants: IV. Reactions of pigmy goats, normal goats, sheep and cattle to acetic acid and quinine hydrochloride, *Jour. Animal Sci.* **31:**373–382.

Hanson, W. R. (1951): Condition classes on mountain range in southwestern Alberta, *Jour. Range Mgt.* **4:**165–170.

Heady, Harold F. (1950): Studies on bluebunch wheatgrass in Montana and height-weight relationships of certain range grasses, *Ecol. Monogr.* **20:**55–81.

Herbel, Carlton H., and Arnold B. Nelson (1966): Species preference of Hereford and Santa Gertrudis cattle on a southern New Mexico range, *Jour. Range Mgt.* **19:**177–181.

Holscher, Clark E., and E. J. Woolfolk (1953): Forage utilization by cattle on northern Great Plains ranges, *U.S. Dept. Agr. Circ.* **918.**

Humphrey, R. R. (1945): Some fundamentals of the classification of range condition, *Jour. Forestry* **43:**646–647.

——— (1949): A proposed reclassification of range forage types, *Jour. Range Mgt.* **2:**70–82.

Hurd, Richard M., and N. A. Kissinger (1953): Estimating utilization of Idaho fescue *(Festuca idahoensis)* on cattle range by percent of plants grazed, *Rocky Mountain Forest and Range Experiment Station Paper 12* (mimeo.).

Launchbaugh, J. L. (1969): Range condition classification based on regressions of herbage yields on summer stocking rates, *Jour. Range Mgt.* **22:**97–101.

Levy, E. B., and E. A. Madden (1933): The point method of pasture analysis, *New Zealand Jour. Agr.* **46:**267–279.

Lommasson, T., and C. Jensen (1938): Grass volume tables for determining range utilization, *Science* **87:**444.

Malechek, John C., and C. L. Leinweber (1972): Forage selectivity by goats on lightly and heavily grazed ranges, *Jour. Range Mgt.* **25:**105–111.

National Academy of Sciences—National Research Council (1962): Basic problems and tecniques in range research, *NAS–NRC Pub.* **890,** Washington.

National Research Council (1970): *Remote Sensing with Special Reference to Agriculture and Forestry,* National Academy of Sciences, Washington.

O'Connor, K. F. (1971): Utilizing tall tussock, *Tussock Grassland and Mountain Lands Institute Review No. 21,* pp. 10–20.

Parker, Kenneth W. (1951): a method for measuring trend in range condition on national forest ranges, U.S. Forest Service, Washington (mimeo.).

Paulsen, Harold A., Jr. (1969): Forage values on a mountain grassland-aspen range in western Colorado, *Jour. Range Mgt.* **22:**102–107.

Pechanec, J. F., and G. D. Pickford (1937): A weight estimate for the determination of range or pasture production, *Jour. Am. Soc. Agronomy* **29:**894–904.

Perry, R. A. (1960): Pasture lands of the Northern Territory, Australia, land research series No. 5, Commonwealth Scientific and Industrial Research Organization, Canberra.

—— (1961): Pasture lands of the Alice Springs area, land research series No. 6. Commonwealth Scientific and Industrial Research Organization, Canberra.

Plice, Max J. (1952): Sugar versus the intuitive choice of foods by livestock, *Jour. Range Mgt.* **5:**69–75.

Roach, Mack E. (1950): Estimating perennial grass utilization on semidesert cattle ranges by percentage of ungrazed plants, *Jour. Range Mgt.* **3:**182–185.

Sampson, A. W. (1919): Plant succession in relation to range management, *U.S. Dept. Agr. Bull. No. 791.*

Sims, Phillip L., and Don D. Dwyer (1965): Pattern of retrogression of native vegetation in north central Oklahoma, *Jour. Range Mgt.* **18:**20–25.

Smith, A. D. (1944): A study of the reliability of range vegetation estimates, *Ecology* **25:**441–448.

—— (1950): Inquiries into differential consumption of juniper by mule deer, *Utah Fish Game Bull.* **8**(5):4.

—— (1953): Consumption of native forage species by captive mule deer during summer, *Jour. Wildlife Mgt.* **6:**30–37

—— (1958): Considerations affecting the place of big game on western ranges, *Proceedings of the Society of American Foresters,* Salt Lake City, pp. 188–192.

——, and Philip J. Urness (1962): Analyses of the twig-length method of determining utilization of browse, *Utah State Department of Fish and Game,* No. 62-9.

—— (1965): Determining common use grazing capacities by application of the key species concept, *Jour. Range Mgt.* **18:**196–201.

Smoliak, S., J. F. Dormaar, and A. Johnston (1972): Long-term grazing effects on *Stipa-Bouteloua* prairie soils, *Jour. Range Mgt.* **25:**246–250.

Springfield, H. W. (1961): The grazed-plant method for judging the utilization of crested wheatgrass, *Jour. Forestry* **59:**666–670.

Standing, A. R. (1933): Ratings of forest species for grazing surveys based on volume produced, *Utah Juniper* **4:**11–14; 40–41.

Stickney, Peter F. (1966): Browse utilization based on percentage of twig numbers browsed, *Jour. Wildlife Mgt.* **30:**204–206.

Stoddart, L. A. (1935): Range capacity determination, *Ecology* **16:**531–533.

—— (1952): Problems in estimating grazing capacity of ranges, *Proceedings of the Sixth International Grassland Congress,* State College, Pa., pp. 1368–1373.

U.S. Department of Agriculture, Forest Service (1969): Range environmental analysis handbook, Intermountain Region.

U.S. Department of Agriculture, Forest Service (1959): Range allotment analysis instructions, revision of Chapters 1 and 2 (mimeo.), October 21, 1959, pp. 13.

West, Neil E. (ed.) (1972): Galleta: Taxonomy, ecology, and management of *Hilaria jamesii* on western rangelands, *Utah Agr. Expt. Sta. Bull.* **487,** Utah State University, Logan.

Chapter 7

Animal Nutrition in Relation to Range Management

A major objective of range management is the production of animals —both wild and domestic. Efficiency of animal production is closely correlated with the nutrient value of the forage available. Consequently, a knowledge of the basic nutritional requirements of animals and the ability of range forage to supply them is of great significance both to the range manager and the rancher whose income depends on the level of production achieved.

The nutrient requirements of animals include *protein* to build and repair animal tissue; *fats* and *carbohydrates* for the production of energy; *minerals* for bone building, cellular formation (especially in soft tissues), regulating pH in body fluids, and enzyme regulation; and *vitamins*, which play a multiple role in the body processes.

MINERAL NUTRIENTS

Numerous minerals are essential for animal life; but most of these normally are present in adequate amounts, and only those likely to be deficient are of immediate importance to the range manager. The following minerals are of most concern to range and pasture management: phosphorus, calcium, iron, copper, cobalt, sodium, chlorine, iodine, potassium, and magnesium.

Others which are essential to animals are sulfur, manganese, and zinc. There is evidence that barium, bromine, fluorine, molybdenum, selenium, vanadium, chromium, and strontium also may be needed, but the required amounts of some of these are so minute that their precise role is difficult to establish. It is of more interest to the range manager that fluorine, molybdenum, and selenium, if taken in more than trace amounts, may be toxic, which is also true of copper and cobalt.

Calcium and phosphorus are essential for bone growth and numerous other body functions. They are by far the most important of the minerals from the standpoint of the amounts present in the body, constituting over 70 percent of the body ash (Maynard and Loosli, 1969). Recommended allowances for these two minerals are usually quite similar, calcium-to-phosphorus ratios commonly being from 2:1 to 1:1. In the case of ruminants, however, the range may be wider than this; there is evidence that calcium-to-phosphorus ratios are not critical for ruminants provided there is a sufficient amount of each (Beeson et al., 1945). There is a close relationship between calcium and phosphorus requirements and vitamin D, although the mechanisms involved are unclear, and calcium-to-phosphorus ratios are less critical when vitamin D is ample. There are many terms used to designate the nutritional failures due to inadequate calcium and phosphorus, but generally they can be limited to *rickets* for younger animals and *osteomalacia* for adult animals.

Sodium, potassium, and chlorine are constituents of body fluids and soft tissues. Of these, potassium is of little concern with range animals, as it is amply present in forage plants. Sodium and chlorine are excreted from the body in considerable quantities and normally must be supplied to animals, usually in the form of common salt, although in arid areas vegetation and saline waters may contain ample supplies. Absence of either of these elements in the diet is attended by loss of appetite and weight, and a decrease in milk supply of lactating females.

Certain of the trace minerals are inadequately present in the soil of some areas of the world. Iodine is deficient in parts of the Western United States, and its absence is characterized by abnormalities in reproduction, impaired growth, and goiter. Selenium is deficient in parts of New England, the Southeastern states, and the Pacific Northwest. Its lack causes white-muscle disease which is characterized by calcium deposits in the muscle (Carter et al., 1970). In arid areas it may reach toxic levels causing what is called *alkali disease.*

Cobalt, iron, and copper are deficient in forage from widely distributed areas throughout the world. Symptoms of deficiencies of any of these three minerals are anemia, emaciation, and eventually death (Madsen, 1942). Iron deficiency first was believed to be the cause of this malady, but later it was shown to be a more complex relationship among the three elements. *Salt sick* in Florida, *coast disease* in Australia, *bush sickness* in New Zealand, *wakuritis* in Kenya, and *pining* in Scotland are names given to diseases of similar nature caused by deficiencies of these three elements, and attest to the wide occurrence of this malady.

Complex mineral imbalances involving divalent elements, such as calcium and magnesium, and monovalent elements, such as potassium and sodium, are believed to cause such sickness as grass tetany, milk fever, and wheat poisoning. High potassium/calcium-plus-magnesium ratios have been found associated with grass tetany (Azevedo and Rendeg, 1972). Improper proportions of these minerals may produce an improper balance in the extracellular fluids and within the cells and result in nervous or glandular disturbances.

VITAMINS

Of the several vitamins, only vitamins A and D appear to be of importance to herbivorous animals (Hart and Guilbert, 1933). Vitamin A is important to the growth of young animals, in maintaining healthy membranes, in preventing night blindness, and in maintaining vigor and normal reproduction in adults. It is formed from carotene, found in fresh green feed, and is likely to be deficient only where animals are grazed on dry feed. When the diet is high in carotene, vitamin A is stored in the body, primarily in the liver and fat. These can be drawn on, and livestock on dry leached forage may not show vitamin A deficiency symptoms for 60 to 180 days. The period of time before deficiencies appear will depend upon the age of the animal and general feed conditions prior to being on dry forage. Since vitamin D is obtained from sunlight, it is unlikely to be deficient under range conditions.

NUTRIENTS FOR ENERGY AND TISSUE BUILDING

The major nutrient need for body functioning in terms of quantity is that of energy. This is provided by carbohydrates (starch, cellulose, hemicellulose, and sugars) and fats. Energy provided by the metabolism of these nutrients is necessary for maintenance of body heat and for work, growth, fattening, and reproduction. Proteins may also provide energy if supplied in excess of the animal's needs for muscle and tissue formation.

Measures of Food Value

Various schemes have been proposed to assess the nutrient value of animal forages. In the traditional method, plant material is divided chemically into protein, crude fiber, ether extract, ash (determined by analysis), and nitrogen-free extract (determined by differences after the other components are determined directly by chemical analysis). The amounts of each of these chemical constituents in the feed ingested and the feces excreted is arrived at by analysis. Digestibility is determined by subtracting from the amounts ingested the amounts found in the feces. The ether-extract fraction is conventionally multiplied by 2.25 (because of the greater caloric value of the fats and oils) and added to the digestible components, excluding ash, and a measure of feed value is obtained called *total digestible nutrients* (TDN). Because protein has a special nutritive function in body building, a further measure, the nutrient ratio, often is added, which is the ratio of digestible protein to all other components. Total digestible nutrient data do not express energy values. By virtue of inconsistencies when applied to feeds greatly different in quality, the TDN system of feed evaluation is rapidly being replaced by others which are based on energy content of feeds.

Measures of Food Energy

Several measures of energy value of forage have been proposed: digestible energy, metabolizable energy, and net energy. Successively, each of these represents a further refinement, and theoretically, each gives a more accurate index to nutrient value than the preceding index.

Digestible energy (DE) is the caloric value of the digestible portion of a food and can be estimated from TDN values by the formula: *TDN* (lbs) $\times$ 2018 = *DE*(kcal) (Swift, 1957). It may also be worked out calometrically by determining the differences between the caloric content of the food ingested and that in fecal matter. Rittenhouse et al. (1971) derived a regression equation by which digestible energy could be estimated either from organic-matter digestibilities or from dry-matter (including ash) digestibilities of forage.

Metabolizable energy (ME) is calculated by subtracting the energy lost through the urine and that due to formation of combustible gases in the rumen (mostly methane) from digestible energy.

Net energy (NE) is derived from metabolizable energy by subtracting heat produced in the animal body and in the rumen.

The Agricultural Research Council (Britain) and the National Research Council (United States) have arrived at differing conversion values between metabolizable energy and net energy, and Burroughs et al.

(1970) developed conversion ratios with beef cattle and reported conversion efficiencies in between United States and British standards. For example, conversion ratios for steers in the feedlot were 52, 56, and 64 percent for NRC, Burroughs, and ARC, respectively. These differences are not surprising, for conversion efficiencies differ with plane of nutrition and the amount of concentrate in the diet. They were higher at the maintenance level than during fattening—75 and 45 percent, respectively—when the ration contained 80 percent concentrate. Reducing the amount of concentrate to 40 percent reduced the conversion ratio to 65 percent at maintenance levels (Burroughs et al., 1970). Such differences preclude extrapolating one energy measure from another under range conditions. The relationships among the different energy measurements are shown in Fig. 7.1.

Metabolizable Energy as a Measure of Forage Value

Metabolizable energy has been used widely to evaluate range forages (Cook et al., 1952); but its determination is not without inaccuracies. Only the energy loss in the urine can be measured. The methane losses usually are estimated from values established under stall feeding and are commonly assumed to be 10 percent, although there is evidence that losses greater than this occur. Thus, Crampton and Lloyd (1959) reported methane losses of 15.5 percent of digestible energy in cattle and 7.7 percent in rabbits. Whether different ruminants differ widely with respect to methane production is not known. It is known that methane production depends greatly on the kind of forage; forages high in easily digested starch and soluble carbohydrates produce more methane. Wallace et al. (1970) estimated methane losses to be 5 to 6 percent on a grass range (Table 7.1).

A good case can be made, however, for using metabolizable energy in range situations, especially where shrubs are an important part of the diet. Shrubs are high in essential oils which appear in the ether-extract fraction, but these oils are largely lost through the urine, yielding no energy usable by the animal. Collection of urine for determination of energy losses is a relatively simple operation even on the range.

Because energy lost in heat increment and gases is not measurable on the range, a measure of nutrient value which considered only the losses from the feces and urine and ignored other losses might be preferable to either metabolizable energy or net energy under range conditions. It would be equally informative since, at present, corrections for gas and heat loss are treated as constants.

Theoretically, net energy establishes the amount of energy available to the animal for productive purposes, all energy inefficiencies, and los-

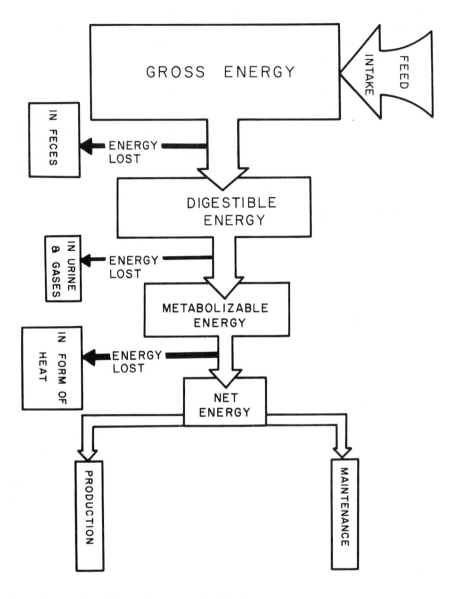

Fig. 7.1 Relationship among conventional feed-energy measurements and sources of loss of energy between ingestion and utilization by a range animal. Urinary losses, heat losses in inclement weather, and work performed to harvest forage are higher on the range than elsewhere which reduces the efficiency of feed conversion.

ses being accounted for; however, the assumptions that must be made with respect to fermentation, heat increment, and methane losses give both net energy and metabolizable energy an appearance of precision not actually achieved. Because of these difficulties, Swift (1957) concluded that digestible energy was the best practical measure of nutrient value, a conclusion in which Garrett et al. (1959) concurred; and Crampton (1959) expressed the view that there is no "workable" descriptive scheme for forages based on chemical analyses which relate well to actual performance.

Swift (1957) gave a ratio of metabolizable energy to digestible energy of 0.793. Based on in vitro methods, losses of energy in percent of energy in the substrate (alfalfa hay) were found to be as follows: methane, 11, heat of fermentation, 5; hydrogen, 2; and microorganisms, 5; this leaves 77 percent of the energy for production and heat increment (Herschberger and Hartsook, 1970). Estimated energy values in a range situation are shown in Table 7.1.

PROTEIN AND ENERGY REQUIREMENTS OF RANGE LIVESTOCK

A mature animal's basic requirement for food to maintain its internal body processes and support normal muscular activity is called the *maintenance requirement*. It is not a useful measure of the food that animals require, for simply maintaining animals yields no product. Additional needs are protein (for growth of the fetus and young animals and the production of milk, hair, horns, and other fibers) and energy for fattening and, in range animals, for the work required to harvest forage.

Table 7.1   Estimated Fate of Energy in Sandhill-Range Forage Selected by Cattle

Data from Wallace et al. (1970)

	Percent of energy in forage			
Energy component	June	July	August	December
Feces	30	38	50	58
Urine	4	4	3	2
Methane	6	6	5	5
Digestible energy	70	62	50	42
Metabolizable energy	60	52	42	35
Energy available for maintenance	44	37	28	22
Energy available for production	31	24	16	11

Thus nutrient requirements vary with the biological status of the animal, although at times range livestock may require little more than that required for maintenance, as in overwintering of mature animals.

Calculating Food Requirements

Weights have frequently been used as a guide to establishing food requirements of animals, and a good deal of research has gone into the relationship of body size to nutrient needs. Although size does affect the need for maintenance of body tissue, heat losses and consequently energy requirements are more directly related to body surface which varies with the two-thirds power of the body weight. Brody (1945) proposed, based on studies of basal metabolism (when an animal is in a fasting and resting state), that food requirements varied as the 0.73 power of the body weight (Brody, 1945). This figure has been widely used, although other investigators have reported higher values (Graham, 1972, suggested 0.9 between sheep and cattle) and 0.75 commonly is accepted as a mean value for all species of animals.

As a practical matter, basal metabolism may not be the best basis for calculating feed requirements. Needs for growth and fattening may not relate to surface area but to size. Practically, it may suffice to use weight as the basis for calculations, if it is remembered that young animals—i.e., smaller animals—have greater needs than mature animals which have no growth requirements to meet. Moreover, under range conditions, there is no control over forage intake, so that percentages are more practical guides to nutrients than are amounts.

Nutrient Standards of Herbivores

Based upon the foregoing theoretical relationships and feeding tests, requirements for different nutrients have been developed for domestic livestock (Table 7.2). These allowances are probably safe, and range livestock may perform adequately even though these standards are only approximately met. Marked departures, however, should be corrected by supplementation.

Inadequate work has been done on nutritional requirements and dietary adequacies of the diets of wild ungulates, although information is accumulating. Ullrey et al. (1969) judged that 160 kilocalories (kcal) per kilogram of body weight raised to the two-third power was an adequate winter maintenance diet for pregnant white-tailed does, and developed a basal diet containing 17 percent protein and 2.27 kcal per gram (Ullrey et al., 1971).

Wild ungulates that shed their horns and antlers each year have

Table 7.2 Recommended Nutrient Allowances for Wintering Mature Beef Cattle and Sheep during Early Pregnancy and Early Lactation

Data from National Research Council (1968; 1970)

	Body weight, kg	Daily feed		Protein, percent	Energy, megacalories*	Calcium, percent	Phosphorus, percent	Carotene, mg
		kg	Percent of body weight					
Beef Cattle								
Pregnant	450	6.8	1.5	5.9	12.4	0.18	0.18	42.0
	500	7.6	1.5	5.9	13.6	0.16	0.16	45.5
Early lactation	450	9.9	2.2	9.2	20.4	0.28	0.22	96.2
	500	10.5	2.1	9.2	21.6	0.27	0.22	105.2
Sheep								
Early gestation	45	1.2	2.6	8.0	2.6	0.27	0.21	1.7
	64	1.5	2.4	8.0	3.4	0.22	0.17	2.4
Early lactation	45	2.1	4.6	8.7	5.4	0.30	0.22	5.8
	64	2.5	3.9	8.0	6.2	0.27	0.20	7.9

*Metabolizable energy for cattle; digestible energy for sheep.

greater requirements than other animals for nutrients, particularly calcium and phosphorus. Magruder et al. (1957) were able to produce stunted antler growth in white-tailed deer by limiting either of these minerals in the diet. They reached the conclusion that a deficiency of either one was better tolerated if the other was present in adequate amounts.

Because of selectivity in eating, such data as are available do not suggest inadequacies in the diets of wild herbivores. Neither white-tailed nor mule deer forages in Arizona gave any indication of nutritional deficiencies when subject to in vitro tests for digestibility, although no determination of forage intakes was made (Urness et al., 1971). Similar indications of forage adequacy were found respecting the diets of pronghorn antelope in Utah (Smith and Malechek, 1974).

DETERMINING THE NUTRIENT VALUE OF RANGE FORAGE

Determining the value of animal foods is a costly and painstaking venture even when done under feedlot conditions. It is even more so under pasture or range conditions. Despite the additional problems which confront the researcher in range situations, considerable progress in range-nutrition research has been made possible through the development of new techniques and devices.

Chemical Analyses for Determining the Value of Forage

Chemical analyses often are employed to give some index to the value of forage. Although they cannot be accepted as totally reliable indicators of the worth of forage to the animal, they have some utility in comparing foodstuffs of similar kinds. Only insofar as plants are equally digestible are they indicative of nutrient value. Chemical analyses have less utility when forage plants differ greatly in kind and condition, as may be true on the range. A mixture of feathers, sawdust, and charcoal can be formed that, on the basis of chemical analyses, would appear to be a satisfactory animal food. This illustrates the need for digestibility determinations before the nutritive value of range forage can be assessed. Digestibility determinations can be done using animals (in vivo) or in the laboratory (in vitro).

Digestion Determination with Animals

Traditionally, digestion-balance trials were made with animals confined in metabolism stalls so that accurate records could be kept of food intake and excreta output. Chemical analyses are made of the food offered, the food rejected, and the feces. The difference in the amount of any constituent in the food offered minus the amounts in rejected feed and that in the feces is regarded as the digestible portion. Digestibility coefficients thus determined are only approximate, however, no account being taken of metabolic fecal matter, catabolic matter from the body of the animal, and products of ruminal digestion. Although digestion-balance trials on range forage using penned sheep equipped with fecal bags were performed prior to 1909 (Kennedy and Dinsmore, 1909), the technique is not well suited to range-forage evaluation. Either plant species must be tested separately, in which case it must be assumed that the digestion of a species is not affected by the presence of other forages, or it is necessary to composite a diet made up of several species and test the mixture.

By equipping animals with bags which collect the feces as it is voided and allowing the animals to graze normally, an accurate composite diet is assured without the need of assuming a dietary mix. Digestibility can then be arrived at in one of two ways. Samples of forage judged to be the same as that eaten by animals are collected, the amount of forage eaten determined, and chemical analyses made of the forage samples and the feces. Calculations then follow the procedures in standard digestion trials, digestibility being calculated from the difference between the nutrients ingested and the nutrients voided in the feces.

Two obstacles are inherent in this procedure: (1) determining the amount of forage consumed, and (2) selecting a forage sample which is equivalent to that actually consumed. The first is subject to errors inherent in range sampling. As for the second problem, hand-picked samples are not nutritionally equivalent to forage actually selected by animals. Hand-picked samples differed both from material collected from esophageal (Kiesling et al., 1969) or rumen fistulas (Galt et al., 1969). The second method uses tracers.

Use of Tracers for Estimating Forage Digestibility

If fistulated animals are used to provide a sample of the forage ingested, forage intake and digestibility can be estimated by the ratio technique. This is accomplished by comparing the changes in proportion of any nutrient in the feed and feces to the changes in proportion of some inert substance. This may be an indigestible substance in plants themselves (internal indicator or tracer) or a substance that is added to the ingesta (external tracer) usually either by means of a fistula or orally. Digestibility of any nutrient component (protein, cellulose, etc.) may then be determined by the following formula:

$$100 - 100 \left(\frac{\% \text{ indicator in food consumed}}{\% \text{ indicator in feces}} \times \frac{\% \text{ nutrient in feces}}{\% \text{ nutrient in feed}} \right)$$

By use of a similar formula, it is possible to determine the amount of feed ingested if the total output of the feces is known (Harris, 1968).

Lignin has frequently been used as an internal tracer (Harris, 1968) but increasingly there is evidence that lignin is not the inert substance it was once thought to be (Wallace and Van Dyne, 1970). Corrections can be made in the lignin-ratio formula for lignin digestibility, although, in view of the wide range of values reported, selection of a suitable correction factor is by no means assured. Chromic oxide may also be added to ingested feed to serve as an external tracer, but the two methods do not necessarily agree (Table 7.3).

Use of Fistulated Animals for Studying Forage Digestibility

A most useful technique that has greatly facilitated range-nutrition research is the use of fistulated animals (Fig. 7.2). A cannula, which has a removable cap, is implanted in an incision in the esophagus or rumen. By removing the cap from an esophageal fistula and placing a bag over the aperture, one can collect a forage sample which accurately represents the diet selected by the animal. Plant material thus obtained may then be introduced into another animal by means of a rumen fistula, or subjected to in vitro or chemical analyses. Forage habits often are studied by examination of material recovered by means of both domestic and wild fistulated animals (Veteto et al., 1972).

Wallace and Denham (1970) used fistulated cattle to collect forage samples which then were subjected to digestion trials using sheep.

Determination of the species of forage plants consumed by grazing

Table 7.3 Digestibility and Dry-Matter Intake of Winter-Range Forages Calculated by the Lignin-Ratio Technique and the Chromogen Method

Data from Harris (1968)

Forage species	Dry-matter intake		Dry-matter digestibility	
	Lignin technique	Chromogen method	Lignin technique	Chromogen method
	(kg per day)		(percent)	
Big sagebrush	1.44	0.20	50.5	−240.0
Shadscale	1.33	1.21	48.8	43.3
Winterfat	1.50	1.23	29.3	13.8
Black sage	1.39	0.72	40.3	−15.3
Squirreltail grass	1.86	3.15	47.9	68.7
Alfalfa	....	....	61.9	60.7

Fig. 7.2 Esophageal fistula in a goat (left) and bag for collecting forage eaten (right). With the cap in place the animal can feed normally. When a forage sample is desired, the cap is removed and food selected collects in the bag. *(Courtesy J. C. Malechek.)*

animals is not without problems. The composition of rumen contents may not agree with the composition of material recovered from esophageal fistulas (Table 7.4). The differences were probably attributable to rapid breakdown of the more soluble plant fraction and became less as the season advanced. In addition, the mineral content of material taken from fistulas is altered by additions from the saliva, and corrections must be made for this factor.

Another technique used in the study of range forage is the in vivo

Table 7.4 Proportions of Contents by Forage Classes Found in Esophageal and Rumen Samples of Sheep

Data from Rice et al. (1971)

	Rumen	Esophagus
	(Percent)	
Grass	95	82
Forbs	4	14
Shrubs	1	4

nylon-bag method. A known quantity of plant material is placed in a bag of nylon (or similar material) and introduced into the rumen of a fistulated animal where it is allowed to remain for a specified time (Van Dyne, 1962). The degree of digestibility is determined by the reduction in weight of dry matter in the sample. Good results have been reported using this technique, although they were not identical to results from conventional digestion trials (Wallace et al., 1972).

Laboratory Methods

Because of the difficulties involved in digestion trials with animals on the range, laboratory techniques have been developed to facilitate range-forage evaluation. This is done by subjecting samples of the forage to be tested to rumen liquor from a ruminant (commonly a sheep or cow) in the laboratory under conditions that simulate those within the rumen. Rumen or esophageal fistulas provide the forage material to be tested, or arbitrarily selected plant samples may be used. Digestibility is determined by the reduction in weight of a sample incurred during subjection to the liquid media. The method can be validated by regressing values found by this technique against values obtained from standard digestion tests, thus making it possible to extrapolate the results to conventional digestibility values. This technique can be used to estimate the efficiency of digestion of wild ruminants by obtaining rumen liquor from wild species.

Another laboratory method not requiring rumen liquor has come into use (Van Soest, 1967). This method involves fractioning plant material into cell contents and cell wall. In turn the cell-wall portion is fractioned into cellulose, hemicellulose, and lignin using designated chemical reagents. The "digestion coefficient" then is determined by summation of the amounts extracted in each step of the analyses. The method shows promise of being a reliable index to forage value (Crampton and Harris, 1969).

In any laboratory approach to forage evaluation, there still remain the problems of getting proper dietary samples and of determining forage intake if more than a qualitative index is desired.

NUTRITIONAL VALUE OF RANGE FORAGE

Range forage varies tremendously in quality from time to time and from place to place. The growing season usually lasts for only a small portion of the year, and it is during the growth stages that plants are most nutritious. Once mature, plants are subject to leaching and reduction in nutritive value, the amount of leaching depending upon the climate, particularly rainfall. Declines in nutrient composition and leaching are especially serious in the case of herbaceous plants, grasses, and forbs. By contrast, shrubs have a longer growing season and maintain their nutritive values longer, and some evergreen shrubs remain highly nutritious year-round. Shrubs may play an important function in meeting the needs for protein and vitamins when herbaceous forage is dormant and lacking in these components.

Effects of Season upon Forage Value

As plants mature, crude protein, the more readily digested carbohydrates, and phosphorus decrease and crude fiber, lignin, and cellulose increase. These changes are caused both by changes in the stem-leaf ratios and by actual changes in the composition within each plant part. The effect of these changes is illustrated by data for three perennial range grasses (Table 7.5). With annual plants even greater reductions take place. Hart et al. (1932) reported that in foothill ranges in California the quality of the forage changed from being comparable to a protein-rich concentrate during the early vegetative stages to that of a poor roughage when dry. Even among grasses, however, decline in nutritive value may not continue throughout the growing season. Kamstra (1973) found higher in vitro digestibility in some grasses in September than he did in August.

Leaching

Much of the change in composition of forage with advancing maturity is caused by leaching of the soluble constituents by rain. This is especially so in the hotter and more humid climates. California studies (Guilbert et al., 1931) indicated that calcium is not affected greatly but showed phosphorus to be distinctly lowered, thus widening the calcium-phosphorus ratio. Nitrogen-free extract decreases with leaching; most of the loss is easily digested sugars. In the same region, as the protein of bur clover (*Medicago hispida*) was leached, an increasingly wide nutritive ratio resulted (Guilbert and Mead, 1931). Watkins (1937) reported both calcium and phosphorus to be leached by rain from mature range grasses.

Effects of Type of Plant upon Forage Value

There are thousands of plant species growing on the range, and many of these are important constituents of the animal diet. Although there are

Table 7.5   Chemical Composition and in vitro Cellulose Digestion of Range
Grasses at Various Stages of Growth during 1959 and 1960

Data from Raleigh (1970)

Growth stage	Air-dry matter	N	Ether extract	Crude fiber	P	Ca	In vitro cellulose digestion
Agropyron spicatum				(Percent dry matter)			
Pre-boot	34	2.6	1.6	21	0.22	0.23	70
Boot	36	2.3	2.1	22	0.23	0.26	69
Head	43	1.8	2.4	25	0.17	0.23	57
Early flower	43	1.6	2.4	26	0.18	0.22	55
Early seed	60	1.2	3.3	27	0.18	0.16	47
Mature	79	0.9	3.5	28	0.18	0.24	41
Koeleria cristata							
Pre-boot	33	3.1	1.8	20	0.26	0.31	74
Early boot	30	2.4	2.9	21	0.25	0.24	77
Early head	33	2.1	2.2	21	0.22	0.28	71
Head	38	1.8	2.8	25	0.23	0.27	74
Flower	40	1.5	2.9	28	0.19	0.25	62
Seed stage	58	1.4	4.9	23	0.23	0.31	61
Mature	80	1.2	4.6	24	0.20	0.24	63
Agropyron desertorum							
Pre-boot	30	3.0	2.1	16	0.22	0.23	75
Boot	34	2.3	2.0	17	0.21	0.21	69
Late boot	34	2.2	2.3	20	0.21	0.28	73
Early head	40	1.7	2.1	19	0.18	0.18	68
Head	42	2.0	5.1	23	0.18	0.24	69
Seed stage	51	1.4	1.9	22	0.14	0.18	53
Mature	65	0.7	3.0	27	0.22	0.26	48

infinite variations in forage value among species, there are some
similarities among various plant groups. The legumes, bearing
nitrogen-fixing bacteria, are inclined toward a high-protein content.
Browse plants, being generally deep-rooted and tending to store food
reserve in the stems rather than in the roots, do not decrease in protein,
vitamin A, and carbohydrates during dry periods or during the winter as
much as do grasses (Fig. 7.3). Forbs generally do not cure well and are

consequently inferior as forage to both grass and browse during the nongrowing season.

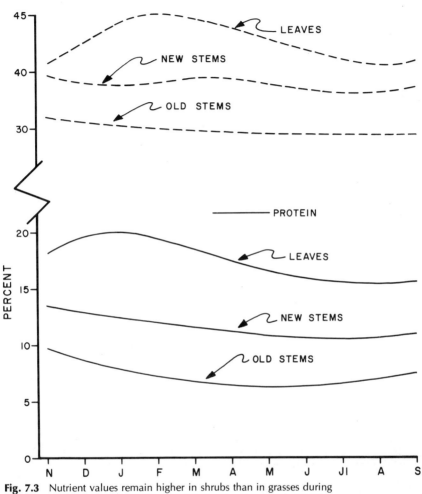

Fig. 7.3 Nutrient values remain higher in shrubs than in grasses during the dormant season as shown by the data for desert saltbush *(Atriplex polycarpa)*. *(From Chatterton et al., 1971.)*

There are some exceptions to this: bur clover maintained crude protein contents as high as 15 percent even when dry (Guilbert and Rochford, 1940). Van Dyne (1965) reported somewhat lower values for bur clover (10 percent), but two species of *Trifolium* exceeded 12 percent protein in midsummer in California.

Shrubs more nearly maintain their peak values throughout the growing season, although reductions in protein and phosphorus and increases in lignin occur; but protein levels are higher than in the grasses (Table 7.6).

The data in Table 7.6 further emphasize the importance of the parts of the plant in determining forage composition. Thus, were animals to remove only the leaves from grasses, the difference among the three classes of forage would be less marked than when the entire plant is taken. As for the actual forage selected by sheep, forbs were highest in crude protein (17.8 percent) followed by browse (13.4 percent) and grass (12.2 percent). In the area where these data were taken, there are many late-growing forbs so that these data do not reflect the seasonal changes in individual forbs.

Browse species in New Mexico are more than three times as high in calcium and 61 percent in phosphorus than grass in the fall (Watkins, 1937). In Texas (Tash, 1941), the short grasses are higher in phosphorus, total mineral, and protein content than tall grasses. Shrubs and valuable forbs, especially legumes, are consistently higher in phosphorus

Table 7.6 Seasonal Changes in Crude Protein and Lignin of Three Classes of Forage in Northern Utah

Data from Cook and Harris (1950a)

	Protein			Lignin		
	Early*	Middle	Late	Early	Middle	Late
Grass						
Stems	5.02	3.62	3.16	10.99	11.16	12.98
Leaves	14.60	11.98	10.21	7.87	8.50	10.30
Heads	13.76	15.20	. . . . .	8.20	8.22	. . . . .
Entire plant	8.22	6.02	4.49	9.95	10.48	12.48
Forbs						
Stems	4.73	4.31	4.48	11.73	11.78	13.59
Leaves	15.66	13.45	13.25	8.05	7.64	9.86
Entire plant	10.62	9.18	8.79	9.74	9.36	11.54
Shrubs						
Stems	6.50	6.00	6.46	22.72	16.03	22.95
Leaves	13.77	12.91	12.45	13.88	11.09	12.01
Current growth	12.26	11.71	10.76	16.20	13.76	15.08

*Early = July 10 to Aug. 4; Middle = Aug. 5 to Aug. 24; Late = Aug. 25 to Sept. 13.

than grasses. These examples illustrate the importance of browse and forbs in improving the diet of range animals (Fig. 7.4).

Vitamin A

The vitamin A value of range forages is of considerable importance, especially as it influences reproduction in livestock. Investigations have shown that carotene content in green plants is exceedingly high (Woods et al., 1932; 1935). All pasture plants have a relatively high carotene content during early summer, with the content decreasing markedly in midsummer and increasing after fall rains when new growth is apparent (Atkeson et al., 1937). The carotene content of blue grama (*Bouteloua gracilis*) is extremely high in August, decreasing to about one-half in September, and to about one/one-hundredth in November when it becomes dry (Smith and Stanley, 1938). Studies on the monthly variation in carotene content of two New Mexico range grasses, mesa dropseed (*Sporobolus flexuosus*) and black grama (*Bouteloua eriopoda*) showed that both grasses were moderately high during the growing season. The dropseed lost all its carotene soon after the end of the growing season, but the grama maintained an amount ample to satisfy the vitamin A requirements of range cattle throughout the winter (Watkins, 1939).

Fig. 7.4 Mixtures of grasses and browse such as this one consisting of wheatgrass *(Agropyron)*, ricegrass *(Oryzopsis)*, and bitterbrush *(Purshia)* provide a better-balanced diet than does a single class of plants.

Research in Utah shows that, in the winter, salt-desert browse species contain an average of 6.8 p.p.m. (parts per million) of carotene, whereas salt-desert grass species contain only about 0.1 p.p.m. carotene. Thus, even in winter most browse species in the Great Basin area are good sources of vitamin A, whereas grasses are very poor sources.

Effects of Habitat upon Forage Value

There are numerous studies dealing with the influence of habitat, chiefly climate and soil, upon the chemical composition of forages. To date the results are conflicting. It has been reported (Daniel and Harper, 1935) that the calcium content of plants decreased and the phosphorus content increased during periods of high rainfall, and that when rainfall was low the reverse occurred. Scott (1929), however, found that the amount of precipitation did not affect the mineral content of forage. Studies in New Zealand (Rigg and Askew, 1929) showed no correlation between soil and grass mineral content, except that soils low in phosphorus generally produced forage low in phosphorus.

In Utah, Cook and Harris (1950b) studied the effect of vegetation type and site upon nutritive value of forage and concluded that environmental factors and soil moisture were more important in determining the nutrient content of range plants than the chemical content of the soil. Conversely, Midgley (1937) reported that plants are affected materially by the soil in which they grow; thus, soils with high mineral content yield forages high in minerals. In addition, the soil type may influence the species composition and the availability of the nutrients. When one principal nutrient is deficient, others are taken up by the plant in abnormal amounts, and an even smaller amount of the deficient nutrient is absorbed than normally (Beeson, 1941). Studies on the effect of season, site, and soil upon *Symphoricarpos rotundifolius* showed that chemical composition varied more with season than with either site or soil (Stoddart, 1941). However, plants growing on highly productive sites contained more protein and less nitrogen-free extract than those on poor sites. Soil type did influence the total ash and phosphorus content of the plant, and possibly affected protein content. Sims et al. (1971) reported substantially greater crude-protein values for the blue grama on some sites than on others. One can only conclude from this evidence that good sites and good growing conditions produce good forage.

DIGESTIBILITY OF RANGE FORAGE

Considerable data are accumulating with respect to digestibilities of range forages. The greatly differing conditions under which studies were performed and the different methods employed have produced widely differing data, but general principles are becoming clear.

Effect of Season upon Forage Digestibility

During early growth stages range grasses are highly digestible (40 to 70 percent), but they decline sharply as the season advances to less than 40 percent (Burzlaff, 1971). Annual grasses in California averaged 47 percent digestible in midsummer when they were dry (Van Dyne, 1965). Declines in digestibility are not due simply to change in chemical composition. Digestibility of all chemical components also declines (Table 7.7).

Data reported from salt-desert ranges of Utah provide comparisons of digestibility of mature grasses and dormant shrubs. Eight species of shrubs averaged 39 percent digestible while nine species of grass were 47 percent digestible. Individual values ranged from 27 to 55 percent for shrubs and 32 to 69 percent for grasses (Cook et al., 1954). The digestibility of crude protein differed markedly from this, however, averaging 58 percent for browse and only 12 percent for grass.

These values for shrubs are lower than those found when deer were used as the subject animal, although, except for *Artemisia,* different species were involved (Table 7.8). Some show surprisingly high digestibility values, considering the woody nature of the material. Others have reported somewhat lower values. Bissell et al. (1955) conducted digestion trials with deer and found generally lower values than those reported from Utah. In the species common to both tests, Bissell's digestion coefficients were slightly better for bitterbrush *(Purshia tridentata)*

Table 7.7 Means of Digestion Coefficients by Sheep Fed Cattle-Grazed Range Forage

Data from Wallace and Denham (1970)

Component	\multicolumn{4}{c}{Digestibility of forage selected by cattle in:}			
	June	July	September	December
	\multicolumn{4}{c}{(Percent)}			
Organic matter	73	68	56	49
Fiber	67	63	52	45
Gross energy	70	63	50	43
Crude protein	69	59	34	2
Ether extract	40	24	23	14

Table 7.8 Digestibility Coefficients for Some Common Browse Species as
Winter Feed for Mule Deer

Data from Smith (1957)

| Species | Digestibility coefficients, percent | | | | Total digestible nutrients, lbs per hundredweight |
	Protein	Ether extract	Crude fiber	Nitrogen-free extract	
Artemisia tridentata	66.6	68.3	51.4	78.0	78.1
Cercocarpus ledifolius	54.3	42.9	35.9	76.3	65.5
Juniperus utahensis	16.8	58.9	33.7	70.4	63.5
Cercocarpus montanus	48.5	37.6	31.8	60.0	49.6
Cowania stansburiana	39.8	47.7	4.4	59.4	47.2
Purshia tridentata	35.7	53.0	18.3	57.3	44.9
Prunus virginiana	48.4	23.3	8.8	56.1	38.9
Quercus gambelii	10.7	38.4	16.6	53.6	36.2

but considerably lower for sagebrush (*Artemisia tridentata*), 54.8 and 55.9
for bitterbrush and sagebrush, respectively. Ullrey et al. (1972) report
dry-matter digestibility of 50 and 60 percent for aspen (*Populus
grandidentata*) and white cedar (*Thuja occidentalis*), respectively.

Some workers have reported that the high content of volatile oils in
sagebrush depresses digestibility by inhibiting the activity of rumen organisms (Nagy et al., 1964). This view is not supported in standard
digestion trials in Oregon with sheep; no detrimental effect was observed when sagebrush was included in the diet at a 50:50 ratio with
alfalfa (Smith et al., 1966). Probably the reduction in forage intake of
plants high in strongly aromatic oils is more important to the failure of
such plants to provide adequate nutrition. Neither juniper nor sagebrush, both important in the winter diets of mule deer in the Western
United States, is eaten in large amounts when either is the sole forage
(Smith, 1959). White cedar is reported to depress in vitro fermentation
in rumen liquor from a steer, but not with deer liquor (Short, 1963).

Effect of Species of Animal on Digestibility

Little is known of comparative efficiencies of different kinds of animals
in digestion of food. Forbes et al. (1937) cited evidence that seemed to
show cattle digest crude fiber more efficiently than sheep unless the
proportion of concentrate to roughage is high; then sheep are more
efficient. Gallagher and Shelton (1972) provide comparative data with
respect to efficiencies of production of sheep and angora goats. Sheep

made much greater weight gains than goats but produced only about half as much fiber. Whether these differences were due to more efficient utilization of food or to differences in the nutrient requirements for wool and mohair is not clear.

There are too few data, arrived at by different methods, and the variation in range forage within species is too great to permit definitive comparisons among other range animals. Cook et al. (1954) reported ranges of 38 to 73 percent in total digestible nutrients for sagebrush in separate trials by sheep; Bissell et al. (1955) reported 40 to 75 percent digestibilities for the same species by deer (Table 7.9). Mean values, too, differed markedly. These variations emphasize the hazards of comparing data reported from different areas without considering the methods used; the ecotype, variety, or subspecies of plants; and the conditions of the experiment.

In controlled digestion tests Cowan et al. (1970) compared the ability of sheep and deer to digest timothy hay by two methods, using a nylon bag in fistulated animals and digestion cages. There were small differences in digestibility between either animals or methods (Table 7.10).

There is a widely held view that game animals have distinctive rumen floras and thus are able to digest plant material high in fiber better than domestic livestock and, conversely, are unable to digest common cultivated forages. Such data as are available do not support this view. Alfalfa hay was fed simultaneously to deer and sheep and very similar digestibility coefficients were found for all components except ether extract. There were almost identical yields of digestible nutrients (Smith, 1952). Bissell et al. (1955) and Dietz et al. (1962) fed alfalfa hay to mule deer in standard digestion trials and found alfalfa hay not to be indigestible; in fact, the latter workers used it in conjunction with other native feeds believing they could not be tested alone.

Although Short (1963) found white-tailed deer digested fibrous material better than did a steer, the rate of digestion in deer was insufficiently rapid to provide adequate energy for their needs.

The ratio between rumen volume and body weight is smaller in deer than in cattle. In consequence, Short concluded that more rapid movement of food through the digestive tract necessitated by their smaller rumen would make deer less able to fully digest fibrous material. Moreover, the lower forage intakes that accompany less digestible food may further limit utilization. For example, food intake of white-tailed fawns at first increased and then decreased as the digestibility of the food declined (Ammann et al., 1973).

Such work as has been done on microflora of the rumen of wild

Table 7.9 Comparison of Mean Digestibility Data for the Same Species Reported by Different Investigators Using Different Methods

| | Digestion cages | | Fecal bags | In vitro | |
| | Deer | | Sheep | Steer inoculum | Elk inoculum |
	(Smith, 1957)	(Bissell et al., 1955)	(Dietz et al., 1962)	(Cook et al., 1954)	(Ward, 1971)
	Total digestible nutrients, percent			Percent digestible	
Bitterbrush	44.9	54.8	52.5	. . . .	26.5	28.6
Sagebrush	78.1	55.9	58.9*	50.7	41.1	54.1	55.1

*Fed with alfalfa hay and determined by difference.

Table 7.10   Apparent Digestibility of Timothy Hay by Deer and Sheep Using Two Techniques

Data from Cowan et al. (1970)

Component	Deer		Sheep	
	Fistulated	Caged	Fistulated	Caged
	(Percent)			
Cellulose	56.8	55.6	54.6	55.6
Crude protein	56.0	45.0	52.9	58.9
Energy	54.5	52.0	53.8	56.1

herbivores does not support the theory of an obligatory role of rumen floras among different animal hosts. Similar organisms are found in wild and domestic ruminants (Pearson, 1965) and alterations in the diets of mule deer between native and cultivated feeds produced only minor percentage changes in ruminal bacteria in deer (Pearson, 1969b). Organisms in the rumen of elk in winter did not differ from those found in cattle when on a poor quality hay (McBee et al., 1969), and the rumen flora from starved pronghorn antelope were normal (Pearson, 1969a). It appears more likely that starvation to the point where the environment of the rumen becomes altered so as to make digestion of any food impossible better explains incidents where wild herbivores have died despite having hay-filled rumens.

NUTRITION OF RANGE HERBIVORES

Sufficient work has been done on nutritional value of range forage to draw some general conclusions with regard to possible nutritional deficiencies, based upon guidelines for domestic livestock developed by the National Research Council (1968; 1970). Considering the great variations from place to place in kind of vegetation, severity of the weather, and the work expended by grazing animals, these can only be accepted as approximate guides. There is still a dearth of data which demonstrates by actual performance the specific insufficiencies in range forage for domestic livestock, and there is even less for wild ungulates.

Factors Affecting Determination of Nutritional Requirements

There are several factors which make the precise determination of nutrient standards of herbivores on the range difficult: level of food intake, nutritional level of food, variations in metabolic rates of the animals, and recycling of nutrients, especially nitrogen.

We have earlier noted that phosphorus-to-calcium ratios may be unimportant in ruminants so long as neither is in short supply. Similarly, if food intake is high the amounts of specific nutrients may be adequate though present in moderate amounts in the feed. It has been found, however, that food intakes are reduced when measures of food quality decline. Weston (1971) found substantially greater forage intake by lambs as the crude-protein content of the forage increased (Fig. 7.5). Fels et al. (1959) believed sheep selected for nitrogen when the content

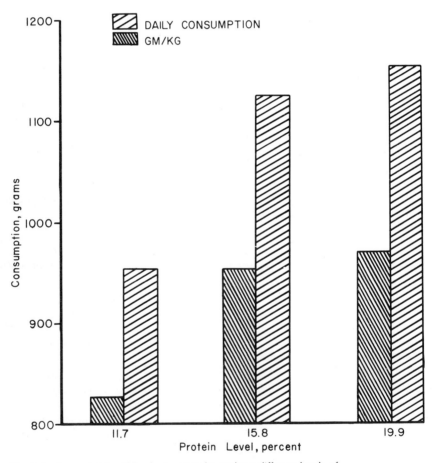

Fig. 7.5 Forage intake of lambs in Australia at three different levels of protein. Adequate voluntary forage intakes were obtained at 18 percent crude protein. *(Data from Weston, 1971.)*

of forage was 2.5 percent (15.6 crude protein) or less, but that there was no selection if nitrogen exceeded 3.5 percent.

Church (1971) cites substantial evidence to support the view that caloric content is more important than crude protein in limiting food intakes. Pressure of gas in the rumen has also been thought to determine intakes. Whatever the specific cause, forage intakes differ seasonally on the range (Table 7.11). It is probable that lowered intake rather than nutritionally deficient forage is the major reason for failure of range animals to perform well on poor-condition range.

The level of nutrients in the forage, and particularly the balance among different components, may determine animal preference and affect the results of attempts to measure forage adequacy. Blaxter and Wainman (1964) added maize to hay rations in increasing proportions and found that the efficiency of utilization of metabolizible energy was increased slightly at a maintenance level, but efficiency more than doubled in fattening rations.

There is some evidence that metabolic rates vary by season in wild animals. Klein (1970) cites evidence that native ungulates in northern regions have lower metabolic rates in winter, thus requiring less forage. Even in the presence of ample forage when in captivity, forage intakes of caribou are voluntarily limited (McEwan and Wood, 1966). Silver et al. (1959), however, could find no difference in the basal metabolism of white-tailed deer between winter and summer, although in a later study deer were observed to reduce their food intakes in winter (Silver et al., 1969).

An interesting, and possibly important, factor in the nitrogen requirements of herbivores, is that of recycling nitrogen supplies in the body in times of a shortage. This is believed by some to explain the success of wild ungulates in Africa in the face of long droughts when

Table 7.11 Intake by Sheep and Seasonal Changes in Chemical Content of Forage on a Forb Range in Montana

Data from Buchanan et al. (1972)

	Intake, kg per day			Chemical composition of fistula samples	
Season	Wet	Dry	Crude protein	Acid-detergent fiber	Acid-detergent lignin
Early	4.6	0.9	16.7	27.4	5.9
Mid	5.0	1.1	11.4	30.4	8.0
Late	3.2	1.1	10.5	32.1	9.0

animals must subsist on low-quality forage. Livingston et al. (1962) measured urea excretion in identical twins of both European and Indian cattle and found that urea excretion declined as the crude protein content of the ration declined. Availability of water also affected urea excretion. Results of experiments on camels and sheep also support this view; injections of urea into the bloodstream of camels were not accounted for in the urine, confirming the utilization of retained urea in protein synthesis (Schmidt-Nielsen et al., 1957; Schmidt-Nielsen and Osaki, 1958). This ability to recycle nitrogen is of survival value only; ultimately the animals must have adequate nitrogen to balance obligatory excretion. So long as nitrogen supply is adequate, the ability to recycle it probably confers no advantage. Further, the ability of domestic livestock to recycle nitrogen appears to be limited (Vercoe, 1969).

Adequacy of Range Forage

The nutritional adequacy of range forage depends upon many factors: season, vegetative type, intensity of use, kind of animals, and status of the animal. From the data thus far developed it appears that adequate nutrients are available during the growing season. Work in Oregon showed digestible nitrogen was adequate for a mature cow until midsummer, but that it was inadequate for a cow and calf by June. Thresholds for digestible energy came later in the season (Fig. 7.6).

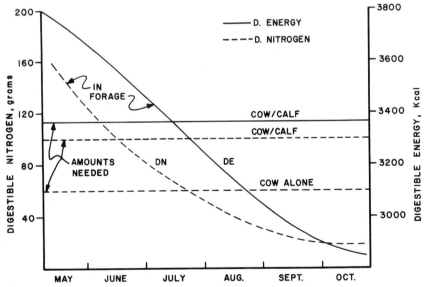

Fig. 7.6 Seasonal nutrients supplied by range forage at Squaw Butte, Oregon, compared to the nutrient requirements of a cow with calf and cow alone. The adequacy of range forage depends upon the kind of animal and its biological status. *(Data from Raleigh, 1970.)*

Animals required to subsist on mature range forage undoubtedly are subjected to some nutritional deficiencies depending upon the kind of forage available. This is of particular concern in more arid tropical-range areas where growth to maturity is rapid, and, throughout most of the year, forage is high in fiber and low in protein, vitamins, and minerals (Pfander, 1971).

On similar ranges and during the growing season, the nutrient level of diets differs with the kind of animal. Sheep diets had a higher percentage of digestible protein than did cattle diets; sheep ingested feed higher in energy in early season than did cattle, but in late season this was reversed (Table 7.12). These differences were probably due to the greater dependence of cattle on grass.

There were nutritional differences in various forage types. Both cattle and sheep diets in the aspen type were superior to those in sagebrush-grass types. The average percent of digestible protein for sheep was 4.5 and 7.2 percent for sagebrush-grass and aspen, respectively, and for cattle 3.6 and 5.4 percent (Cook and Harris, 1968).

Despite the fact that much of the forage is mature and dormant, salt-desert forage in the Great Basin is adequate or nearly so in producing the amount of energy required for winter grazing. Most of the grasses are excellent sources of energy and provide a third to a half more than is required normally. However, most salt-desert browse plants are borderline or deficient in energy and must be supplemented in order to furnish adequate energy in the absence of ample grass to supply it.

Conversely, all grass species on salt-desert range in the intermoun-

Table 7.12 Seasonal Changes in Digestibility of Forage Eaten by Sheep and Cattle on a Mountain Summer Range

Data from Cook and Harris (1968)

Animal	Periods*	Digestible protein, percent	TDN Percent	Digestible energy kcal/lb
Sheep	Early	6.9	50.2	1026
	Mid	4.8	46.0	949
	Late	4.2	40.9	752
Cattle	Early	4.9	49.5	983
	Mid	4.0	48.8	968
	Late	3.0	46.6	814

*Early, June 10 to July 15; mid, July 16 to Aug. 9; late, Aug. 10 to Sept. 15.

tain area are materially deficient in protein during winter, whereas browse species are borderline or only slightly deficient (Cook and Harris, 1950a). The adequacy of winter ranges, then, varies with the different forage types and the proportions of the different kinds of forage present. Shadscale is low in protein, and sheep grazing this type do not receive adequate protein, obtaining only 80 percent of their energy needs (Fig. 7.7).

Energy Budgets on the Range

One approach to estimating forage requirements of range livestock is to set up energy budgets based upon the particular physiological status of the animal. The basic energy requirement for maintenance animals is derived from the metabolic weight relationship: Kcals required = 70 × $W^{0.75}$, where W is the weight of the animal in kilograms. This relationship would closely estimate forage requirements for mature animals

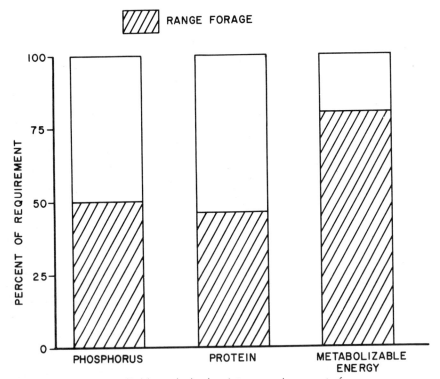

Fig. 7.7 Nutrients supplied by a shadscale winter range in percent of requirements for a sheep. Nutrient value of range forage depends upon the vegetative type and the season. *(From Harris, 1968.)*

Fig. 7.8 Sheep equipped with fecal bags and urinals make it possible to determine digestible energy of range forage, but it is not possible to measure losses due to gases and heat increment which are needed for complete energy budgets.

over winter, since no weight gains are anticipated. Additions must be made for fetal growth in brood stock and wool growth in fiber-producing animals. At other seasons additions must be made for weight gains, lactation, and at all seasons, for work performed in travel incident to harvesting forage and obtaining water. By making certain assumptions, sometimes heroic, each of these functions can be expressed in terms of energy requirements. When fecal, urinary, and gaseous losses are considered, the energy yield of the range can be partitioned among all elements in the energy matrix (Fig. 7.8).

Given the present state of our knowledge, the utility of such an exercise is open to question. There are such wide discrepancies in the value of constants going into separate steps of the energy matrix that the results at best are approximations. Rather than a practical tool, this approach probably serves a better function as a theoretical model upon which present knowledge can be assembled and evaluated and the need for new areas of knowledge and research isolated and identified.

More direct approaches may be more utilitarian, in which the relationship between forage intake, measured as digestible matter, is equated to actual output of animal products (Pearson, 1972).

Energy budgets, though rough estimates, have demonstrated the low efficiency of energy transfer (see Chapter 5 and Table 7.1). It has been estimated, for example, that only about 1 percent of the sun's energy reaching a given area is converted into forage (Thomas, 1973) and as little as 1/40,000 reaches the food product, meat (Love, 1970). Estimates indicate that only about 10 percent of the energy in the feed consumed by animals in a Western cow-calf operation can be sold and removed from the ranch (Cook, 1970).

CORRECTING NUTRITIONAL INADEQUACIES

From the foregoing discussion it is evident that range forage is inadequate nutritionally at certain times throughout the year. Experience has shown that production is increased by recognizing these deficiencies and providing supplements. This will be discussed in Chapter 10.

BIBLIOGRAPHY

Ammann, Alan P., Robert L. Cowan, Charles L. Mothershead, and B. R. Baumgardt (1973): Dry matter and energy intake in relation to digestibility in white-tailed deer, *Jour. Wildlife Mgt.* **37:**195–201.

Atkeson, F. W., W. J. Peterson, and A. E. Aldous (1937): Observations on the carotene content of some typical pasture plants, *Jour. Dairy Sci.* **20:**557–562.

Azevado, J., and V. V. Rendig (1972): Chemical composition and fertilizer response of two range plants in relation to grass tetany, *Jour. Range Mgt.* **25:**24–27.

Beeson, K. C. (1941): The mineral composition of crops with particular reference to the soils in which they were grown, *U.S. Dept. Agr. Misc. Pub.* **369.**

————, C. H. Hickman, and R. F. Johnson (1945): Calcium and phosphorus for breeding ewes, *Flour & Feeds*, **46:**12–13.

Bissell, Harold D., Bruce Harris, Helen Strong, and Frank James (1955): The digestibility of certain natural and artificial foods eaten by deer in California, *Calif. Fish and Game* **41:**57–78.

Blaxter, K. L., and F. W. Wainman (1964): The utilization of the energy of different rations by sheep and cattle for maintenance and for fattening, *Jour. Agri. Sci.* **63:**113–128.

Bredon, R. M., D. T. Torell, and B. Marshall (1967): Measurement of selective grazing of tropical pastures using esophageal fistulated steers, *Jour. Range Mgt.* **20:**317–320.

Brody, Samuel (1945): *Bioenergetics and Growth*, Reinhold, New York.

Brown, E. M. (1939): Some effects of temperature on the growth and chemical composition of certain pasture grasses, *Mo. Agr. Exp. Sta. Res. Bull.* **299.**

Buchanan, Hayle, W. A. Laycock, and D. A. Price (1972): Botanical and nutritive content of the summer diet of sheep on a tall forb range in southwestern Montana, *Jour. Anim. Sci.* **35:**423–430.

Burroughs, Wise, M. A. Fowler, and S. A. Adeyanju (1970): Net energy evaluation for beef cattle rations compared with evaluation by metabolizable energy measurements, *Jour. Anim. Sci.* **30**:450–454.

Burzlaff, D. F. (1971): Seasonal variations of in vitro dry-matter digestibility of three sandhill grasses, *Jour. Range Mgt.* **24**:60–63.

Carter, D. L., C. W. Robbins, and M. J. Brown (1970): Selenium concentrations on some high northwestern ranges, *Jour. Range Mgt.* **23**:234–238.

Chatterton, N. J., et al. (1971): Monthly variation in the chemical composition of desert saltbush, *Jour. Range Mgt.* **24**:37–40.

Church, D. C. (1971): Digestive physiology and nutrition of ruminants, Vol. 2, *Nutrition,* Oregon State University Bookstore, Corvallis.

Cook, C. Wayne (1970): Energy budget of range and range livestock, *Colo. State Univ. Expt. Sta. Bull.* TB **109**.

————, and Lorin E. Harris (1950a): The nutritive content of the grazing sheep's diet on the summer and winter ranges of Utah, *Utah Agr. Expt. Sta. Bull.* **342**.

————, and ———— (1950b): The nutritive value of range forage as affected by vegetation, type, site, and stage of maturity, *Utah Agr. Expt. Sta. Bull.* **344**.

————, and ———— (1968): Nutritive value of seasonal ranges, *Utah State Univ. Agr. Expt. Sta. Bull.* **472**.

————, L. A. Stoddart, and Lorin E. Harris (1952): Determining the digestibility and metabolizable energy of winter range plants by sheep, *Jour. Animal Sci.* **11**:578–590.

————, ————, and ———— (1954): The nutritive value of winter range plants in the Great Basin, *Utah State Agr. Expt. Sta. Bull.* **372**.

Cowan, R. L., J. S. Jordan, J. L. Grimes, and J. D. Dill (1970): Comparative nutritive values of forage species, in range and wildlife habitat evaluation—a research symposium, *U.S. Dept. Agr. Misc. Pub.* **1147**:48–56.

Crampton, E. W. (1959): Interrelations between digestible nutrient and energy content, voluntary dry matter intake, and overall feeding value of forages, in Howard B. Sprague (ed.), *Grasslands,* Publication No. 53, American Association for the Advancement of Science, Washington, pp. 205–212.

————, and L. E. Harris (1969): *Applied Animal Nutrition: The Use of Feedstuffs in the Formulation of Livestock Rations,* 2d ed., W. H. Freeman and Co., San Francisco.

————, and L. E. Lloyd (1959): *Fundamentals of Nutrition,* W. H. Freeman and Co., San Francisco.

Daniel, H. A., and H. J. Harper (1935): The relation between effective rainfall and total calcium and phosphorus in alfalfa and prairie hay, *Jour. Am. Soc. Agronomy,* **27**:644–652.

Dietz, Donald R., Robert H. Udall, and Lee E. Yeager (1962): Chemical composition and digestibility of selected forage species, Cache La Poudre Range, Colorado, *Col. Dept. Game Fish and Tech. Pub.* **14**.

Fels, H. E., J. R. Moir, and R. C. Rossiter (1959): Herbage intake of grazing sheep in south-western Australia, *Austral. Jour. Agri. Res.* **10**:237–247.

Forbes, E. B., J. W. Bratzler, A. Black, and W. W. Braman (1937): The digestibility of rations by cattle and sheep, *Pa. Agr. Expt. Sta. Bull.* **339.**

Gallagher, J. R., and Maurice Shelton (1972): Efficiencies of conversion of feed to fiber of Angora goats and Rambouillet sheep, *Jour. Animal Sci.* **34:**319–321.

Galt, H. D., et al. (1969): Botanical composition of diet of steers grazing a desert grassland range, *Jour. Range Mgt.* **22:**14–19.

Garrett, W. N., J. H. Meyer, and G. P. Lofgreen (1959): The comparative energy requirements of sheep and cattle for maintenance and gain, *Jour. Animal Sci.* **18:**528–547.

Graham, N. McC. (1972): Units of metabolic body size for comparisons amongst adult sheep and cattle, *Proc. Austral. Soc. Animal Prod.* **9:**352–359.

Guilbert, H. R., and S. W. Mead (1931): Digestibility of bur clover as affected by exposure to sunlight and rain, *Hilgardia* **6:**1–12.

———, ———, and A. C. Jackson (1931): The effect of leaching on the nutritive value of forage plants, *Hilgardia* **6:**13–26.

———, and L. H. Rochford (1940): Beef production in Califorinia, *Calif. Agr. Ext. Serv. Circ.* **115.**

Harris, Lorin E. (1968): Range nutrition in an arid region, Honor Lecture 36, Utah State University, Logan.

Hart, G. H., and H. R. Guilbert (1933): Vitamin A deficiency as related to reproduction in range cattle, *Calif. Agr. Expt. Sta. Bull.* **560.**

———, ———, and H. Goss (1932): Seasonal changes in the chemical composition of range forage and their relation to nutrition of animals, *Calif. Agr. Expt. Sta. Bull.* **543.**

Herschberger, Truman V., and E. W. Hartsook (1970): In vitro rumen fermentation of alfalfa hay; carbon dioxide, methane and heat production, *Jour. Animal Sci.* **30:**257–261.

Kamstra, L. D. (1973): Seasonal changes in quality of some important range grasses, *Jour. Range Mgt.* **26:**289–291.

Kennedy, P. P., and S. C. Dinsmore (1909): Digestion experiments on the range, *Nev. Agr. Expt. Sta. Bull.* **71.**

Kiesling, H. E., A. B. Nelson, and C. H. Herbel (1969): Chemical composition of tobosa grass collected by hand-plucking and esophageal-fistulated steers, *Jour. Range Mgt.* **22:**155–159.

Klein, David R. (1970): Tundra ranges north of the boreal forest, *Jour. Range Mgt.* **23:**8–14.

Livingston, H. G., J. A. Payne, and M. T. Friend (1962): Urea excretion in ruminants, *Nature* **194:**1057–1058.

Love, Merton (1970): The rangelands of the western U.S., *Scientific American* **222**(2):89–96.

Madsen, Louis L. (1942): Nutritional diseases of farm animals, in U.S. Department of Agriculture Yearbook of Agriculture, *Keeping Livestock Healthy,* pp. 323–353.

Magruder, N. D., C. E. French, H. L. McEwan, and R. W. Swift (1957): Nutritional requirements of white-tailed deer for growth and antler development II, *Pa. Agr. Expt. Sta. Bull.* **628.**

Maynard, Leonard A., and John K. Loosli (1969): *Animal Nutrition*, 6th ed., McGraw-Hill, New York.

McBee, Richard H., John L. Johnson, and Marvin P. Bryant (1969): Ruminal microorganisms from elk, *Jour. Wildlife Mgt.* **33**:181–186.

McEwan, E. H., and A. J. Wood (1966): Growth and development of the barren ground caribou; I, heart, girth, hind foot lengths, and body weight relationships, *Can. Jour. Zoology* **44**:401–411.

Midgley, A. R. (1937): Modification of the chemical composition of pasture plants by soil, *Jour. Am. Soc. Agronomy* **29**:498–503.

Nagy, Julius G., Harold W. Steinhoff, and Gerald M. Ward (1964): Effects of essential oils of sagebrush on deer rumen microbial function, *Jour. Wildlife Mgt.* **28**:785–790.

National Research Council (1968): Nutrient requirements of domestic animals; No. 5, *Sheep*, 4th rev. ed., National Academy of Science, Washington.

―――― (1970): Nutrient requirements of beef cattle No. 4, *Beef Cattle* 4th rev. ed., National Academy of Science, Washington.

Pearson, Henry A. (1965): Rumen organisms in white-tailed deer from south Texas, *Jour. Wildlife Mgt.* **29**:493–496.

―――― (1969a): Starvation in antelope with stomachs full of feed, *Rocky Mt. Forest and Range Expt. Sta. Res. Note* **148**.

―――― (1969b): Rumen microbial ecology in mule deer, *Applied Microbiol.* **17**:819–824.

―――― (1972): Estimating cattle gains from consumption of digestible forage on ponderosa pine range, *Jour. Range Mgt.* **25**:18–20.

Pfander, W. H. (1971): Animal nutrition in the tropics—problems and solutions, *Jour. Animal Sci.* **33**:843–854.

Raleigh, R. J. (1970): Symposium on pasture methods for maximum production in beef cattle; manipulation of both livestock and forage management to give optimum production, *Jour. Animal Sci.* **30**:108–114.

Rice, R. W., D. R. Cundy, and P. R. Weyerts (1971): Botanical and chemical composition of esophageal and rumen fistula samples of sheep, *Jour. Range Mgt.* **24**:121–124.

Rigg, T., and H. O. Askew (1929): Mineral contents of some typical pastures in Waimea County, *New Zealand Dept. Agr. Jour.* **38**:304–316.

Rittenhouse, L. R., C. L. Streeter, and D. C. Clanton (1971): Estimating digestible energy from digestible dry and organic matter in diets of grazing cattle, *Jour. Range Mgt.* **24**:73–75.

Schmidt-Nielsen, Bodil, and Humio Osaki (1958): Renal response to changes in nitrogen metabolism in sheep, *Amer. Jour. Physiology* **193**:657–661.

―――, Knut Schmidt-Nielsen, T. R. Houpt, and S. A. Jarnum (1957): Urea excretion in the camel, *Amer. Jour. Physiology* **188**:477–484.

Scott, S. G. (1929): Phosphorus deficiency in forage feeds of range cattle, *Jour. Agr. Research* **38**:113–130.

Short, Henry L. (1963): Rumen fermentations and energy relationships in white-tailed deer, *Jour. Wildlife Mgt.* **27**:184–195.

―――― (1971): Forage digestibility and diet of deer on southern upland range, *Jour. Wildlife Mgt.* **35**:698–706.

Silver, Helenette, N. F. Colovos, and Haven H. Hayes (1959): Basal metabolism of white-tailed deer—a pilot study, *Jour. Wildlife Mgt.* **23**:434–438.

———, ———, J. B. Holter, and H. H. Hayes (1969): Fasting metabolism of white-tailed deer, *Jour. Wildlife Mgt.* **33**:490–498.

———, J. B. Holter, N. F. Colovos, H. H. Hayes (1971): Effect of falling temperature on heat production in fasting white-tailed deer, *Jour. Wildlife Mgt.* **35**:37–46.

Sims, Phillip L., Gilbert R. Lovell, and Donald F. Hervey (1971): Seasonal trends in herbage and nutrient production of important sandhill grasses, *Jour. Range Mgt.* **24**:55–59.

Smith, Arthur D. (1952): Digestibility of some native forages for mule deer, *Jour. Wildlife Mgt.* **16**:309–312.

——— (1957): Nutritive value of some browse plants in winter, *Jour. Range Mgt.* **10**:162–164.

——— (1959): Adequacy of some important browse species in over-wintering of mule deer, *Jour. Range Mgt.* **12**:8–13.

———, and John M. Malechek (1974): Nutritional quality of summer diets of pronghorn antelope in Utah. *Jour. Wildlife Mgt.* 38(4).

———, Robert B. Turner, and Grant A. Harris (1956): The apparent digestibility of lignin by mule deer, *Jour. Range Mgt.* **9**:142–145.

Smith, G. E., D. C. Church, J. E. Oldfield, and W. C. Lightfoot (1966): Effect of sagebrush on forage digestibility of lambs, Proceedings Western section, American Society of Animal Science, New Mexico State University, University Park, pp. 373–378.

Smith, M. G., and E. B. Stanley (1938): The vitamin A value of blue grama range grass at different stages of maturity, *Jour. Agr. Research* **56**:69–71.

Stoddart, L. A. (1941): Chemical composition of *Symphoricarpos rotundifolius* as influenced by soil, site, and date of collection, *Jour. Agr. Research* **63**:727–739.

Swift, R. W. (1957): The nutritive evaluation of forages, *Pa. Agr. Expt. Sta. Bull.* **615.**

Tash, L. H. (1941): Mineral deficiency investigations in range cattle production in south Texas, *Cattleman* **27**:67–70.

Thomas, Gerald W. (1973): Agriculture faces the energy crisis, New Mexico State University, Las Cruces.

Tilley, J. M. A., and R. A. Terry (1963): A two-stage technique for the in vitro digestion of forage crops, *Jour. British Grassland Soc.* **18**:104–111.

Torell, D. T., R. M. Bredon, and B. Marshall (1967): Variation of esophageal fistula samples between animals and days on tropical grasslands, *Jour. Range Mgt.* **20**:314–316.

Ullrey, D. E., et al. (1969): Digestible energy requirements for winter maintenance of Michigan white-tailed does, *Jour. Wildlife Mgt.* **33**:482–490.

———, et al. (1971): A basal diet for deer nutrition research, *Jour. Wildlife Mgt.* **35**:57–62.

———, et al. (1972): Digestibility and estimated metabolizability of aspen browse for white-tailed deer, *Jour. Wildlife Mgt.* **36**:885–891.

Urness, Philip J., Win Green, and Ross K. Watkins (1971): Nutrient intake of deer in Arizona chaparral and desert habitats, *Jour. Wildlife Mgt.* **35**:469–475.

Van Dyne, George M. (1962): Micro-methods for nutritive evaluation of range forages, *Jour. Range Mgt.* **15**:303–314.

―――― (1965): Chemical composition and digestibility of plants from annual range and from pure-stand plots, *Jour. Range Mgt.* **18**:332–337.

Van Soest, P. J. (1967): Development of a comprehensive system of feed analyses and its application to forages, *Jour. Animal Sci.* **26**:119–128.

Vercoe, J. E. (1969): The transfer of nitrogen from the blood to the rumen in cattle, *Austral. Jour. Agr. Res.* **20**:191–197.

Veteto, George, C. E. Davis, Ray Hart, and R. M. Robinson (1972): An esophageal cannula for white-tailed deer, *Jour. Wildlife Mgt.* **36**:906–912.

Wallace, Joe D., and A. H. Denham (1970): Digestion of range forage by sheep collected by esophageal fistulated cattle, *Jour. Animal Sci.* **30**:605–608.

―――――, J. C. Free, and A. H. Denham (1972): Seasonal changes in herbage and cattle diets on sand-hill grassland, *Jour. Range Mgt.* **25**:100–104.

―――――, K. L. Knox, and D. N. Hyder (1970): Energy and nitrogen value of sandhill range forage selected by cattle, *Jour. Animal Sci.* **31**:398–403.

―――――, and George M. Van Dyne (1970): Precision of indirect methods of estimating digestibility of forage consumed by grazing cattle, *Jour. Range Mgt.* **23**:424–430.

Ward, A. Lorin (1971): In vitro digestibility of elk winter forage in southern Wyoming, *Jour. Wildlife Mgt.* **35**:681–688.

Watkins, W. E. (1937): The calcium and phosphorous content of important New Mexico range forages. *New Mexico Agr. Expt. Stat. Tech. Bull.* **246.**

―――――, (1939): Monthly variation in carotene content of two important range grasses, *Sporobolus flexuosus* and *Bouteloua eriopoda*, *Jour. Agr. Res.* **58**:695–699.

Weston, R. H. (1971): Factors limiting the intake of feed by sheep; V, Feed intake and the productive performance of the ruminant lamb in relation to the quantity of crude protein digested in the intestines, *Austral. Jour. Agr. Res.* **22**:307–320.

Woods, Ella, et al. (1935): Vitamin A content of pasture plants; IV, White blossom sweet clover (*Melilotus alba* Desux.), orchard grass (*Dactylis glomerata* L.), and meadow fescue (*Festuca elatior* L.) under pasturage conditions and fed green, *Jour. Dairy Sci.* **18**:639–645.

―――――, A. O. Shaw, F. W. Atkeson, and R. F. Johnson (1932): Vitamin A content of pasture plants; I, White clover (*Trifolium repens*) and Kentucky blue grass (*Poa pratensis*) under pasturage conditions and fed green, *Jour. Dairy Sci.* **15**:475–479.

Chapter 8

Management for Proper Range Use

Maximum production from a given range unit is dependent upon proper management and use of the resources. Of fundamental importance are (1) grazing the range with the proper kind of animal, (2) balancing numbers of animals with forage resources, (3) grazing at the correct season of year, and (4) obtaining proper distribution of livestock over the range. The importance of these and means of accomplishing them are the subjects of this chapter.

KIND OF ANIMAL

Kind of animal denotes species as distinct from breed or class, whether domestic or wild. Occasionally the term *class* is used synonymously or, especially, is applied to cattle, horses, and mules as one class and to sheep and goats as another. It is better used, however, to apply to particular ages, sexes, or types, such as steers, dry ewes, feeders, or breeders. The various kinds of grazing animals, both domestic and wild, have certain characteristics that make them differentially adapted to ranges of various sorts. These differences are reflected both in the influence of the animal upon the range and in the influence of the range upon the animal.

An important consideration in determining stock adaptation is the vegetation. Different animals do not have the same forage preferences and so do not choose the same plants when given choice. Observation of these preferences makes possible a wiser use of the range by grazing the kind of stock that makes best use of the existing forage.

Forage Preference of Livestock

Browse ranges are best adapted to sheep and goats. Of the two, perhaps goats can make better use of shrubs. Since browse often grows in thick stands, larger animals may be prevented from working through and fully utilizing the forage present. Furthermore, sheep and goats prefer the forage produced on shrubby plants more than do cattle or horses. Though sheep eat grass in large quantities, most grasses must be young and green to be fully used. Other factors being equal, browse ranges should be stocked with sheep or goats for best results. Camels are even better adapted to utilizing coarse shrubs, since the morphology of their mouth, tongue, and lips enables them to remove small leaves from thorny trees and shrubs (Fig. 8.1).

Sheep utilize forbs more fully than any other kind of livestock using larger quantities of them and a greater number of species. Where forbs make up a large part of the forage, most satisfactory results are obtained from sheep.

Horses are the most selective of the domestic animals. They are primarily grass eaters and utilize relatively small amounts of other forage. The horse has an ability to utilize coarse grass unsurpassed by any other kind of stock.

Cattle are chiefly grass grazers, although they consume many shrubby species readily and obtain some forage from the broad-leaved herbs. Grass, browse, and forbs are preferred by cattle in the order named, the volume of the last two being small in relation to that of grass (Culley, 1938).

Despite the preferences of animals, any kind of grazing animal can thrive on virtually any kind of forage and will accept a variety of species outside the preferred class.

Forage preferences of animals are important in preventing livestock poisoning. Whereas a given plant may be a constant threat to one kind of livestock, another may graze the same range with impunity (see Chapter 10).

Fig. 8.1 Camels in India. Camels are unbelievably adept at removing forage even from well-armed shrubs. *(Courtesy F.A.O.)*

Topography

Various kinds of animals respond quite differently to the physiography of the land. Cattle graze level land to best advantage, although rolling country can be fully utilized by them. Where the range is rough, cattle congregate on more level areas, such as valley bottoms or ridge tops, leaving the steeper portions unused or only partly used (Table 8.1). This has direct bearing upon range-grazing capacity. (See also Chapter 10.)

It is not unusual on mountain range to see overgrazing and range deterioration on the best part of the range, even though the total forage is adequate. Under such conditions, both the range and the livestock are affected adversely. Many mountain ranges are mistreated because of failure to consider this factor. Forage on steep slopes must be disregarded when steep land is grazed by cattle. Usable, not total, forage must be the basis for stocking-capacity calculations.

Sheep are better adapted to grazing steep topography than other domestic animals, with the possible exception of the goat. Their smaller size, climbing instinct, and sure-footedness enable them to negotiate steep areas with less difficulty than larger animals, and they are ordinarily under the control of a herder and can be made to graze steep slopes. Overuse of the valley bottoms can thus be avoided.

Rocky areas and poorly accessible mesas have important influences upon grazing. Studies of cattle ranges in the Southwest revealed that rocky areas were not used so heavily as those free from stones (Culley, 1938). Cattle cannot travel as well or so far over rocky terrain, and poorer utilization results.

Common Use

Since each kind of stock grazes most heavily on certain plant species and certain types of topography, most efficient range use likely can be attained by grazing more than one kind of livestock on the same range. Common use can result in more uniform utilization of both species and areas than is obtained by single use, provided the combined numbers of each kind of animal are commensurate with forage production. This does not mean adding another kind of stock to those already there; rather, they should replace some of the present stock. Common use does not imply double use.

The more kinds of animals grazing, the more likely that every species will contribute its share to the total forage consumption. Likewise, many kinds of animals will cover the range more thoroughly and make full use of steep and less accessible areas. This fact is recognized in the Edwards Plateau area of Texas where deer, cattle, sheep, and goats are commonly grazed together.

Table 8.1   Relative Time Cattle Spend at Different Distances Up Slopes of Various Steepness

Data from Mueggler (1965)

Distance upslope, yards	Slope, percent							
	0	10	20	30	40	50	60	70
	percent							
0	100	100	100	100	100	100	100	100
100	78	42	29	22	18	15	13	11
200	70	38	26	20	16	13	11	10
300	63	34	23	18	14	12	10	0
400	57	31	21	16	13	11	0	
600	47	25	17	13	11	0		
800	38	21	14	11	0			
1,000	32	17	12	0				
1,400	21	12	0					

If the vegetation is particularly suited to one kind of animal, or is liked equally by more than one animal, little increase in grazing capacity results from common use. Where there are a great many forage species some of which are used by one kind of animal such as cattle, and an equally large number of species used only by another animal—deer —common use can materially increase grazing. The amount of the increase depends upon the mix of animals and vegetation. (See "Grazing Capacities under Common Use," Chapter 11.)

It is sometimes thought that sheep and cattle cannot graze together because cattle will not use a range that has been grazed by sheep. This is not true if each so grazes the range as to leave forage for the other. Cattle and sheep will graze a range concurrently or successively, and each after the other. This is demonstrated by nomadic pastoralists who achieve common use by successively grazing camels, cattle, sheep, and goats.

Exchange Ratios

The volume of forage consumed by various kinds of animals is related to body weight and body surface (see Chapter 7). Commonly, five sheep are considered equivalent to one cow when changing from one to the other on the range, although six to one is likely more proportionate to their body weights. This ratio, however, is profoundly influenced by type of range, and by age, sex, and breed of animal.

On federal ranges, animals under 6 months of age are not counted in determining permitted numbers. Since lambs on summer range are proportionately larger, mature faster, and are more abundant percentagewise than calves, a proportionately larger food supply is necessary for ewes with lambs than for cows with calves.

In actual practice, the suitability of the range is considered in determining stocking ratios. A proportionately larger number of sheep can graze on rough range, poorly watered range, ranges high in forbs and low in grasses, and ranges having plants poisonous to cattle but less so to sheep. The standard exchange or conversion ratio based upon forage consumption is modified locally according to these factors.

As has been pointed out, many ranges are more efficiently used by cattle and sheep together. The conversion ratio, therefore, would be influenced. On a range now grazed entirely by cattle, some might be replaced by sheep on a very wide ratio. On a range now grazed by both sheep and cattle, the cattle would be entirely replaced by sheep only at a very narrow ratio.

Pests, Diseases, and Predators

Of importance under some circumstances in determining the kind of stock for a given range is the presence of various pests and diseases. In certain areas, stock have been precluded because of the presence of bloodsucking insects. Occasionally, on high mountain meadows the distress inflicted upon cattle will prevent their feeding normally and making satisfactory gains. It is better to graze with sheep, since their wool gives them greater protection. Other insects, such as the cattle botfly or ox warble fly, are specific to a certain kind of animal and do not interfere with the grazing of another. Game farming has been advocated as a means of utilizing rangelands unsuited to domestic livestock because of the tsetse fly in Africa (Skovlin, 1971).

Certain predators might determine the best-adapted stock. Unusual concentration of coyotes makes sheep production very difficult, though they are of minor importance to cattle. Similarly, diseases often affect sheep only or cattle only.

Range Damage by Different Kinds of Stock

Comparative range damage by various kinds of livestock has received little scientific investigation, and opinions are based upon early observations. These were, commonly, that sheep were the most detrimental, and many early-day corrective measures applied to unsatisfactory ranges involved their removal. These opinions were based upon damage that followed increased sheep numbers, which caused overuse, rather than by any feature inherent in sheep grazing. Actually, because they are under the control of a herder, it is possible to give the range greater protection when grazing with sheep. Overgrazed spots may be avoided, and the less heavily used areas may be made to support the stock. Conversely, cattle congregate in choice grazing areas such as meadows and canyon bottoms, where local overgrazing may cause serious erosion. Although this problem is not serious on nearly level plains, it is so serious on some rough, mountainous ranges as to raise the question of whether cattle should be grazed.

The ability of sheep to crop forage closely, together with intense trampling when they are permitted to bunch together, may result in serious damage. Sheep, being smaller than cattle, probably cause less damage to the range by trampling, provided that they are herded properly. Cattle can cause great soil disturbance on wet areas and hillsides.

Because the cow and the horse are large and graze with a pulling motion and the sheep merely nibbles, the cow and horse can pull more plants from the soil. Because sheep prefer the leaves of grasses, they do less damage to the stems and hence to the seeds than do cattle or horses. Sheep are more harmful to timber reproduction than are cattle because of their forage preferences (see Chapter 12).

Horses do not tend to congregate but constantly seek areas of fresh feed. Having both upper and lower incisors they are able to crop forage closely and for that reason may cause great injury to individual plants. If horses are confined and made to crop closely, no other animal can match their impact.

Goats are much like sheep, though hardier and capable of existing under very unfavorable conditions, which, undoubtedly, has given them a bad name. When forage is scarce, goats do great damage by debarking woody plants. Because they are inclined to climb onto shrubs, breakage results. Where good forage is present, however, they are fastidious and graze less destructively than sheep (Semple et al., 1940), although they travel more than either sheep or cattle (Cory, 1927). Goats often are kept for milk and are herded to a central headquarters each night. Such a practice results in concentration which destroys the range in the vicinity of the ranch headquarters.

The animal most suitable to the range biologically may not be acceptable because of economic or cultural reasons. For instance, much of the desert-shrub vegetation in North America could probably best be utilized by camels and donkeys, but since no market exists for these, sheep and cattle are grazed. The steady shift from sheep to cattle production by ranchers in America and in Australia more practically demonstrates how biological considerations may not determine choice of livestock. The ranges may be better suited to sheep, but cattle are more profitable. This trend is regrettable for many rangelands are poorly suited to cattle and some cannot be grazed by them at all. The consequence will be less livestock production and, possibly, deterioration of the range resource.

IMPORTANCE OF CORRECT LIVESTOCK NUMBERS

Correct numbers are important for the perpetuation of the range, the well-being of the livestock, and the economic stability of the operator. Continued overoptimism regarding the capacity of rangeland is certain to result in a failure of the livestock enterprise and an impaired range resource. When overuse takes place, the forage cover is reduced and less meat and wool are produced from an acre, reducing the capital value of the land. Such declines in income and ranch value cannot be permitted if the operation is to be conducted on a permanent basis. The livestock operator should not place too much importance upon numbers and not enough upon livestock quality and long-term production.

Unfortunately, range forage can be neither weighed nor measured accurately. Unlike hay and concentrates, range forage cannot be used in its entirety. When the haystack is gone, the fact is evident. When the harvestable portion of the range forage is gone, however, there is a residue that must be left if the range is to continue normal production.

If forced by economic emergency, drought, or sometimes by the mere ignorance or greed of their manager, livestock can survive for a long time and in apparent good health on this residue. It is unfortunate that there exists no precise measuring stick for determining full range use. Good judgment on the part of the experienced manager is still indispensable to good range management.

Range ecosystems have limits to their ability to withstand grazing and support grazing animals. This limit depends upon physical and climatic factors and the kind of vegetation each area supports. It varies also with the season of use and the kind of animal being grazed and may be modified by management practices which lessen the impact of the grazing animal on vegetation. When this limit is reached, changes in vegetation cease or become very slow, establishing a balance not unlike the natural climax (see Chapter 5). The distinctive task of the range practitioner is to detect vegetational changes and achieve a balance that maintains an acceptable level of production both of animals and other rangeland products (Stoddart, 1960).

Effects of Different Intensities of Grazing upon Vegetation

Vegetative changes have frequently been documented under different levels of stocking, usually arbitrary, and commonly labeled as light, moderate, and heavy. Often an ungrazed area is maintained for purposes of comparison, a measuring stick against which the grazed areas can be equated. These studies have universally shown that heavy grazing reduces the capacity of the range and causes undesirable vegetation changes. As between protection and less heavy grazing, results vary largely in consequence of the fact that no uniform grazing intensity is represented by these use categories and the measure of vegetation that is used. For example, 19 years of continuous grazing in the Great Plains in Alberta resulted in greater plant cover on grazed than ungrazed plots (Smoliak et al., 1972). Most of the increased cover came from plants of low forage value or low productivity; the ground cover of the taller-growing, more palatable plants was much less under heavy grazing. Cover changes and differences under light and moderate grazing varied with individual plant species.

The differential effects upon species of different clipping schedules simulating grazing intensities are demonstrated by 6-year clipping and production data in South Dakota (Table 8.2). The yield of buffalo grass was lower when clipped only at season end than when clipped more frequently throughout the season. Yields of all grasses were lowered by frequent clippings, due primarily to lower yields of western wheatgrass.

Similar reduction in palatable forage, although cover was not re-

Table 8.2   Air-Dry Weight of Vegetation Clipped from 1-M Quadrats when Clipped at Different Intervals, 1927–1932. (clipping begun about May 21 each year)

Data from Black et al. (1937)

Species	Clipping interval				
	10 days	20 days	30 days	40 days	End of season*
	(Grams)				
Buffalo grass	25.9	15.8	25.1	28.4	17.2
Blue grama	4.8	10.2	5.6	8.3	6.0
Western wheatgrass	13.0	24.0	28.1	51.1	64.3
Plains bluegrass	3.2	4.8	8.5	5.4	3.6
Total grass	46.9	54.8	67.3	93.2	91.1
Total other plants	4.2	10.3	17.9	13.5	4.0
Total all plants	51.1	65.1	85.2	106.7	95.1

*Buffalo grass and blue grama could not readily be distinguished at this season.

duced, was caused by heavy sheep grazing in the sagebrush type in Idaho during the spring and fall seasons. Heavily used pastures after 8 years of grazing produced only 83 percent of the average yearly production for the 8 years of observation, while moderately used pastures produced 114 percent (Craddock and Forsling, 1938). As determined by ground cover, browse, most of which was sagebrush which is little used, increased under heavy use and grass cover declined. Moderate use resulted in small changes, unpalatable species decreasing and palatable species increasing.

During times of subnormal precipitation as is common in more arid range areas such as the Southwestern United States, cover of forage plants is reduced greatly irrespective of level of stocking. Once precipitation returned to normal or above, however, production from the more lightly grazed areas exceeded that from heavily grazed or protected areas (Paulsen and Ares, 1961).

Effect of Heavy Grazing on Forage Yield and Quality

Decreased animal production under heavy grazing is related directly to reduced quantity and reduced nutritive quality of forage. Pasture studies (Johnstone-Wallace and Kennedy, 1944) have shown that on dense stands of forage (1,120 kg dry matter per hectare), cattle ate 14.5

kg of dry matter per day; with continued grazing and decrease in forage (560 kg per hectare), consumption decreased to 9.5 kg; with further grazing (280 kg per hectare), consumption reached only 4.5 kg per day.

Continued heavy grazing may so reduce forage production that grazing periods must be shortened. After 6 years of grazing in Kansas, heavily grazed pastures provided an average of only 121 days grazing over the next 4 years while moderately and lightly grazed pastures sustained the original 180 days of use (Launchbaugh, 1957). Drought accentuated the effects. Conversely, Sharp (1970) in southern Idaho could find no difference due to grazing in production of *Agropyron cristatum* in years of low precipitation, but in favorable years heavily grazed pastures produced considerably less than those more lightly grazed.

Experiments with sheep on salt-desert range in Utah (Cook et al., 1953) showed that as grazing continued during the nongrowing season there was a progressive decline in nutritive value of forage consumed (Table 8.3). In contrast to this, Kamstra et al. (1968) found that western wheatgrass was more digestible on heavily grazed than lightly grazed pastures. This was possibly due to the increases in vigor and stem production by the more vigorous, lightly grazed plants.

Trampling may result in losses of forage. Bryant et al. (1972) found maximum reductions in yield of over 60 percent in June; seasonal reductions varied from none to 65 percent depending upon the month, height of vegetation, and the travel performed per cow. Distance animals travel varies with grazing intensity—2.4 km per day at light intensity and 3.3 km per day at higher intensities (Quinn and Hervey, 1970). Forage losses were from 2 to 19 percent depending upon grazing intensity and the season.

Effects of Numbers of Animals on Soil

The effects of severe overgrazing upon soils—bare and unstable soils, gullies, and evident loss of topsoil—are familiar to all and have been well documented (see Chapter 12). More subtle changes in soil characteristics at different stocking rates have less frequently been reported. Soil compaction by grazing animals has repeatedly been demonstrated, but this does not occur everywhere. Smoliak, et al. (1972) found no evidence of soil compaction by sheep in the northern plains. Other soil effects were found which, though not apparent, can have great consequences in the long run (Table 8.4).

A possible indirect soil effect arising from vegetative changes following heavy grazing relates to nitrification by bacteria. There is evidence that roots of increasing or invading plants inhibit nitrogen fixation. By

Table 8.3 Effect of Intensity of Utilization upon Chemical Composition of the Diet and Digestibility of Nutrients on *Artemisia nova* Range when Grazed by Sheep

Data from Cook et al. (1953)

Utilization percent	Dry matter consumed per day, lbs	Total protein percent	Protein digest-ibility, percent	Cellulose percent	Cellulose digest-ibility, percent	Other carbo-hydrates, percent	Other carbo-hydrates, digestibility percent	Gross energy, cal/kg	Gross-energy digest-ibility percent	Phos-phorus, percent	Total digestible nutrients, percent
0–30	2.9	8.5	54.5	25.6	35.9	33.9	58.2	5,172	43.6	0.16	48.3
31–55	2.4	7.8	53.9	23.7	24.5	33.7	55.3	4,977	35.9	0.12	39.3

Table 8.4 Effects of Long-Time Grazing at Different Intensities on Certain Soil Characteristics of Two Sites in Alberta, Canada

Cattle data from Johnston et al. (1971)

Sheep data from Smoliak et al. (1972)

	Soil moisture percent		pH		Carbon/nitrogen ratio	Soil temperature, C
	Sheep	Cattle	Sheep	Cattle	Sheep	Cattle
Ungrazed	19.0	.....	6.4	.....	9.1	.....
Light	15.1*	40.0	6.3*	5.7	9.9*	13
Moderate	15.0*	37.0*	6.0*	5.8*	10.4*	15*
Heavy	11.1*†	31.0*	5.8*	6.0	10.3*	15*
Very heavy	.....	24.0*†	.....	6.2*	.....	17*†

*Value differs significantly from ungrazed (sheep) or light grazing (cattle).
†Value differs significantly from all other grazing intensities.

contrast no inhibition was found among six dominant grasses (Neal, 1969).

INDICATORS OF PROPER LIVESTOCK NUMBERS

Although long-term production records are indexes to correctness of range stocking, records may not be available and one must rely on range condition and indicators. Deteriorated range conditions may precede reduced livestock production; hence unsatisfactory situations may be discovered by careful ecological study before production is seriously impaired.

The use of vegetation as an indicator is based upon the ecological premise that vegetation is the product of its environment; the product, therefore, can be used as an indicator of the causal relationships. The most accurate indexes of overgrazing are the early changes that take place in the vegetation as a result of plant succession. Grazing gradually reduces the more desirable plants and makes available soil nutrients and moisture for less desirable plants. The result is a change in the composition (Fig. 8.2). Early recognition of such changes is of the greatest value in determining the adequacy of existing management procedures.

Just as it is important to the physician to recognize and diagnose disease before it has progressed so far that recovery is impossible, so it is important to the range manager and rancher to recognize range ills in their developmental stage.

Fig. 8.2 A short-grass range which is badly invaded by cactus *(Opuntia)* as a result of heavy grazing. In such large quantities, this plant is a reliable indicator of misuse. *(Photograph by J. E. Weaver.)*

Unsatisfactory range condition may result from habitat factors other than the biological factor of grazing. Drought and fire, especially, should be studied as causative factors. Much confusion has resulted from interpreting degeneration from drought to be a result of excess grazing. Erosion resulting from abnormal weather conditions often erroneously is attributed to mismanagement. Many of the so-called grazing indicators are in reality indicative of poor growing conditions or disturbances from other causes. Poor soil may support weedy or stunted species that superficially appear to be misused. Before a decision is made it is necessary to learn all that is possible of the history of settlement, cultivation, grazing, climate, recent weather conditions, and normal vegetation. It is well to remember that an indicator is only an indicator and that one, alone, may be unreliable. All possible clues should be considered. (See range-condition indicators in Chapter 6.)

Plant Indicators of Too Heavy Grazing

There are countless plants that are useful and well-known indicators of overuse, but few such plants by their occurrence alone categorically confirm overuse. Many are normal constituents of climax or near-climax vegetation and mere presence does not indicate unsatisfactory conditions. It is when they occur in unnaturally large proportions that they indicate deterioration in range conditions. Invaders more clearly indicate overuse of the range but these, too, may have been present in ecologically impoverished areas from which they spread as range conditions worsened and additional niches were provided for them. A knowledge of each range type and site is indispensable to the range manager if he is to correctly interpret range conditions from evidence on the ground, often spoken of as "reading the range." Nevertheless, some guidelines are possible. The following are conditions that, generally, are cause for concern:

Preponderance of plants of low palatability
Vegetation dominated by a few species
Presence of a high percentage of annual plants, particularly forbs
Palatable shrubs hedged and with dead branch stubs (Fig. 8.3)

Fig. 8.3 Bitterbrush *(Purshia)* plant showing effects of too-heavy utilization. Note stubs of large branches and broomlike clustering of smaller twigs due to cropping and regrowth from basal buds. Palatable shrubs can be used as guides to range use.

The range manager must become familiar with the vegetation and the potential for each range area so that he can recognize and identify plants useful as indicators and be able to detect a "sick" range before animal-production potential declines.

Disturbances Indicating Heavy Grazing

Many indicators of the past use of rangeland are associated with disturbance of the soil. Soil changes, like vegetation changes, should be recognized early, for by the time that they are obvious, much damage has been done. That concentrations of animals can have a profound effect upon soil is evident from terracing of steep slopes by animals moving constantly back and forth (Fig. 8.4). This may be so severe as to utilize a major part of the area in trails.

Other signs of overuse are (1) widespread networks of gullies, (2) steep gully banks in major drainage channels, (3) pedestaled plants resulting from removal of soil from the base of plants, and (4) hummocks where soil has been deposited under plants by wind. Unfortunately, by the time these signs are evident range deterioration is in an advanced stage.

Fig. 8.4 Evident terracing of a hillside resulting from trampling by livestock. Forage production is prevented on such trails, and the soils are too compacted to permit infiltration.

Livestock Condition as a Grazing Indicator

If their animals are vigorous and thrifty, stockmen sometimes consider range condition to be satisfactory. Research has shown, however, that range deterioration may progress for a considerable time before the change is reflected in livestock condition. Forage condition is a more sensitive indicator.

Use of forage condition as the criterion of proper grazing does not imply that meat production is ignored. Continued and dependable meat production is contingent upon maintaining the forage production at its maximum. Forage production is a means to that goal.

IMPORTANCE OF CORRECT NUMBERS TO ANIMAL PRODUCTION

Most ranchers understand that ranges are harmed by misuse, but some do not understand how directly this affects their income. That they do not practice better management is attributable to insufficient understanding of the problems confronting them. Range management is a business of no mean complexity. Often, its intricacies are secondary in the minds of men more concerned with meeting bills and paying taxes. Management procedure must be coldly practical and economically sound to be acceptable. The technical advisor should not become so obsessed with the desire to improve range conditions that he loses sight of the economic limitations to range-management theory. Nor should the rancher be so concerned with immediate income that he loses sight of the far-reaching influence of good husbandry upon his economic welfare.

In a business largely controlled by economic necessity, any management practice must pay its way if it is to be acceptable. Fortunately, the objectives of sound range management can be made consonant with the more impelling economic forces. In no other way can good management be generally acceptable. If it is not economically sound, there is need for improvement in the practice or for reordering of the economic structure so that ranchers can adopt what to them, all too often, seems highly theoretical objectives.

Proper grazing has an important bearing upon the success of the ranching business. Through improved grazing management, the calf and lamb crop may be increased, the death loss minimized, more livestock produced and marketed, and animals sold at higher prices. Good range conditions increase the number of calves and lambs produced. High calf and lamb percentage is likely the most reliable index to success of a commercial breeding-livestock operation. By maintaining animals in better vigor, they are more able to withstand the vicissitudes of bad weather, diseases, and other adverse factors to which they are subjected; hence death loss is decreased.

Determining Effects of Stocking Rates on Production

Grazing intensities can be evaluated from their effect on individual animals (weight gains, weaning weights, calf or lamb crops, and wool yields) or per unit area. These do not lead to the same conclusion. Wherever conducted, results have shown high gains per animal at low stocking rates and high gains per unit of range at heavy stocking rates. When a few animals are stocked on a range, their diet is not limited and gains are limited only by diseases, social behavior, or factors other than nutrition. Under such circumstances, their genetic potential for gain may be approximated if proper husbandry practices are applied. As the stocking rate is increased, forage becomes scarce and possibly less nutritive, and less productivity results.

Lowered productivity may express itself in a lower percentage calf or lamb crop, less wool produced, or less gain on market animals. In all cases productivity declines *per animal unit* as stocking rate increases. Productivity *per unit area,* on the other hand, usually increases with increased stocking rate (Fig. 8.5). One cannot have maximum animal gains and maximum area yield concurrently. Animals may feed longer and travel more, and food intake may be lower as grazing intensities increase, as shown by merino sheep in Australia (Leigh et al., 1968).

Grazing Intensities and Turnoff

The classic experiment designed to measure the effects of grazing intensity on production was conducted in North Dakota on mixed short- and mid-grass range (Sarvis, 1941). Stocking was about 4, 2.8, 2.0, and 1.2 ha per steer over a 5-month season. Gains per animal were successively smaller as stocking intensity was increased, and gains per hectare were successively larger. Similar results have more recently been reported from Kansas (Table 8.5).

Yield of animal products, called *turnoff,* over long periods determines the success of livestock operations. Hereford cows grazed on Montana short-grass ranges at the rates of 9.4, 12.3, and 15.7 ha per cow-year produced over a 12-year period an average weaned-calf weight per cow of 126.8, 146.4, and 148.6 kg, respectively (Hurtt, 1946).

Cows were grazed for 11 years on California annual grass foothill ranges for a 6-month season at rates of about 4.0, 6.1, and 8.1 ha per head (Bentley and Talbot, 1951). Cows gained 65.0, 101.8, and 109.1 kg,

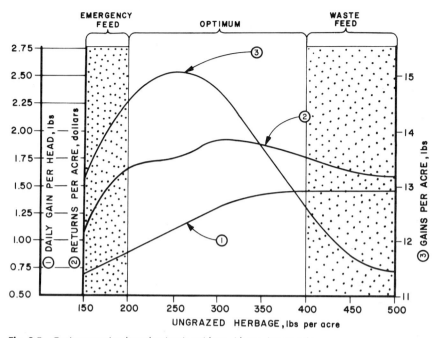

Fig. 8.5 Optimum animal production is neither with maximum gains per unit area nor per head. Moderate stocking provides most returns in the short run and assures continued high yields of forage. *(Adapted from Bement, 1969.)*

and weaning weights of calves were 171.8, 196.4, and 205.5 kg, respectively.

Open pine ranges in central Colorado mountains were grazed by yearling Hereford heifers from June 1 to October 31 for a 6-year period at three intensities giving utilization of 10 to 20 percent, 20 to 40 percent, and 50 percent or more (Johnson, 1953). Average gains were

Table 8.5   Rates of Stocking, Gain per Head, and Gain per Acre of Steers Grazed on Native Pastures, Average for 11 Years

Data from Launchbaugh (1957)

Acres per head 6-month season	Season gain per head, lbs	Season gains per acre, lbs
5.1	217	43
3.4	188	55
2.0	122	61

107.3, 100.9, and 82.3 kg per head and 9.5, 17.9, and 16.6 kg per hectare, respectively.

On mixed prairie vegetation in South Dakota, Hereford cows were grazed during the 7 summer months for a 5-year period at rates as follows: light, 1.3 ha per head per month; moderate, 0.9 ha; and heavy, 0.6 ha. Average utilization of forage was 29, 34, and 54 percent, respectively. For the following 3 years, hectares per head were decreased to 1.0, 0.7, and 0.4 to give utilizations of 48, 60, and 74 percent, respectively (Johnson et al., 1951). Calf weights at weaning were 175.9, 170.5, and 164.1 kg in the first 5-year period and 170.5, 160.9, and 158.6 kg in the subsequent 3 years.

Hereford steers were grazed yearlong at three rates of stocking in the southern Great Plains in Oklahoma for 10 years. Animals were allowed 2.6, 3.9, and 5.2 ha per head and gained an average of 164.1, 174.5, and 181.8 kg, respectively, per head (McIlvain, 1953). Per-hectare gains, however, were 62.7, 44.8, and 34.7 kg.

With one exception (Oklahoma), range deterioration or reduced forage yield accompanied heavy use in the instances cited above. Moderate grazing was regarded as most desirable.

On salt-desert range in Utah (Hutchings and Stewart, 1953), sheep were grazed over an 11-year period for about 5 winter months at two intensities, one about one-fourth heavier than the other. Ewes under moderate grazing weighed 1.8 to 8.2 kg more than those under heavy grazing at the end of the season, and produced 0.45 kg more wool and an 11 percent higher lamb crop. The moderate stocking resulted in improved range condition and increased herbage production.

Economic Returns

The many experiments on intensity of grazing are inconclusive because (1) they did not sample sufficient levels of grazing intensity, (2) they were not continued long enough to determine vegetation and soil responses, and (3) they were not analyzed in terms of true economic effect upon the operator or sociological costs to the nation. For immediate maximum production of meat and wool, heavy grazing may be profitable. Overgrazing, however, can only lead to reduced herbage production and low profit if continued indefinitely. In the short run, an individual may benefit from deliberate overgrazing to secure quick income, provided the land will respond rapidly to good management following overgrazing. By thus *mining* the land, he is ahead, even though his range declines in productivity and value. If soil is not eroded by misuse, improved practices or seeding may bring the land back to its original productivity at reasonable cost. Klipple and Bement (1961) cite evidence that light grazing is an economical means of improving ranges that have not been too greatly deteriorated. Of course, the social hazard of deliberate misuse of land, especially if it results in land abandonment or expensive publicly financed range-improvement projects, should not be ignored.

Light or moderate grazing may have special economic significance on ranges that are grazed during the growing season. Ample forage is usually available early in the season even on heavily grazed areas, and there is little difference in gain between stocking rates. As the season progresses, the animals on moderately grazed pastures continue to gain, while animals on heavily grazed pastures actually lose weight (Fig. 8.6).

The advantage of different grazing intensities may be neither with maximum gains per head or per area and consequently, with light or heavy grazing. Due to differing investments and operating costs most profits can be expected from some intermediate stocking. Hutchings and Stewart (1953) concluded that moderate grazing of sheep during winter was most profitable, and Johnson (1953) reported lower income from heavy grazing by cattle (Table 8.6). Pearson (1973) concluded that maximum profit on ponderosa pine rangeland came from moderate use. Conversely, higher incomes through heavy grazing were reported by McIlvain (1953) on the southern Great Plains.

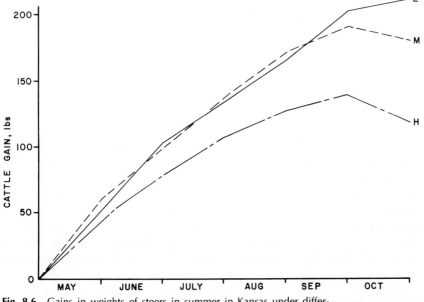

Fig. 8.6 Gains in weights of steers in summer in Kansas under different stocking rates. Under heavy use weight losses occur at the end of the season. *(From Launchbaugh, 1957.)*

Table 8.6 Production and Estimated Return per Section from Pine-Bunchgrass Range in Colorado Grazed at Three Intensities, 1946–1947

Data from Johnson (1953)

	Heavy grazing	Moderate grazing	Light grazing
Animals per section	53	47	27
Gain per head (5 months), pounds	172	211	231
Gross return per section, dollars	661	1,027	724
Costs (death loss, grazing fee,* interest, etc.), dollars	188	163	97
Net return per section, dollars	473	864	627

*Land costs used were the cost of standard Forest Service fees. This results in the same cost per head regardless of stocking intensity. Actually, of course, the landowner by heavy stocking reduces the forage cost per head since interest and taxes chargeable to land are independent of stocking intensity.

Bement (1969) used data from 19-year grazing experiments in Colorado and by means of budget analyses determined the dollar returns based on then current beef prices. The results clearly show that heavy or light grazing gives lower profits than stocking somewhere in between (Fig. 8.5). He prepared a utilization guide suggesting that leaving 224 to 448 kg of forage per hectare gave satisfactory returns, and leaving 336 kg per hectare gave maximum returns.

Light or moderate stocking is especially important during drought. Studies in Texas following the drought of the 1950s and in Australia (Heathcote, 1969) showed that ranch operators who stocked their ranges lightly came through drought periods with more assets and less financial loss than those who stocked heavily.

Over a long-time period conservative stocking will pay dividends. When the cost of (1) extra investment in animals and the accompanying extra work, extra salt, and extra equipment, (2) reduced calf or lamb crops, (3) reduced gains, (4) reduced price per pound for poorer stock, and (5) increased supplemental feed are considered and weighed against the cost of more land to supply needed forage, the rancher can appreciate that overstocking does not make for high income on a sustained basis. It not only decreases the meat yield but ultimately it greatly injures the range. Conservative stocking results in a healthy, productive range.

IMPORTANCE OF SEASON OF GRAZING

From the discussion in Chapter 4 it is evident that an animal-unit month of grazing may affect the range quite differently depending upon season of the year. The start of the growing season is the most critical period for plant growth and it is this season that requires special consideration. This applies to seasonally used ranges as well as to those grazed year-long.

Physical Determinants of Seasonal Use

Over much of the range area animals migrate from one range to another during different times of the year. These seasonal ranges are mainly of two types: (1) arid ranges where seasonal migration occurs because of rainfall, and (2) mountainous ranges where migration occurs because of low temperature and snow cover. Most seasonal ranges in the tropics are associated with monsoonal climates. Animals are grazed in remote areas during the wet season when surface water is available, and flocks are returned to areas with permanent water during the dry period. Migrations may also take place because of disease-bearing organisms such as the tsetse fly that inhabits watercourses during the wet season or because soils are muddy.

Seasonal migration in mountain areas is regulated by many factors such as snow, quantity of forage, quantity of water, condition of stock, loss of nutriment in drying forage, growth state of spring forage plants, soil-moisture conditions, and occasionally, insects or parasites. The seasonal use often is not obligatory but merely convenient.

In high latitudes, snow prohibits winter grazing on ranges of high elevation in all but the summer, a period of 4 months or less. During winter months, grazing is confined to ranges of lower elevation or those kept free of snow by aridity and winds. Lack of water is a factor in their use also, and such snow as falls permits use of otherwise waterless areas.

The winter grazing season generally starts when animals are forced by snow from the fall ranges and ends when shortage of water forces the animals to leave or when the beginning of plant growth in the spring or shortage of feed makes further use unwise. Ranges at intermediate elevations are used in spring and in the fall.

On the plains and southern desert areas where topographical features are comparatively uniform, migrations are unknown and livestock are grazed yearlong, although not always on the same part of the range. Yearlong grazing is the common practice in Australia, Argentina, and South Africa.

Vegetational Determinants of Seasonal Use

In addition to physical factors, the type of vegetation must be considered. Arid browse ranges are best suited to winter use. Nutrient values in shrubs remain higher than in grasses in the dormant season. The spring growing season is usually short and, if plants are used then, soil moisture is insufficient to permit regrowth. Shrubs are more dependa-

ble sources of forage where snow accumulates, since, in contrast to herbaceous vegetation, they protrude from the snow.

Areas supporting forbs are more productive if grazed during the
growing season because many are short-lived, quickly disappearing from
the scene, and lose nutrients rapidly upon maturing. There are exceptions to this, and in vegetation of Mediterranean climates, annual
legumes cure well and remain nutritive in the dry stage. In high elevations in the Western United States, forbs are an important part of the
forage crop.

Grass ranges are best grazed during the growing season, for they, of
all vegetation, are most adapted to regrow after cropping, and they are
most nutritious then. They can be grazed at any season, although if no
other forage is present they are poor nutritionally being at best a
maintenance diet in the mature state.

Irrespective of whether ranges are grazed seasonally or yearlong,
the early growing season is the critical one both from the standpoint of
vegetation and the grazing animal. If it is at all possible, grazing should
not take place at this time so that forage plants could recover from
dormancy and provide sufficient forage for the grazing animal. This
means that on spring and summer ranges livestock should not be placed
on the range too early; on winter ranges livestock should be removed
before the growing season begins. Livestock do not do well while green
growth is insufficient for their needs, for once green feed appears they
avoid mature forage. In yearlong ranges, there may be no recourse
other than feeding livestock during the early growing period. Grazing
systems which vary the time a range unit is used from year to year are
thus especially important on yearlong ranges.

Indicators of Range Readiness

Spring grazing generally commences when soil is firm after winter snow
and when plants have had opportunity to make good growth. This is
known as the period of *range readiness*. Rapid growth of plants in spring
may temporarily deplete food reserves. Deferring grazing until the plant
has had opportunity to begin restoration of these food supplies is advisable. Examples of plants that should be in bloom before grazing begins
in the Western United States are *Dicentra uniflora, Erythronium grandif-
lorum, Claytonia lanceolata, Hydrophyllum capitatum, Ranunculus glaberrimus,
Orogenia linearifolia, Physaria newberryi,* and *Viola linguaefolia.* These generally are unimportant from the standpoint of production of forage, but
they are useful as indicators of the stage of forage growth (Sampson and
Malmsten, 1926).

The proper time for grazing also may be based upon forage grasses.
Early-maturing species such as *Koeleria cristata, Poa fendleriana, P. sec-*

unda, and *Bromus tectorum* usually will have started to produce flower heads by the time stock should go on the range. Grasses that mature later, such as *Agropyron* spp., *Bromus* spp., and *Stipa* spp., will not have shown any flower heads, but the foliage will usually have reached a height of 6 in. or more.

Where the majority of the forage is made up of browse, these plants should be observed closely. Usually the main browse species will have leaves that are one-half to three-fourths developed by the time grazing should begin, and early-flowering shrubs will have started blossoming.

Range-readiness problems are quite distinct on ranges where annual plants are the major forage species. Here, heavy use during the growing season is the usual practice. Most annuals, even under heavy use, are able to produce sufficient seed to ensure a subsequent crop.

Variation in Date of Readiness

There are great differences from year to year in the time at which comparable growth conditions are reached. The same phenological stages varied as much as 47 days from year to year in central Utah (Costello and Price, 1939), and in southeastern Idaho the date of beginning of growth of the most important grass varied from March 20 to April 24 during a 9-year period. This is of great importance to the rancher, for he must feed livestock for a much longer period in extremely late years.

Loss of Nutrient from Too-Late Grazing

Plants, especially herbaceous species, as they pass maturity lose much of their succulence, decrease in digestibility, become less nutritious, and are less palatable to animals. The degree of these changes depends upon the species. Frequently, much of the forage cannot be harvested if the season is delayed unreasonably. Mature grass, for example, may be almost unused by sheep. Early-growing forbs do not cure well and may not even remain if grazing is long delayed. Shrubs store nutriments in the smaller stems, which are eaten by grazing animals, and maturity detracts little from their grazing value. Annual-vegetation types lose value rapidly.

Nutrient losses in tropical areas are of special concern. Soils usually are leached and low in nutrients. Plants grown on them may be marginal for good livestock growth even when they are at their best. Soon after the monsoonal rains stop, the sun bleaches the standing crop and nutrient levels decline. Under such conditions a yearling steer may gain 140 kg during the "wet" and lose 100 kg during the following "dry," a dramatic difference directly related to forage quality.

Animal gains decrease as maturing vegetation becomes less palatable and nutritious. Of course, animals are usually thin when they are placed on green forage in spring and hence naturally gain rapidly. But, eliminating this factor of animal condition, the early growing season still is the time of greatest animal gains (Table 8.7).

In the mountain and intermountain sections of the American West the spring and fall range is the most difficult to fit into the grazing system. Expansive valleys and plains provide winter grazing, or stock are wintered on harvested feeds. If the ranch is one upon which winter feeding is practiced, economic pressure encourages use of the ranges at the earliest possible time, for feeding is more costly. Under such circumstances it is natural that the rancher desires to get onto the range as quickly as possible. However, winter ranges are almost always at lower elevations; hence, plant growth begins earlier than on spring ranges, and spring range is not ready to graze as early as animals must leave winter range. Consequently, administration of federal ranges is complicated and deterioration of spring-fall range is widespread.

EFFECT OF DISTRIBUTION UPON GRAZING CAPACITY

Overgrazing on a range is not dependent entirely upon the number of animals; all the attendant results can be realized locally if stock are not distributed properly. Animals naturally congregate at certain points. On cattle ranges, the most accessible areas such as valley bottoms (Fig. 8.7), low saddles between drainages, areas around waterholes, and level mesas are utilized first. Steep areas and areas far from water are less well-utilized or even unutilized. In some areas steep slopes are unable to stand heavy use, whereas valley bottoms are less injured by concentrated grazing (Holscher and Woolfolk, 1953). Therefore, this habit of cattle may not be altogether undesirable. Unherded sheep will overutilize ridge tops, where they return each night to bed. Windward sides of large paddocks in Australia are more heavily used due to sheep moving into the wind (Squires, 1973). Many apparently overstocked ranges will be able to improve without number reductions if management is adjusted to secure more uniform utilization.

Several factors affect the choice of grazing grounds by animals: (1) topography, including steepness and length of slopes; (2) distribution of water; (3) vegetation; (4) prevailing winds; and (5) kind of livestock. Even breeds of livestock may be important. For example, merino sheep in Australia walked shorter distances than Border Leicesters, 3.2 km as compared to 4.8 km (Squires and Wilson, 1971). Devices for improving distribution are salting, fencing, herding, trail building, water developing, fertilizing (see Chapter 14), or changing the kind of livestock.

Table 8.7   Daily Gains of Livestock in Pound per Head, Showing Effect of
Season (animals are all of Hereford breed except as designated)

Cattle and location	May	June	July	Aug.	Sept.	Oct.	Source
Steers, Oklahoma prairie	2.0	2.0	1.8	1.4	1.0	1.0	McIlvain, 1953
Large steers, Utah mountains	...	...	2.16	1.92	1.11	0.68	Stoddart, 1944
Small steers, Utah mountains	...	...	1.96	1.93	1.31	0.91	Stoddart, 1944
Two-year-old steers, North Dakota plains	4.2	3.5	2.3	1.8	1.3	0.0	Sarvis, 1941
Yearling steers, North Dakota plains	4.0	2.3	1.7	1.6	1.0	0.7	Sarvis, 1941
Heifers, Colorado mountains	...	2.40	1.74	1.50	1.37	-0.03	Johnson, 1953
Holstein heifers, Utah mountains	...	...	1.03	1.57	1.02	0.59	Stoddart, 1944
Cows, South Dakota plains	2.6	1.8	1.0	-0.1	-0.7	-2.2	Johnson et al., 1951
Calves, South Dakota plains	1.5	1.5	1.7	1.8	1.5	0.9	Johnson et al., 1951
Cows, Utah mountains	...	1.7	2.0	1.2	0.9	-0.2	
Calves, Utah mountains	...	1.5	1.6	1.5	1.9	0.8	
Cows, Colorado plains	1.7	1.9	1.6	0.0	-0.2	-1.2	Klipple, 1953
Calves, Colorado plains	1.7	1.7	2.0	2.0	1.7	1.0	Klipple, 1953

Salting

In areas other than salt-desert types, salt provides an economical means
of improving livestock distribution. Salt is less costly than other means
and, since salt usually must be provided, proper placement adds little to
the expense of providing it. On mountain range, a study must be made
of the proper time to graze different parts of the range. If salt is placed
in all parts of an unfenced range in the spring, cattle will move quickly
over the entire area, reaching the higher ranges before they should be
grazed. Much can be accomplished in the way of preventing this if salt is
placed only on the lower ranges in the early season. Localized overgrazing may be minimized by proper placement. If a range supplies enough
forage for 100 head of cattle for 2 months in the early season, by placing
only enough salt there for 200 animal-months, the salt will be exhausted

Fig. 8.7 Cattle normally tend to congregate on meadows and stream bottoms, especially in the fall. Obtaining good distribution through fencing, herding, or salting is an important problem of range management and is essential in avoiding local overgrazing.

at the same time as is the forage. Too much salt will tend to hold stock in the vicinity after forage is fully used. Although other factors influence livestock distribution more than does salt, grazing capacity may be increased as much as 19 percent by salting properly (Cook, 1967).

In locating salt grounds, areas in proximity to water and low saddles between drainages should be avoided, since the use on these areas will be complete without the added attraction of salt. Some parts of the range customarily are utilized poorly. Placing salt in such places increases use there, though heavy-use areas may not be benefited materially (Martin and Ward, 1973).

No exact and generally applicable rules can be made for the location of salt grounds, but certain principles can be followed. Locate salt grounds on ridges, knolls, benches, openings in timber or brush, and gentle slopes. These areas should be easily accessible and should be sufficiently level so that the stock may move around the salt ground without difficulty.

Water

Since livestock are as dependent upon water as upon food, lack of water may prevent proper utilization of forage. Cattle, and especially sheep, occasionally travel long distances to water and go for long periods without it, particularly when snow is present. However, poor water distribution is probably the chief cause of poor distribution of livestock on the range.

In arid regions of the world, water is difficult to obtain and its paucity and sporadic distribution cause many range utilization problems, and result in denuded sacrifice areas around watering points (Fig. 8.8). Where available watering points are infrequent, the vast numbers of animals using a single water point lead to widespread erosion. In the Northern Territory of Australia, statutes prohibit the development of water points closer than 10 mi apart to prevent overuse of the range. Such conditions lead to severely denuded areas around water and poor animal yield.

Studies on eastern Montana plains (Holscher and Woolfolk, 1953) showed that, in winter, forage utilization reached virtually 100 percent around water but declined gradually as distance from water increased (Fig. 8.9).

Two means of obtaining proper utilization of the range while avoiding numerous costly permanent facilities are (1) changing access to water from one water facility to another, and (2) hauling water. Martin and Ward (1970) found that by making water available to cattle at only one point at a time and changing this throughout the season, more even utilization of the range could be obtained (Table 8.8). Utilization at some distance from water was increased, and use of the range near water was decreased by rotating water sources. Adding water on a permanent basis

Fig. 8.8 Areas near to watering points become denuded and vegetation is heavily used for some distance away. Proper numbers and spacing of watering facilities and control of livestock are required to minimize range deterioration.

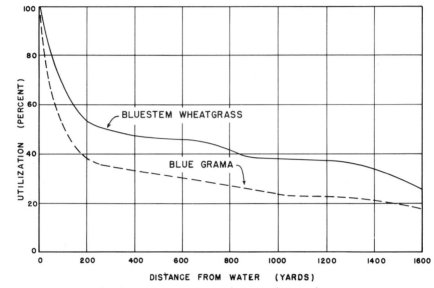

Fig. 8.9 Relationship between percentage utilization of major forage species and distance from water on winter range in eastern Montana plains. *(After Holscher and Woolfolk, 1953.)*

Table 8.8 Utilization on a Range with Water Available Yearlong Compared to One Where Water Was Available Part of the Time

Data from Martin and Ward (1970)

Year July 1– June 30	Use away from water		Use near water		Seasons when part-time water was open
	Year long	Part-time	Year long	Part-time	
	(Percent)				
1959-60	34	44	73	65*	None
1960-61	48	57	78	73	Summer-fall
1961-62	20	32	63	45*	Spring
1962-63	67	77	80	79	None
1963-64	24	62	63	73	Summer
1964-65	33	52	69	71	Summer-spring
1965-66	41	42	75	62*	Fall
Average	38	52	71	67	

*Years when difference in use near water between yearlong and part-time water was significant at 95 percent level.

frequently does not achieve proper use; it merely increases the area of overuse. Jensen and Schumacher (1969) report that adding a third water facility to a prairie pasture of about 470 ha, while it resulted in slight improvement in the good and fair portion of the range, deteriorated the excellent portions.

In parts of the Western United States, hauling water to range livestock is an accepted practice, especially on the salt-desert winter range. Sheep which were supplied water by trucks gained several pounds more than those trailing to water, and their range was more uniformly grazed (Hutchings, 1946).

Herding

Cattle are not herded on ranges in the Americas, Australia, Europe, and South Africa, but are tended by a herder in much of Africa and Asia. However, many ranchers have found that, on rough or poorly watered range, a rider used to guide cattle pays dividends. Calf crops are increased by keeping bulls and cows distributed to proper ratio. Ranges are utilized better by pushing cattle from bottomland and onto underused steeper ranges.

Sheep customarily are herded, except in Australia and New Zealand, for protection against predators and to avoid fencing costs. Occasionally in the United States, sheep have been allowed to roam unherded on unfenced public ranges. Jones and Paddock (1966) reported this practice gave higher lamb gains in one instance in Colorado; Strasia et al. (1970), however, could find no differences in performances between herded and unherded bands in Wyoming. Unherded sheep are somewhat better than cattle in distributing themselves over mountainous range, but they, too, tend to trample out parts of the range by repeated use, by bedding on the same location night after night (Bowns, 1971). Travel to and from these bedgrounds may further increase trampling with resulting range deterioration.

Sheep-Herding Methods

Open herding of sheep, as opposed to close herding is an effective means of conserving the range. Close herding keeps the animals bunched, and much forage is trampled into the ground. Better results are obtained by herding from in front rather than from behind, i.e., by guiding the movement of the lead animals and by avoiding the excessive use of dogs. This allows the sheep to feed quietly and move only enough to secure feed.

By following a 1-night bedding system, excess travel is eliminated.

At night, the band is bedded down on a convenient knoll or other suitable place wherever the animals finish grazing for the day. In the morning, ample forage is near, and the animals can begin grazing as soon as they leave camp; thus the period of trailing is eliminated which would be necessary if they were brought into a central camp (Fig. 8.10). Continued bedding in the same area causes vegetation to be trampled out, and the forage in the vicinity of the camp is utilized closely making it necessary for the sheep to move a great distance before good grazing is secured. Again at night, after grazing has terminated, it is necessary for the herd to retravel the distance to the bedding ground, trailing over parts of the range where feed is not obtainable.

BIBLIOGRAPHY

Albertson, F. W., and J. E. Weaver (1944): Effects of drought, dust, and intensity of grazing on cover and yield of short-grass pastures, *Ecol. Monogr.* **14**:1–29.

Bement, R. E. (1969): A stocking-rate guide for beef production on blue-grama range, *Jour. Range Mgt.* **22**:83–86.

Bentley, J. R., and M. W. Talbot (1951): Efficient use of annual plants on cattle ranges in the California foothills, *U.S. Dept. Agr. Circ.* **870.**

Fig. 8.10 Bedding where there is ample forage enables sheep to begin grazing immediately and saves trailing to forage. Such trailing is ruinous to the range and, also, harmful to animals.

Black, W. H., A. L. Baker, V. I. Clark, and O. R. Mathews (1937): Effect of different methods of grazing on native vegetation and gains of steers in Northern Great Plains, *U.S. Dept. Agr. Tech. Bull.* **547.**

Bowns, James E. (1971): Sheep behavior under unherded conditions on mountain summer ranges, *Jour. Range Mgt.* **24:** 105–109.

Bryant, H. T., R. E. Blaser, and J. R. Peterson (1972): Effect of trampling by cattle on bluegrass yield and soil compaction of a Meadowville loam, *Agron. Jour.* **64:**331–334.

Chamrad, Albert D., and Thadis W. Box (1965): Drought-associated mortality of range grasses in south Texas, *Ecology* **46:**780–785.

Cook, C. Wayne (1967): Increased capacity through better distribution on mountain ranges, *Utah Science* **28:**39–42.

———, L. A. Stoddart, and Lorin E. Harris (1953): Effects of grazing intensity upon the nutritive value of range forage, *Jour. Range Mgt.* **6:**51–54.

Cory, V. L. (1927): Activities of livestock on the range, *Tex. Agr. Expt. Sta. Bull.* **367.**

Costello, David F., and Raymond Price (1939): Weather and plant-development data as determinants of grazing periods on mountain range, *U.S. Dept. Agr. Tech. Bull.* **686.**

Craddock, G. W., and C. L. Forsling (1938): The influence of climate and grazing on spring-fall sheep range in southern Idaho, *U.S. Dept. Agr. Tech. Bull.* **600.**

Culley, Matt (1938): Grazing habits of range cattle, *Am. Cattle Producer* **19:**3–4, 16–17.

Heady, Harold F. (1961): Continuous vs. specialized grazing systems: a review and application to the California annual type, *Jour. Range Mgt.* **14:**182–193.

Heathcote, R. L. (1969): Drought in Australia: a problem of perception, *Geographical Review* **59:**175–194.

Holscher, Clark E., and E. J. Woolfolk (1953): Forage utilization by cattle on Northern Great Plains ranges, *U.S. Dept. Agr. Circ.* **918.**

Hurtt, L. C. (1946): Penalties of heavy range use, *Am. Hereford Jour.*, July.

Hutchings, Selar S. (1946): Drive the water to the sheep, *Nat. Wool Grower* **36**(4):10–11, 48.

———, and George Stewart (1953): Increasing forage yields and sheep production on intermountain winter ranges, *U.S. Dept. Agr. Circ.* **925.**

Jensen, Peter N., and C. M. Schumacher (1969): Changes in prairie plant composition, *Jour. Range Mgt.* **22:**57–60.

Johnson, Leslie E., Leslie R. Albee, R. O. Smith, and Alvin L. Moxon (1951): Cows, calves, and grass, *S. Dak. Expt. Sta. Bull.* **412.**

Johnson, W. M. (1953): Effect of grazing intensity upon vegetation and cattle gains on ponderosa pine-bunchgrass ranges of the Front Range of Colorado, *U.S. Dept. Agr. Circ.* **929.**

Johnston, A. , J. F. Dormaar, and S. Smoliak (1971): Long-term grazing effects on fescue grassland soils, *Jour. Range Mgt.* **24:**185–188.

Johnstone-Wallace, D. B., and Keith Kennedy (1944): Grazing management practices and their relationship to the behaviour and grazing habits of cattle, *Jour. Agr. Sci.* **34**:190–197.

Jones, Dale A., and Raymond Paddock (1966): You can't turn 'em loose—or can you?, *Jour. Range Mgt.* **19**:96–98.

Kamstra, L. D., D. L. Schentzel, J. K. Lewis, and R. L. Elderkin (1968): Maturity studies with western wheatgrass, *Jour. Range Mgt.* **21**:235–239.

Klipple, G. E. (1953): Weight gains made by range cattle while grazing summer ranges, Rocky Mountain Forest and Range Experiment Station Research. Note 12 (mimeo.).

———, and R. E. Bement (1961): Light grazing—is it economically feasible as a range-improvement practice, *Jour. Range Mgt.* **14**:57–62.

———, and David F. Costello (1960): Vegetation and cattle responses to different intensities of grazing on short-grass ranges on the central Great Plains, *U.S. Dept. Agr. Tech. Bull.* **1216.**

Launchbaugh, J. L. (1957): The effect of stocking rate on cattle gains and on native shortgrass vegetation in west-central Kansas, *Kan. Agr. Expt. Sta. Bull.* **394.**

Leigh, J. H., A. D. Wilson, and W. E. Mulham (1968): A study of merino sheep grazing a cotton-bush (*Kochia aphylla*)—grassland (*Stipa variabilis-Danthonia caespitosa*) community on the Riverine Plain, *Austral. Jour. Agr. Res.* **19**:947–961.

Leopold, A. (1939): A biotic view of the land, *Jour. Forestry* **37**:727–730.

Martin, S. Clark, and Donald E. Ward (1970): Rotating access to water to improve semidesert cattle range near water, *Jour. Range Mgt.* **23**:22–26.

———, and ——— (1973): Salt and meal-salt help distribute cattle use on semidesert range, *Jour. Range Mgt.* **26**:94–97.

McIlvain, E. H. (1953): Seventeen-year summary of range improvement studies at the U.S. Southern Great Plains Field Station, Woodward, Okla. (mimeo.).

Mueggler, Walter F. (1965): Cattle distribution on steep slopes, *Jour. Range Mgt.* **18**:255–257.

Neal, J. L., Jr. (1969): Inhibition of nitrifying bacteria by grass and forb root extracts, *Can. Jour. Microbiol.* **15**:633–635.

Nelson, Enoch W. (1934): The influence of precipitation and grazing upon black grama grass range, *U.S. Dept. Agr. Tech. Bull.* **409.**

Paulsen, Harold A., Jr., and Fred N. Ares (1961): Trends in carrying capacity and vegetation on an arid southwestern range, *Jour. Range Mgt.* **14**:78–83.

Pearson, Henry A. (1973): Calculating grazing intensity for maximum profit on ponderosa pine range in northern Arizona, *Jour. Range Mgt.* **26**:277–279.

Quinn, James A., and Donald F. Hervey (1970): Trampling losses and travel by cattle on sandhills range, *Jour. Range Mgt.* **23**:50–55.

Sampson, Arthur W., and Harry E. Malmsten (1926): Grazing periods and forage production on the national forests, *U.S. Dept. Agr. Dept. Bull.* **1405.**

Sarvis, J. T. (1941): Grazing investigations on the Northern Great Plains, *N. Dak. Agr. Expt. Sta. Bull.* **308.**

Semple, A. T., H. N. Vinall, C. R. Enlow, and T. E. Woodward (1940): A pasture handbook, *U.S. Dept. Agr. Misc. Pub.* **194.**

Sharp, Lee A. (1970): Suggested management programs for grazing crested wheatgrass, *Univ. Idaho Forest, Wildlife and Range Expt. Sta. Bull.* **4.**

Skovlin, Jon M. (1971): Ranching in East Africa; a case study, *Jour. Range Mgt.* **24:**263–270.

Smith, Jared G. (1895): Forage conditions of the prairie region, *U.S. Dept. Agr. Yearbook,* pp. 309–324.

Smoliak, S., J. F. Dormaar, and A. J. Johnston (1972): Long-term grazing effects on *Stipa-Bouteloua* prairie soils, *Jour. Range Mgt.* **25:**246–250.

Squires, Victor R. (1973): Role of livestock behavior in the utilization of range-lands, Ph.D. diss., Utah State University, Logan.

——, and A. D. Wilson (1971): Distance between food and water supply and its effect on drinking frequency, and food and water intake of Merino and Border Leicester sheep, *Austral. Jour. Agr. Res.* **22:**283–290.

Stoddart, L. A. (1944): Gains made by cattle on summer range in northern Utah, *Utah Agr. Expt. Sta. Bull.* **314.**

—— (1960): Determining correct stocking rate on range land, *Jour. Range Mgt.* **13:**251–255.

Strasia, C. A., M. Thorn, R. W. Rice, and D. R. Smith (1970): Grazing habits, diet, and performance of sheep on alpine ranges, *Jour. Range Mgt.* **23:**201–208.

Workman, John P., and Donald W. MacPherson (1973): Calculating yearlong grazing capacity—an algebraic approach, *Jour. Range Mgt.* **26:**174–277.

Chapter 9

Planning Grazing Use
of the Range

Range forage is one of the most important resources for meeting the red meat requirements of the world's human population. In the past it has been exploited through heavy, uncontrolled grazing. Today there are principles of scientific management that can be applied to improve the range resource and insure a sustained yield of goods and services from rangeland. In order to apply these principles, grazing use must be planned and the plan executed. Several planned-grazing systems are available to improve range productivity.

PLANNED-GRAZING SYSTEMS

The first consideration in planning range use is to ensure that the basic plant and soil resources are used in such a way that they continue to be productive under the grazing system employed. The selection of a particular system will depend upon the kind of vegetation, the physiography of the range, the kind of animals, and the management objectives of the operator.

Continuous grazing wherein livestock are placed on the range and allowed to remain yearlong or throughout the grazing season has been shown to result in undesirable successional changes in range forage. To prevent this, specialized systems of grazing management have been used widely. Although differing greatly in details, they have two features in common, a period of rest to allow forage plants to grow unmolested and a systematic grazing schedule among different parts of the range (Fig. 9.1.

The objectives sought are (1) restoring vigor of forage plants, (2) allowing plants to produce seed, (3) attaining heavier and more uniform utilization, and (4) increasing animal production.

Many terms have been used to describe grazing systems and there are inconsistencies in their use. Despite differences in terminology and variations in details all can be considered as belonging to a few basic types.

Deferred Grazing

Deferred grazing means delayed grazing. The longer the beginning of grazing on a range unit can be delayed, the better opportunity exists for new plants to become established and for old plants to gain vigor. Thus any delay in grazing constitutes deferment, but *deferred grazing* has come to mean delaying grazing until after the most important range plants have set seed, although with plants that reproduce vegetatively, seed maturity may have less significance. Sampson (1923) outlined a three-pasture, deferred rotation system thus:

Year	Unit 1	Unit 2	Unit 3
1	Spring	Summer	Fall
2	Spring	Summer	Fall
3	Summer	Fall	Spring
4	Summer	Fall	Spring
5	Fall	Spring	Summer
6	Fall	Spring	Summer

Deferred grazing has certain theoretical advantages. If grazing can be deferred every few years, then forage plants have better opportunity to reproduce. Grazing after seed maturity injures plants less and is believed to be beneficial, since animals scatter and trample the seeds into the soil promoting seedling establishment. By allowing important forage plants to grow unhindered during the period most favorable for their growth, they are enabled to produce a greater quantity of seed. Nearly equal advantages result from deferring grazing on plants that reproduce

Fig. 9.1 Season of use affects plant composition of rangelands. Here both season and kind of animal have produced profound differences within a few meters due to a fence. The area on top was used heavily by domestic livestock in spring resulting in a stand of sagebrush *(Artemisia)*. The area below was used only by mule deer in winter which resulted in loss of shrubs and dense stands of grass and *Balsamorhiza. (Photograph by Lee Kay.)*

vegetatively. Rhizome production is decreased greatly with continued heavy grazing; in fact, there may be a total absence of production.

Rotation Grazing

Rotation grazing, or alternate grazing, involves subdividing the range into units and grazing one range unit, then another, in regular succession. The rotation system of grazing is based upon the assumptions that animals in large numbers make a more uniform use of the forage, and that a rest from grazing is beneficial to the plant, even though it must support a greater number of animals during the shorter time during which it is grazed. Certainly, proper rotation grazing results in more uniform utilization. Larger numbers of animals in smaller units are forced to spread over the entire area and to use the available forage more uniformly. Trampling is reduced because animals are held on small areas where feed is more abundant and, hence, where less travel is necessary. Another advantage lies in a reduction of forage loss from dung. Pasture studies in New Zealand indicate that, where sheep are allowed to spend only 1 day and night in each range unit and do not return for 1 to 2 weeks the dung is spread and has dried to the extent that it is no longer objectionable to grazing animals (Peren et al., 1938). This latter point is of special importance on small and intensively used pastures, such as under the *hurdle* system in which animals are held upon small units by means of movable panels.

Deferred-Rotation Grazing

The ideas of deferred grazing and rotation grazing frequently are combined into a *deferred-rotation system.* Under such a plan, grazing on one part of the range may be deferred during one or more years; then by rotation, other areas are successively given the benefit of deferment until all have been deferred. It was first proposed by Sampson (1913) following a study of mountain-range plants.

A change in rotational order not sooner than after each 2 years generally is considered desirable; thus the seeds that are allowed to mature the first year will germinate the second year, and the young plants are given protection from grazing while they are becoming established. Longer periods may be advisable if growth conditions are poor or if the range is in poor condition.

In its simplest form, the deferred-rotation system consists in dividing a given range into two subunits, A and B. Animals are placed on unit A for the first half of the season to allow seed to mature on unit B. Unit B is grazed only during the last half of the grazing season. During the third year, unit B is grazed during the first half of the season and unit A is allowed to mature seed. Again, after 2 years, the original order is followed. In practice, three pastures are most common, but there may be as many as ten.

Rest-Rotation Grazing

Rest-rotation grazing is a system wherein the deferred part of the range is given complete rest for an entire year. It is similar to deferred-rotation grazing, differing mainly in a longer rest period and heavier use of the grazed portion, since, unlike deferred grazing, the rested portion is not grazed at all. The system was developed on the ponderosa pine ranges in California by the U.S. Forest Service (Hormay and Evanko, 1958). It has been widely accepted in the temperate latitudes of the United States where seasonal grazing is practiced and cool-season grasses make up most of the vegetation (Hormay, 1956; Johnson, 1965).

Other Grazing Systems

Several other grazing systems are commonly in use and have given varying degrees of success. The South African *switchback system* involves rotating animals between two pastures to enable each pasture to rest during a portion of the year. This and other two-pasture rotation systems have had sporadic success and are not widely accepted.

Short-duration grazing used in parts of Africa is an example of grazing in which individual range units are many and small so that heavy use results, a plan that is especially adapted to ranges where the herbaceous vegetation is abundant and rank (Goodloe, 1969). *Mob stocking* as practiced in New Zealand is meant to achieve similar heavy use. These are essentially rotation systems.

A modified grazing system, the *Merrill Four-Pasture System,* was developed at Sonora, Texas (Merrill and Young, 1952; Merrill, 1954). This system used four pastures of equal carrying capacity. All animals normally carried on the four pastures are divided equally among three pastures and grazed for 12 months. The fourth pasture is rested for 4 months (Table 9.1). This system works well on yearlong ranges in climates where rainfall is spread throughout the year and where the mixture of vegetation is such that some species respond and produce seed at any season of the year. Thus, deferment for some species is achieved regardless of when the rest period falls.

A simple, practical system developed in New Mexico shows promise for large, arid areas. The animals grazing a property are put in one pasture or grazing unit; all others are deferred. When the animals have used 50 percent of the forage, they are moved to the best pasture available. When it is grazed to 50 percent utilization, the animals again are

Table 9.1   Sequence of Grazing in the Merrill Four-
Pasture Rotation System*

Adapted From Merrill (1954)

Year	Period	Pastures			
		1	2	3	4
	Mar.-June	Rest	Graze	Graze	Graze
1	July-Oct.	Graze	Rest	Graze	Graze
	Nov.-Feb.	Graze	Graze	Rest	Graze
	Mar.-June	Graze	Graze	Graze	Rest
2	July-Oct.	Rest	Graze	Graze	Graze
	Nov.-Feb.	Graze	Rest	Graze	Graze

*Each pasture is grazed continuously for 12 months and rested
for 4 months, causing successive rest periods to fall at different
seasons.

moved to the best available pasture even if it is the one they were in
previously. This system involves continuous evaluation of pastures and
subjective judgment of what is the "best" pasture. It works best in vast,
dry areas where rainfall is sporadic and erratically distributed. It is not
unlike the natural rotation of pastures by wildlife and that practiced by
nomadic peoples.

A modification of rotation grazing involves rotating of kind of stock
rather than time of use. Where an operator grazes both sheep and cattle,
one area may be grazed by cattle for a few years and then by sheep for a
few years. Since the kind of vegetation grazed by the different kinds of
stock differs, this method allows plants highly preferred by cattle some
rest while the area is grazed by sheep, and vice versa. Continuous grazing by sheep may change a mountain range into a pure grass type,
excellent for cattle but poor for sheep. The use of a browsing animal
such as a goat in the rotation has been shown to reduce brush invasion
(Huss, 1972; Cassady, 1972) (Fig. 9.2).

Advantages of Special Grazing Systems

One of the justifications for the deferred-rotation system was that it was
a means of harvesting the forage and still securing proper regeneration.
In comparing similar areas, some grazed yearlong, others ungrazed, and
still others grazed by the deferred-rotation system, it was found that
more seedlings were established under deferment than on unused
ranges (Sampson and Malmsten, 1926). This was explained on the basis
that grazing following the casting of seed served to plant the seed.

Fig. 9.2 Rotating kind of animals in a planned grazing scheme can aid in controlling the composition of vegetation. Goats are especially useful in controlling brush as demonstrated by these angoras. *(Courtesy of John C. Malechek.)*

One of the major values of rotation grazing over continuous grazing on the range is better livestock distribution. Animals tend to overgraze certain parts of a range, grazing and regrazing the same areas to utilize the tender young growth, while other parts may be untouched. Rotation

Fig. 9.3 Well-planned grazing systems minimize damage to vegetation around water holes as shown here in western Montana.

involves putting a large population on a small portion of the range; hence, complete use is achieved over a short period of time. Thereby, animals can be forced to utilize all parts of the area and all species consistent with good range use (Fig. 9.3).

Rotation of season of use on ranges unquestionably has advantages. Plants vary greatly in their season of palatability. Under rotation grazing, different plants will be grazed at one season then another resulting in all being more equally utilized. For example, balsamroot is highly palatable to sheep in spring. Continuous spring grazing can eliminate it, but by alternating spring and fall grazing it can be maintained.

Martin (1973) found that a spring-summer rest period 2 years in succession gave significantly greater yields of perennial grasses than any other rest period on a semidesert grass range. Rest-rotation grazing is especially helpful in minimizing the effects of drought (Ratliff and Rader, 1962; Woolfolk, 1960).

Among different systems, there is no clear-cut evidence for one's superiority over another or over continuous grazing. This may be due to the fact that in actual practice, grazing systems used do not always clearly fit into a particular type. There is a good deal of inconsistency in terminology. In addition, grazing intensities undoubtedly have varied from one experiment to another thereby obscuring the results.

There is little doubt that grazing systems result in better distribution of livestock and more uniform utilization of the range. Their beneficial effect on vegetation is less clear. Early experimental results seemed to indicate marked improvements, and there are many instances in which forage production has been increased upon their adoption. At Sonora, Texas, a four-pasture rotation has been credited with a 25 percent increase in grazing capacity (Keng and Merrill, 1960).

Insofar as benefits to livestock are concerned, the results too are variable. Smoliak (1960) reports lower cattle gains under rotation grazing than under continuous grazing at the same intensities in Alberta. In Texas, Kothmann et al. (1971) obtained greater weaning weights of calves under a rotation system than under continuous grazing, but the highest production per acre resulted from continuous heavy grazing.

From the data at hand it cannot be categorically stated that a rotational-grazing system will invariably improve the range or give greater livestock production than moderate, continuous seasonal grazing. Existing evidence is contradictory. The answer may possibly lie with the conditions of the range on which the system is imposed. When ranchers adopt management systems, it is usually because of unsatisfactory range conditions; hence, range improvement follows. If the range already is good, there is little to be gained. The fact that no grazing system can be implemented without improved management practices—fencing to control livestock, better distribution of water, and careful herding—greatly influences the results. Moreover, and possibly more importantly, the rancher who adopts a grazing system is a more alert and observing manager which is reflected in other phases of his operation. However, when poor results are obtained from a grazing system, some basic rule in range management has been broken. Hickey reviewed 115 papers reporting results of grazing-system trials. He began his conclusions thus: " *'The maximum number of cattle that can safely be carried on any square mile of territory is the number that the land will support during a poor season.* Whenever this rule is ignored there is bound to be loss.' (Smith, 1895). This "rule" is of utmost importance in grazing management."

Difficulties of Special Grazing Systems.

There are certain obstacles to rotation-grazing systems. Fencing is almost a necessity in the case of cattle and is often prohibitively expensive, though the cost may be diminished by the use of drift fences, natural barriers, and movable electric fences. Salting and herding may be used in certain instances, but they are only partially effective. As is frequently the case in the practical application of grazing principles, a compromise must be made, sacrificing some effectiveness of the grazing plan to secure decreased costs. The cost of fences is not a major problem in deferred-rotation grazing of herded sheep, for they can be guided to a certain unit at any desired time.

In many range areas, water is not evenly distributed; hence subdividing the range is difficult. Further, the water must be a season-long source, for during a rotation a given unit might be needed to support stock at any season. Unless water can be provided at all seasons, a range is not amenable to the use of deferred-rotation systems.

Deferred-rotation grazing originally was designed to apply to ranges where plants cure well on the ground. In parts of the prairie region, it has been found that the grasses become too coarse to be effectively utilized toward the end of the season and, if not grazed early, forage becomes unusable.

Rest-rotation systems derive their advantage from complete utilization of range forage during the year of use. Where much of the range vegetation is made up of unpalatable shrubs, they may be the benefactors rather than the desirable forage plants.

Despite their disadvantages, proper deferred-rotation and rest-rotation schemes offer the range manager one of the most important tools in obtaining sustained productivity from rangelands. They must be properly designed and artfully applied to obtain the desired results. Improper application can lead to overgrazing, soil loss, and reduced yield. The following are some of the characteristics of a good grazing system:

1. It is based on the physiology and life history of the plants.
2. It is suited to the kind of plant present.
3. It is adapted to soil conditions, and erosion, puddling, etc., will not result during heavy grazing.
4. It will move plant succession toward higher productivity by favoring the desired plants.
5. It is not detrimental to animal gain.
6. Its implementation is practical in a ranching operation.

If a system has the characteristics listed above, it should improve range productivity; if any are missing, it will likely fail.

EFFECT OF TENURE SYSTEMS ON RANGE PLANNING

The way land ownership is held or land use is delegated greatly affects the way rangeland is managed. Different decision-making criteria are used for private, freehold ownership, private tribal ownership, or public ownership. In some cultures rangeland is held in common and grazing rights belong to those who can control the land by power, position, or favorable location. Under other conditions land may be held by private individuals with all uses—whether grazing or mineral extraction——controlled by a single individual. American range management has developed under a mixture of private and public ownership. In some states almost all rangelands are privately owned. In many Western states only the *base property* is owned and grazing on public land is by permit. The balance between privately owned land and the use of public lands is important in the range-planning process, since it determines who is involved in making decisions and what constraints there are to making them.

Common Ownership

In most cultures where pastoralism is the dominant pattern of land use, individual rights to grazing land are not recognized. Land is held in common, except for arable land where both individual and clan rights are recognized. To some extent in the past even rangelands have been reserved for individual clan or lineage groups. Where a tribe or clan could control the range, special systems of management evolved. For instance, in Somalia, a country of arid conditions and sporadic rainfall, herdsmen have evolved a system of nomadic rotation of their rangeland that is in harmony with the harsh environment. Their migration across the range fits in very well with both the needs of the range vegetation and the needs of the animal (Box, 1971).

In the yearly use cycle, most clans have a fairly regular movement over the range. In the dry season, they are concentrated near their home wells over which they have primary rights. Where wells are adequate, clans share water and forage with friendly clans of other lineages. During the dry seasons, the lineage groups are normally at their thickest concentration, and the ranges near permanent water are stocked most heavily. The degree of separation between groups with camels and those with other animals is most marked because of the differences in water requirements of the two species (Lewis, 1961).

In the wet seasons, the herds leave their home wells and scatter out over the range where water is temporarily available. This practice provides rest to those areas nearer to the water sources, and, in effect, provides a system of rotation management. The plants on the range evolved under such a system of seasonal rest throughout the year, and range condition declined rapidly when subjected to yearlong grazing.

Attempts to alter traditional use patterns based on tribal or clan rights and to introduce management systems from developed countries have met with little success in Africa or Asia, because no means of controlling grazing has been devised. Many of the world's most severe range problems are caused by land-tenure systems wherein livestock cannot be controlled.

Leasehold Systems

Title to rangeland in many parts of the world is held by the government, and uses are leased to private individuals. The best example of this type of tenure is in Australia where less than 1 percent of the area designated as rangeland is in private ownership or alienated (Heathcote, 1969). Lands are held by the "crown" and leased to private ranchers under various state land laws. Mineral rights are reserved to the crown as they are on privately owned land in some cases. Under leasehold arrangement, land is leased for a specific purpose and time period, usually for 50 to 99 years.

Leasehold tenures offer some definite advantages. Conservation clauses can be put into the leases and use regulated. Unfortunately, leases have been issued to "develop" remote areas of rangeland and conservation clauses are all too often lacking or are ignored. The major disadvantage of leasehold is that, unless leases are for long periods, uncertain tenure may result in minimal investment in range improvements and opportunistic use of the forage resource. With leases of sufficient length to encourage investment and proper monitoring of conservation clauses, the leasehold system offers considerable promise for efficient management of rangelands. Management by enforced regulations is possible, though to date this has seldom been done (Fig. 9.4).

Fig. 9.4 Ranchers operating under public-land permits in America and grazing leases in Australia may have small investments in private holdings consisting only of ranch headquarters and a few acres of pastureland for saddle and breeding stock such as this one in Arizona.

Permit Systems

Public lands in the United States are grazed under a permit system which allows for occupancy of a specified area, usually by a stated number of animals for a stated period. Permittees may be required under terms of the permit to follow specific practices, such as using a particular part of the allotted area between certain dates, and following certain livestock handling procedures. In addition, they may be subject to departures from planned procedures in any grazing season when forage conditions depart from normal. Permits are commonly for either 1 or 10 years. To be eligible for permits, users must own "base property," which may either consist of grazing land or cropland sufficient to provide forage for the livestock for that part of the year when animals are not on public lands, or "own" water facilities which control use of the public lands. This system provides for strict control of numbers and kind of animal, and season and manner of use, but it limits decisions by the range user. Investments on the land by the user, though permitted, are limited by administrative policy or discouraged by the limitations of tenure. Where lands are important for many uses the system may best provide for exploitation of all uses concurrently (Fig. 9.5).

Fig. 9.5 In planning public-range use, the range manager must consider all resources. Wild animals such as this mule deer herd must be considered in management plans. *(Photograph by Lee Kay.)*

Private Ownership

Much of the world's rangeland is in private ownership, title being held by individuals, institutions, or corporations. Land-use decisions are made by the owners, usually on the basis of anticipated economic return. Regulation of range use for the common good is difficult except through legal instruments or manipulation of the economic system. For instance, price ceilings or subsidies on meat products can change range use quickly. Use of private rangeland can be regulated through zoning, easements, and other legal restrictions, though little use has thus far been made of them.

Effects of Tenure on Financial Structure

Grazing leases can greatly reduce the capital investment in land and permit a much greater part of the investment in livestock, which can be as high as 80 percent in western Australia (Waring, 1969). Theoretically, the permit system as practiced in the United States provides similar opportunities. In practice this may not be so, for the privilege of grazing on public lands is commonly purchased from the previous owner when ranch properties change hands. The costs of acquiring the grazing privilege is often far greater than the capitalized value of the differential between the grazing fees charged on public lands, which are lower than fees charged on private lands, and the costs of forage from other sources. Martin (1966) could find no economic justification for the prices paid for the privilege of grazing public lands in southern Arizona and concluded that ranchers were "simply paying for the privilege of being ranchers." Despite this practice there is evidence that grazing on public lands confers an economic advantage to ranchers using them. Cost savings of nearly $2.50 per animal-unit month (AUM) were estimated to accrue to public-land users in plains and mountain sections of the Western United States; cost savings on intermountain and Southwest ranches were less than $1.00 per AUM (Public Land Law Review Commission, 1970). These relationships may change greatly as higher fees are imposed on public-land users.

RANGE-USE PLANNING

A plan is simply a predetermined course of action (LeBrerton and Hennig, 1961). Plans may be strategic or operational (Miller, 1971). A strategic plan might weigh the relative merits of a cow-calf operation versus a sheep ranch on a desert range, while an operational plan may provide the means for obtaining efficient use of resources once the type of ranch has been determined. Plans may be made for a specific ranch unit, for a public-range allotment, or for an entire developing country (Gupta, 1971). Plans for private ranches must include use of any public lands associated with the ranch operation. A range-management plan for a public-land allotment may involve the use of many individual ranches.

A good plan will include (1) definition of goals, (2) inventory of the resources, (3) analysis of how the resources presently are used, and (4) suggested adjustments for more efficient use.

Goals in Rangeland Planning

It should be evident from the discussion that a dichotomy exists between the goals of the range user on one hand and the goals of society on the other. To the rancher, who operates a business of great complexity, the prime goals are economic stability and high profits. His decisions are made on the basis of maximizing income, minimizing losses, and attaining status in the community. He is concerned with resource use only to the extent that attainment of these goals is affected by it (Boykin and Hildreth, 1958).

The public-land manager, representing society, must consider not only the yield of animal products, but also the effects of his decisions upon other rangeland products. He cannot maximize returns from livestock at the expense of some other segment of the using public. His major duty is the conservation of the basic resources—soil, water, vegetation, native animals, and esthetics. Consequently, grazing plans on public range must integrate many uses. Whether the land is publicly or privately owned, desert or rain forest, plain or mountain, its maximum contribution to society can be achieved only if use of resources is carefully planned. The plan must include a detailed inventory of present and potential productivity based on firm ecological principles. It will include suggested management adjustments necessary for reaching the potential productivity, and a timetable for implementation of needed adjustments. Irrespective of the goals conceived, the basis for a plan is a physical inventory (Fig. 9.6).

The Physical Inventory

The ranch planner must know the resource with which he is to work. This requires a thorough inventory of the physical conditions and resources, for the entire operation of a ranch depends upon these resources, and complete understanding of them is essential to planned operation. The following outline covers the kind of data necessary for proper decision making:

Fig. 9.6 Maintaining flow of good quality water and preserving the esthetic quality of range settings are integral parts of public rangeland inventories and planning.

A. Land resources available
 1. Owned range
    a. Geographical location
    b. Area size and boundaries
    c. Elevation
    d. Topography
    e. Soils
    f. Erosion conditions
 2. Rented range
 3. Public Range
 4. Cultivated land
B. Climate
 1. Rain and snow
    a. Expected quantity
    b. Variability influencing forage production
    c. Distribution of precipitation
    d. Snow season and expected depths
    e. Influence on grazing dates
 2. Evaporation
 3. Temperature
 4. Wind
C. Forage
 1. Vegetation types and description, original and present
 2. Grazing capacity of range by seasonal units
 3. Customary use of range by units
 4. Cost per animal-month
    a. Rent of leased land
    b. Grazing fees on federal range
    c. Interest for investment on owned land
 5. Condition of range by units
 6. Kind of stock to which adapted
 7. Season usable and growing season
 8. Poisonous plants
 9. Cultivated crops produced and yields
    a. Kinds produced
    b. Yields
    c. Cropping schedule
D. Water
 1. Developed resources
    a. Location
    b. Type of development
    c. Quantity correlated with range capacity
    d. Distribution or spacing
    e. Season of availability
 2. Undeveloped Resources
E. Improvements

1. Buildings and sheds
2. Fences
3. Corrals and chutes
4. General equipment

These data must be obtained from a number of sources. Land ownership, water developments, and improvement data can often be found in the ranch records or obtained in an interview with the ranch operator. Soil surveys may be obtained from the U.S. Soil Conservation Service, climatic data from the state climatologist, and economic data from the ranch operator. Often, the range planner will have to do the forage survey himself, using techniques described in Chapter 6. Although use of secondary sources is necessary, all data should be validated by the planner in a careful on-site inspection (Fig. 9.7).

Forest Service or Bureau of Land Management allotments are integral parts of many Western ranch operations. Much of the data necessary for ranch planning may be included in the allotment plans of the U.S. Forest Service or the Bureau of Land Management if the rancher uses public lands. The allotment analyses are consequently an important part of the overall ranch-management plan. The ranch plan must reflect

Fig. 9.7 Range-management plans often are made by interdisciplinary teams of specialists. Here an FAO field party in Kenya plans a new ranching scheme.

Fig. 9.8 The total AUMs of feed from meadows and pastures must be balanced against range AUMs when making a ranch plan. Here a Montana hay meadow is evaluated for winter feed.

the total number of AUMs of feed available and the time during which they can be used. The allotment analysis of his particular grazing unit will not only give the rancher this information, but it will suggest the stability of the forage supply (Fig. 9.8).

On Bureau of Land Management land an *allotment-management plan,* though not required, provides a vehicle for planning use of a federal allotment and integrating it with the use of land owned by a rancher. The plan is a written action program designed cooperatively by the livestock operator and the government to reach specific management goals of both parties. It has the following components: (1) statement of forage resources, (2) objectives of livestock management, (3) grazing-management system, (4) proposed range-improvement practices, and (5) evaluation of the effectiveness of the grazing system (Public Land Law Review Commission, 1970).

Physical data from all resources should be marked on a map with an attachment containing detailed descriptions of the various plan components. An aerial photo mosaic often is used as a base map with physical improvements, range sites, and other semipermanent components drawn on the photo in ink, although other maps may be used. Range condition and proposed adjustments are shown in pencil or by overlays to the base map. This physical inventory and map become the instruments from which the management plan is made.

Operational Inventory

The current operation of the ranch must be analyzed before any plan can be effected. This management inventory then is related to the basic resource data to determine if changes in operation can make a more efficient business of the ranch. A ranch plan will include certain livestock-handling techniques not of concern to the administrator of public land. Likewise, a public-land-administration plan will include some details not applicable to ranch operation. Inappropriate sections are, of course, omitted in using this outline.

A. Livestock inventory
   1. Kind of stock
   2. Numbers
   3. Age
   4. Grade
B. Seasonal distribution and movement
   1. Approximate dates of movement
   2. Trailing or transporting program
C. Supplementary feeding
   1. Feeds used
   2. Source of feed
   3. Method of feeding
   4. Location of feed ground
   5. Usual season of feeding for each animal type
   6. Cost of feeding per animal-month
D. Livestock-handling practices
   1. Breeding
     a. Season
     b. Sex ratio
     c. Breeding program and genetical aims
   2. Calving or lambing
     a. Location
     b. Season
     c. Supervision and management program
     d. Calf- and lamb-crop percentage
   3. Branding, castrating, docking, etc.
   4. Weaning
   5. Disease control
   6. Marketing
     a. Season
     b. Age of marketed animals
     c. Culling practices .
     d. Location of market
     e. Transportation to market
     f. Method of marketing

7. Purchasing
    a. Age of replacement
    b. Numbers needed for replacement
    c. Bull or ram source
    d. Cow or ewe source
E. Interrelated demands upon land
    1. Farming
    2. Mining
    3. Wildlife
    4. Recreation
    5. Timber
    6. Watershed

Most of the data on ranch operation can be obtained from an interview with the owner and examination of ranch records. A detailed narrative of the ranch operation should be attached to the map and physical inventory (Fig. 9.9).

Fig. 9.9 Cattle on a Texas ranch are sorted for sale. Marketing practices and management skills are important factors in ranch planning.

Economic Inventory

Range-livestock production differs in several important details from other agricultural enterprises: (1) The size of the investment is large. (2) Income comes mostly from a single product, except for sheep ranches. (3) Productivity of the land is low and variations from year to year are large. (4) There may be great seasonal imbalance in forage supplies. (5) In many cases, dependence upon public range limits decision making. A ranch is a high-risk, high-investment business and depends for success upon the most efficient management techniques.

If one is considering going into ranching as a business, the business management practices of the ranch should be examined in detail. This can best be done by an analysis of records kept for income tax purposes and an interview with the ranch operator. The inventories discussed above should have identified the durable and fixed capital resources. The next step is to identify all other capital assets and relate them to the use and development of the ranch. In some plans this relationship will be cursory and superficial, and in others detailed, and sophisticated management tools such as linear programming and operations research may be used (Goodall, 1971). In any case, the plan should be detailed enough to analyze the ranch as a business and identify areas where profits can be increased with changes in management.

Relative Advantage of Kind of Stock

Whether one is acquiring a ranch or planning the operation of one already owned, he must consider the relative advantage conferred by different kinds and classes of livestock. In some cases physical factors may dictate a single kind of livestock. If the ranch is equally suited physically for either sheep or cattle, then economic considerations will dictate the choice. There are advantages and disadvantages to each.

The following general factors favor sheep production: (1) sheep are well suited to arid conditions; (2) they produce two products, lamb and wool; (3) returns from sheep come more quickly following production decisions; (4) wool is stored and shipped easily; and (5) sheep are able to utilize many plants that are of low palatability to cattle.

There are, however, some major disadvantages to sheep production: (1) wool and lamb prices are politically sensitive; (2) sheep are vulnerable to predation from dogs, coyotes, and other carnivores; (3) sheep are more susceptible to parasites than are cattle; (4) good labor is difficult to obtain; (5) there is constant competition from synthetic fibers; and (6) the demand for lamb is, in many countries, secondary to beef.

Following are listed some of the major advantages of cattle production: (1) cattle operations are more flexible than sheep operations because of high value for beef of all ages; (2) cattle prices are not as subject to government regulation; (3) the demand for beef is greater than for lamb; (4) competition from imports is not as great with cattle as with sheep; (5) labor requirements are less than for sheep; (6) cattle are more hardy under adversity; and (7) there are more operational options available such as cow-calf, cow-steer, or steer ranches.

Some major disadvantages of the cattle industry are that (1) returns are slow due to the long gestation period following management decision, (2) only one product is marketed, and (3) range water requirements are greater.

Adjustments in Ranch Operations

No plan is complete without suggested adjustments and a schedule for their implementation. These adjustments should be spelled out in detail. Each adjustment should be preceded by a cost-benefit analysis that will allow the manager to set priorities between suggested changes in operation. Since few ranches will ever be able to completely reorganize at any given time, the setting of priorities and the timing for implementation of adjustments are critical to an operational plan. It is well in proposing adjustments under the management plan to indicate (1) immediate adjustments, (2) gradual or delayed adjustments, and (3) the final goal toward which the plan aims.

The success of such a plan depends not alone upon the care and foresight with which it is developed, but also upon the zeal with which it is followed by both administrator and livestock operator. The following outline may be used as a checklist for considering changes in the ranching business.

I. Proposed physical adjustments
    A. Range improvement
        1. Water development
        2. Fence and corral construction
        3. Trail and driveway construction
    B. Forage improvement
        1. Seeding
        2. Brush control
        3. Poisonous-plant control
    C. Soil protection and stabilization
        1. Contour furrows
        2. Water spreaders
        3. Check dams

    D. Rodent control
    E. Insect control
    F. Predator control
II. Proposed use adjustments
    A. Livestock
        1. Changes in kind, age, or grade of stock
        2. Changes in stock numbers
            a. Plan of adjustment
            b. Effect upon economy
        3. Changes in sex ratio
        4. Changes in breeding practices
        5. Adjustment in range practices
            a. Grazing systems
            b. Salting
            c. Trailing
            d. Herding or riding
        6. Adjustments in marketing and purchasing
        7. Wildlife adjustments and management
    B. Land and forage
        1. Adjustment in range use
        2. Purchases, sales, exchanges
        3. Supplemental forage
            a. Production or purchase
            b. Season of use
        4. Public-land allotment use
            a. Boundary lines
            b. Seasonal units
    C. Cooperative relationships
        1. Associations
        2. Cooperative marketing
        3. Advisory boards
        4. Provision for interrelated demands upon land

Keeping the Plan Current

No plan can fully predict new demands or anticipate new uses. Any good plan must provide for future changes—like a good constitution, it must provide ways in which it can be amended. Therefore, all plans should be reviewed periodically and adjusted according to the changing goals of the rancher and the society he serves.

BIBLIOGRAPHY

Box, Thadis W. (1971): Nomadism and land use in Somalia, *Econ. Dev. and Cultural Change* **19**:222–228.

Boykin, Cal, and R. J. Hildreth (1958): Management aspects of range management, *Jour. Range Mgt.* **11**:173–176.

Cassady, John (1972): Completion report of FAO Range Management Project, Nairobi, Kenya (mimeo.).

Goodall, D. W. (1971): Extensive grazing systems, in J. B. Dent and J. R. Anderson (eds.), *Systems Analysis in Agricultural Management,* John Wiley & Sons, Australasia, Sydney, pp. 173–187.

Goodloe, Sid (1969): Short duration grazing in Rhodesia, *Jour. Range Mgt.* **22**:369–373.

Gupta, R. K. (1971): *Planning Natural Resources,* Navayug Trailers, New Delhi.

Heathcote, R. L. (1969): Land tenure systems: past and present, in R. O. Slatyer and R. A. Perry (eds.), *Arid Lands of Australia,* Australian National University Press, Canberra, pp. 185–208.

Hickey, Wayne C. (n.d.): A discussion of grazing management systems and some pertinent literature, U.S. Forest Service, Denver, Colo. (multilith.).

Hormay, August L. (1956): How livestock grazing habits and growth requirements of range plants determine sound grazing management, *Jour. Range Mgt.* **9**:161–164.

———, and A. B. Evanko (1958): Rest-rotation grazing—a management system for bunchgrass ranges, U.S. Department of Agriculture, Forest Service, California Forest and Range Experiment Station Miscellaneous Paper 27.

———, and M. W. Talbot (1961): Rest-rotation grazing . . . a new management system for perennial bunchgrass ranges, U.S. Department of Agriculture, Forest Service, Production Research Report No. 51.

Huss, Donald L. (1972): Goat response to use of shrubs as forage, in Wildland shrubs, their biology and utilization, *U.S. Dept. Agr. Forest Service Gen. Tech. Rept.* INT-1 1972, pp. 331–338.

Jeffries, Ned W. (1970): Planned grazing for Montana ranges, *Jour. Range Mgt.* **23**:373–376.

Johnson, W. M. (1965): Rotation, rest-rotation, and season-long grazing on a mountain range in Wyoming, U.S. Department of Agriculture, Forest Service, Rocky Mountain Forest and Range Experiment Station Research Paper RM-14.

Keng, E. B., and Leo B. Merrill (1960): Deferred rotation grazing does pay dividends, *Sheep and Goat Raiser,* June, pp. 12–13.

Killough, John R. (1962): Managing our rangeland resources, *Our Public Lands* **11**(4):13–16.

Kothmann, Merwyn M., Gary W. Mathis, and William J. Waldrip (1971): Cow-calf response to stocking rates and grazing systems on native range, *Jour. Range Mgt.* **24**:100–105.

LeBrerton, Preston P., and Dale A. Hennig (1961): *Planning Theory,* Prentice-Hall, Englewood Cliffs, N. J.

Lewis, I. M. (1961): *A Pastoral Democracy,* Oxford University Press, London.

Martin, S. Clark (1973): Responses of semidesert grasses to seasonal rest, *Jour. Range Mgt.* **26**:165–170.

Martin, William E. (1966): Relating ranch prices and grazing permit values to ranch productivity, *Jour. Range Mgt.* **19**:248–252.

Merrill, Leo B. (1954): A variation of deferred rotation grazing for use under southwest range conditions, *Jour. Range Mgt.* **7**:152–154.

———, and V. A. Young (1952): Range management studies on the ranch experiment station, *Tex. Agr. Expt. Sta. Progress Rep.* **1449.**

Miller, Ernest C. (1971): Advanced techniques for strategic planning, *Am. Mgt. Assoc. Res. Study* **104.**

Peren, G. S. et al. (1938): The economics of the intensive rotational grazing of sheep, *Massey Agr. Coll., Union, New Zealand, Bull.* **9.**

Public Land Law Review Commission (1970): The forage resource, Vol. 2, Report for the P.L.L.R.C. by University of Idaho with Pacific Consultants, U.S. Department of Commerce, Clearing-house for Federal Technical Information, Springfield, Va. P.B. 189–250.

Ratliff, Raymond D., and Lynn Rader (1962): Drought hurts less with rest-rotation management, U.S. Department of Agriculture, Forest Service, Pacific Southwest Forest and Range Experiment Station Research Note 196.

Sampson, Arthur W. (1913): Range improvement by deferred and rotation grazing, *U.S. Dept. Agr. Bull.* **34.**

———, (1923): *Range and Pasture Management,* John Wiley & Sons, New York.

———, and Harry E. Malmsten (1926): Grazing periods and forage production on the national forests, *U.S. Dept. Agr. Dept. Bull.* **1405.**

Smith, J. G. (1895): Forage conditions of the prairie regions, U.S. Department of Agriculture Yearbook 1895, p. 332.

Smoliak, S. (1960): Effects of deferred-rotation and continuous grazing on yearling steer gains and shortgrass prairie vegetation of southeastern Alberta, *Jour. Range Mgt.* **13**:239–243.

Soil Conservation Service (1970): *National Handbook for Resource Conservation Planning,* U.S. Department of Agriculture, Washington.

U.S. Department of the Interior (1955): The Taylor grazing act, U.S. Government Printing Office, Washington.

Waring, E. J. (1969): Some economic aspects of the pastoral industry in Australia, in R. O. Slatyer and R. A. Perry (eds.), *Arid Lands of Australia,* Australian National University Press, Canberra.

Woolfolk, E. J. (1960): Rest-rotation management minimizes effects of drought, U.S. Department of Agriculture, Forest Service, Pacific Southwest Forest Range Experiment Station Research Note 144.

Chapter 10

Range Management
for Livestock
Production

The production of forage from rangeland is not of itself the major goal of the range manager. A properly vegetated range best assures optimum yield of the products sought from rangeland, stable watersheds, clear water, wildlife, amenities, and food and fiber. In the production of edible red meat the range manager must seek to attain the most efficient transfer of energy in the forage to animals eaten by man. The forage plant is the basic producing unit, and the animal that eats it is the harvesting machine. Successful production of red meat depends upon careful handling of both and upon the greatest possible harmony between the two. Livestock producers should not be concerned with the animals at the expense of the· range, nor should range managers confine their attention to the range and ignore the livestock.

Scientific range research and the experience of ranchers have shown that it is possible to harvest and use the products of rangelands without jeopardizing their future productivity. Proper balance between range and livestock can be obtained only by an understanding of the requirements and characteristics of various grazing animals as well as the plants.

ECOLOGICAL ADAPTATION AMONG RANGE ANIMALS

The relation of kind of animals to efficient range utilization was discussed in Chapter 8. However, factors other than forage preferences are important in selecting animals for most efficient production, including (1) animal tolerance to climatic factors; (2) adaptability to scarcity of water and to steep terrain; (3) efficient use of sparse and rough herbage; (4) tolerance to disease, parasites, predators, poisonous plants, and other livestock hazards; and (5) consumer demand for the products. Animals that have evolved for many generations under a particular set of environmental conditions are usually best ecologically adapted to an area. They make efficient use of the forage and are tolerant of adverse environmental factors. They may not, however, be the kinds of animals best suited for the market or those that will best satisfy the consumer's demand for red meat. Consequently, breeds of livestock have been produced through selective breeding to better utilize particular range environments.

Physiological Traits of Range Animals

The ability of a range animal to produce efficiently in a given environment depends upon its physiological adaptations to local conditions. Use of properly adapted animals is one of the major methods available for increasing meat production, but care must be taken to ensure that adaptability to survive has not lowered the productive potential of a given animal. For instance, some endemic African zebu cattle are ideally suited to the fluctuating forage conditions and paucity of water in arid lands, but they calve only every other year. Their reproductive behavior ensures survival, and perhaps maximum production under extensive management of desert regions, but it is unsuited to intensive management.

There is little doubt that camels, donkeys, and goats are biologically best suited to the utilization of poorer rangelands. However, these animals are not well accepted as meat providers by Western cultures and have little potential, at present, of finding markets in the developed nations. Even among the readily accepted kinds of animals—sheep and cattle—there are significant variations that affect efficient use of the range.

Heat Tolerance   Animals differ in their physiological adaptations to heat and aridity. An examination of their traits helps explain why

European cattle—Herefords, Shorthorns, Charolais, Angus, etc.—are well adapted to temperate and cold climates, but zebu cattle are best adapted to hot environments. Domestic cattle probably originated in the wet tropics of Indochina from an animal similar to the native Guar. One line of development led to *Bos indicus* types in Asia and Africa and another to the cold-adapted *Bos taurus* types (Macfarlane, 1972). Compared to other animals, cattle—and European cattle especially—have high metabolic rates and, often, a coat that absorbs heat and transfers it to the skin, raising body temperatures. Zebus, however, have become better adapted to the dry, tropical regions of the world. They have coats that reflect solar energy, lower body-surface temperatures during hot periods, and lower energy turnover, and their body heat is dissipated by sweating and panting.

Physiologically, sheep are a better desert animal in many respects than cattle. Although, in still air in arid Australia, surface temperature of the wool reaches 92°C, very little heat penetrates the wool and much that does is dissipated by panting. They are superior to cattle in energy and water turnover and in other respects (Table 10.1).

Traveling Ability   The difference in ecological adaptations of domestic animals determines their ability to walk distances necessary to utilize sparse rangelands. Grazing studies in an arid range in New Mexico showed that Santa Gertrudis cattle spent less time grazing and more time walking than Herefords (Herbel and Nelson, 1966). However, they distributed themselves better over the range than did Herefords (Herbel et al., 1967).

Similar behavioral differences were found among Merino, Dorset Horn, and Border Leicester sheep in Australia where the distance be-

Table 10.1   Physiological Trait Differences in Merino Sheep and Shorthorn and Zebu Cattle in Arid and Tropical Environments

Data from Macfarlane (1972)

	Merino	Shorthorn	Zebu
Frequency of parturition	Yearly	Alternate years	Alternate years
Tolerance to salt in water, percent	1.3	0.9	1.0
Energy turnover, cal/kg$^{0.75}$, per day	55	90	80
Water turnover, ml/kg$^{0.82}$, per day	220	500	420
Survival without water in summer, days	6–8	3–4	4–5

tween water and pasturage was controlled. When the distance was 4 km, all watered twice daily; but when the distance was increased to 4.8 km, the aridity-adapted merinos watered only once daily and walked only 9.8 km per day. Border Leicester continued to water twice daily and traveled 17.8 km per day (Squires et al., 1972). Bowns (1971) reports daily travel distances of 4.7, 3.9, and 3.1 km for Rambouillet, Targhee, and Columbia sheep, respectively, in mountain range in southern Utah.

Physiological Needs for Water   There are marked differences in the physiological water requirements of different kinds of animals (Table 10.2), showing a definite hierarchy of adaptability of domestic animals to

Table 10.2   Hierarchy of Water Turnover among Bovids, Antelopes, Camel, Sheep, and Dasyurids

Data from Macfarlane (1972)

Breed species or genus	Body weight, kg	Body solids, percent	Water turnover/24 hr		Environment
			ml/kg	ml/kg$^{0.82}$	
Boran cattle	197	23	135	347	Equatorial
Ogaden sheep	31	32	107	197	desert,
Somali goats	40	31	96	185	Kenya NFD,
Somali camels	520	30	61	188	lat. 3°N
Boran cattle	417	28	76	224	Dry
Eland	247	20	78	213	grasslands,
Wildebeest	175	27	53	137	Kenya,
Kongoni	88	15	52	116	lat. 1°S
Oryx	136	30	29	70	
Karakul sheep	31	24	111	205	Tropical
Merino	38	29	94	180	grasslands,
Dorper	42	31	88	170	Kenya, lat. 1°S
Buffalo	354	21	212	535	Wet
Shorthorn	322	26	168	461	tropics,
Santa Gertrudis	523	37	141	373	Arnhem land,
Banteng	372	23	132	348	summer,
Zebu	532	39	121	350	lat. 12°S
Reindeer	100	23	128	293	Taiga,
Moose	186	23	111	284	winter,
Goats	70	33	52	112	Alaska,
Musk oxen	324	34	35	99	lat. 65°N
Dasyurid marsupials					
Sminthopsis	0.017	28	463	224	Laboratory,
Dasycercus	0.087	30	134	87	25°C

arid range conditions. The order of adaptability to increasingly dry climates is as follows: buffalo, European cattle, zebu cattle, wool sheep, hair sheep, goats, and camels. Animals with low water turnover rates excrete less urea, enhancing their nitrogen efficiency and allowing for better use of the dry, low-protein forage of arid areas.

All cattle have high water turnovers which are best provided in moist environments with assured water supplies. Zebus, however, have a greater water-absorption capacity in the colon than do European cattle although neither approaches the water economy of camels or sheep (Macfarlane, 1964). Zebus thus have lower requirements and higher salt tolerance; consequently, they can go for longer periods without water and produce more efficiently in arid environments than can European breeds of livestock. Water efficiency of a number of cattle breeds has been reported by Siebert and Macfarlane (1969).

Wild animals have a wide range of physiological water requirements ranging from the desert-adapted oryx to the swamp-dwelling buffalo, each developed to fit its own habitat. The muskox is physiologically a cold-desert "camel," his low-water requirement reflecting the nonavailability of free water in the Arctic throughout most of the year (Macfarlane, 1972).

The apparent efficiency for meat production from wild ungulates has attracted much attention in recent years. Their relative resistance to local diseases and parasites makes them an attractive possibility for new domestic species. Certainly, the water efficiency of some antelope—for example, the oryx seldom needs to drink—is much superior to that of domestic animals, but others like the eland offer no advantage.

Water efficiency does not necessarily mean an increase in meat production. Many times an animal that has evolved for survival in harsh conditions may not be the most efficient producer. Limited studies in Africa (Macfarlane, 1972) indicate that nitrogen retention and digestibility of savanna-grassland forage may be greater in eland than in cattle. Russian work with domestic eland shows it to be a promising meat animal (Treus and Kravchenko, 1968). However, the limited work with domestication of wild animals to increase the efficiency of animal production does not offer much hope for a major increase in meat production by his means.

WATER REQUIREMENTS OF RANGE LIVESTOCK

The water requirements of livestock depend upon (1) the kind of stock, (2) the nature of forage, and (3) weather conditions. If the forage is green and succulent, the amount of water needed will be much less than if the feed is dry. Studies in Nebraska (Stoddart, 1935) have shown that range plants in the spring contain over 80 percent water, whereas the same species by midsummer contain 40 percent or less. A cow consuming 29 kg of succulent forage per day, dry weight, receives 23 to 45 liters of water. Considerably more water is required by livestock when the temperature is high and the humidity low, because both these factors cause water losses from the body to be high.

Studies with milk cows under controlled temperatures showed that the average daily water consumption increased by 6, 17, and 50 percent over that consumed at 16°C when temperatures were raised to 21°C, 27°C, and 32°C, respectively (Thompson et al., 1949), although there is considerable variation in data on this point (Ragsdale et al., 1950).

Quantities of Water Required by Livestock

In view of the various conditions affecting water usage, precise recommendations are hazardous. The consumption records shown in Table 10.3 serve as approximate guides upon which plans can be based, but needs vary widely from place to place and season to season. Cows with calves averaged only 23.8 liters per day, ranging from 9.8 liters in the winter to 43.5 liters in the summer (Stanley, 1938).

Experiments on salt-desert winter ranges in Utah showed sheep to drink an average of 2.7 liters per day. On especially dry feed, consumption was 5.7 liters; on very salty feed, consumption was 6.8 to 8.3 liters; and in warmer spring months, consumption averaged 3.3 compared with 2.4 liters in colder winter months (Hutchings, 1946).

Mule deer in Utah have been found to consume 1.58 liters daily during summer (Smith, 1954). This is somewhat lower than was observed in Arizona (Nichol, 1938).

Table 10.3 Average Water Requirements of Horses, Cattle, and Sheep per Day

Horses, gal.	Cattle, gal.	Sheep, gal.	Authority
10-12	10	0.25-1.50	Henry and Morrison (1928)
10	10	1	Talbot (1926)
—	6.3	—	Stanley (1938)
—	7.75	—	Talbot*
—	—	0.36-0.375	Ingram (1930)
—	—	0.75-0.84	U.S. Sheep Expt. Station (Anon., 1926)
—	—	0.72	Hutchings (1946)

*Personal correspondence. Data from Burgess Spring experimental range in northern California, a 6,000-ft plateau in the ponderosa pine type.

Frequency of Watering

In nomadic pastoral grazing, cattle are watered every other day, sheep and goats may go 2 to 3 days without water, and camels 5 to 6 days. Intensified production of European cattle on rangelands, such as occurs in North America, demands that water must be provided at frequent intervals. American experience indicates that cattle require water regularly and should be watered every day for best results. Horses and sheep show satisfactory progress under less favorable water conditions. Frequently, horses water only at 3-day intervals when the vegetation is succulent. Under summer conditions, sheep are watered at least every second or third day, but daily watering is preferable.

Grazing sheep for relatively long periods without access to water is possible and is a common practice where forage is succulent. In Montana (Jardine, 1915), sheep were grazed with absolutely no watering for the 3 summer months. The lambs gained 0.1 kg per day, which compared favorably with those produced on well-watered range. Similar results were obtained on high mountain ranges in central Utah. The following conclusions were reached as a result of these studies: (1) On succulent weed ranges in high mountains, sheep can be grazed without water with gains comparable with those from well-watered ranges. (2) On nonsucculent grass ranges in high mountains, sheep can do well if they get a limited amount of moisture in the form of dew, fog, or rain. (3) Rarely is it necessary to drive sheep long distances to water on mountainous summer ranges more often than every third day. (4) Where water is inadequate, quiet open grazing and shading up during the hot midday is imperative.

Grazing sheep on winter ranges without water is not uncommon where the animals have access to snow. Large areas of range with no water development whatever are used in this manner, though in winters of deficient snow it becomes hazardous.

Water intake and frequency of watering of animals during the dry season in Africa differ somewhat from American experience (Table 10.4).

Number of Watering Places Needed

The number of watering places necessary is variable and will depend upon local conditions. Various kinds of livestock differ in the distances that they can travel from water and these distances depend, in turn, upon the topography. Sheep and goats may need only half the number of water developments that are required for cattle. Not only can they go

Table 10.4   Daily Intake and Frequency of Watering of
Range Animals in the Dry Season

Data from Baudelaire (1972)

Animal	Daily intake, liters	Frequency of drinking
Sheep	4-5	Once every 2 days
Goats	4-5	Once a day
Asses and donkeys	10-15	Once a day
Horses	20-30	Once or twice a day
Bovines	30-40	Once a day or once every 2 days
Camels	60-80	Once in 4 or 5 days

longer without water but, because they are herded, they can be made to graze farther from the source of water without damaging the range. Actually, cattle, and especially horses, can move farther out from water within a given time than can sheep, but they show a greater reluctance to move away from the water after drinking.

In excessively steep and rough country, cattle should not be forced to go more than 1 km for water, though in more level areas this can be measurably increased. Even in flat country they should not be expected to travel more than 4 km. This would require water holes 1½ to 8 km apart and one source might serve 1 to 20 sq mi of land, depending upon the topography. These distances would apply equally well to sheep when they are watered daily but would not apply when longer periods elapse (Fig. 10.1).

Importance of Water to Production

Although range animals can adjust to water deficits, this is done at the expense of production. Wilson (1970) observed that forage intakes of sheep in Australia declined the second day after watering, though there were some differences among breeds. Probably both forage intake and work of travel were responsible for differences in lamb gains reported by Squires (1970). Penned lambs gained 219 gm per day; and lambs forced to travel gradually increasing distances from 0.8 to 4.0 km per day gained 126 gm.

Sheep on salt-desert range in Utah gained 1.5 kg when watered daily for 40 days, and 0.36 kg when watered each second day, but when watered each third day, they lost 2.7 kg (Hutchings, 1946). Sheep grazed more quietly, ate a greater variety of feed including dry material, and utilized the range more uniformly when watered daily.

Fig. 10.1 Having ample, well-spaced watering facilities increases animal gains and reduces range deterioration. This metal trough stores enough water to accommodate a band of sheep even though the flow is not large.

BREEDS OF LIVESTOCK

An important means of increasing production from ranching is using high-quality livestock. Increasing the quality of stock means higher returns per animal. This permits a greater value of salable products from the same or fewer number of animals. Improved quality results, therefore, in improved range conditions through less intensive use. This may not mean that weight gain per unit of feed consumed is greater in better-bred animals; but a larger percentage of the carcass is usable, and also the animals bring a higher price (Fig. 10.2).

Straightbred Animals

Animals having fixed characteristics of a given breed are called *straightbred*. Although their genetic makeup may be essentially "pure" for the breed, the term *purebred* is reserved for pedigreed animals registered in an official breed registry. Most range animals are straightbreds.

Certain breeds of animals have been widely used on rangelands. British cattle breeds dominate most commercial range operations. Shorthorns are the most common cattle breed worldwide, ranking first in Australia, South America, and Africa. However, in North America they rank behind Hereford and Angus (Edwards, 1973). Currently, a number of European continental breeds, originally developed for milk or as dual-purpose animals, have found favor as range animals. Charolais, Brown Swiss, Freisan, and Simmental are all used to some extent. Rouse (1970) gives a country-by-country treatment of cattle breeds of the world.

Fig. 10.2 Good-quality animals are basic to range-livestock production, but they must have good-quality range if they are to produce efficiently. Herefords are the most popular cattle breed in the United States.

Tropical ranges are dominated by zebu breeds. Brahman (mostly Guzerrat) breeds are popular in the hot regions of North America and Australia. In Africa several endemic zebus occur, but many of the local herds are giving way to improved Boran or Sahewal herds (Fig. 10.3).

The demand for fine wool and the merino's adaptability to dry conditions have made it the most popular range sheep breed. Specific merino strains have developed in Australia, New Zealand, Argentina, and Russia. In North America the Rambouillet (French Merino) is the standard range breed, with some Delaine Merinos in the Southwest. British mutton breeds of sheep (Suffolk, Hampshire, Border Leicester, Dorset, etc.) are kept as straightbreds on some ranges, mainly to supply rams for crossbreeding.

Fat-tailed and fat-rumped sheep are the most popular breeds on most African and Asian ranges, the major deserts of the world. These hardy animals are suited to converting forage to meat production under arid conditions.

New Breeds

A number of new breeds of livestock have been developed for specific range conditions. These usually are produced by crossing straightbred animals and stabilizing the cross. Several new breeds have been established to achieve greater heat tolerance by introducing zebu blood into

European cattle or fat-tailed sheep blood into wool sheep. Santa Gertrudis, Beefmaster, and Africander cattle represent stable zebu x European cattle bred for heat and disease resistance. The Dorper sheep is a cross between Dorsets and Somali Blackheads. The Columbia and Targhee breeds are crosses between British and European breeds, long- and fine-wooled sheep, bred for Western range conditions.

Crossbreeding

Quality of livestock does not necessarily mean the maintenance of straightbred animals with fixed breed characteristics. The greatest production from rangelands may come from crossbreeding, thereby taking advantage of hybrid vigor. Crossbreeding can be accomplished by (1) maintaining straightbred female stock and using males of another breed, or (2) by mating purebred males to successive generations of crossbred females. The effects of heterosis (hybrid vigor) are significant for most economically important traits in beef cattle. Early work at Miles City, Montana, resulted in greater weaning weights of crossbred steer calves through the third generation of increasingly mixed-blood heifers resulting from using Shorthorn, Angus, and Hereford bulls in succession. Heifers at the outset were Herefords and comparisons were made only to purebred Herefords (Knapp et al., 1949).

The advantages may depend upon what measure of success is used—birth weights, weaning weights, gains to maturity, gains in feed-

Fig. 10.3 Brahman or zebu cattle from Burke Brothers Ranch, Corsicana, Texas. These types are used widely in tropical or subtropical areas either as straightbreds or for crossbreeding with European breeds because of their greater heat and disease tolerance.

lots, or carcass quality. Angus-Hereford crosses were superior to purebreds in some respects but not in others (Gerlaugh et al., 1951). Moreover, the manner of crossing affected the results. Hereford sires on Angus heifers produced steers that gained more rapidly to weaning; steers of Angus sires and Hereford heifers performed better in the feedlot and produced better carcasses. Gregory et al. (1966b) reported better gains and feed efficiency from Hereford x Angus and Hereford x Shorthorn steers than from Shorthorn x Angus crosses.

A Nebraska study showed no significant difference from crossing Hereford, Angus, and Shorthorn and straightbreds for percent of calves born, but a 3 percent greater calf crop was weaned because of differences in postnatal survival (Wiltbank et al., 1967). Crossbred calves were 4.6 percent heavier at weaning than straightbred calves (Gregory et al., 1965). Increased survival and growth rate combined gave the crossbreds an 8.5 percent greater weaned-calf weight for each cow exposed over straightbreds (Cundiff, 1973). Crossbred heifers have also shown better growth at low levels of nutrition (Gregory et al., 1966a), better fertility of females at puberty (Wiltbank et al., 1966), and better mothering ability of females (Cundiff, 1973).

The heterotic effect is inversely proportional to the genetic similarity in the animals crossed. When European breeds are crossed with British breeds (Furr, 1968) or when zebu and European breeds are crossed, greater responses are observed than when British beef breeds are crossed (Fig. 10.4). Spectacular increases in performance have been reported from crossing straightbred British breeds with traditional milk breeds such as Brown Swiss and Holstein. In controlled experiments in Africa, Sahewal—a milk breed—has been very effective in increasing the turnoff from endemic zebu herds (Allen, 1973).

Crossbreeding is the usual practice among sheepmen in the Western United States. Commonly, rams of breeds with better mutton qualities (Suffolks and Hampshire are popular) are bred to Rambouillet ewes, or ewes of Rambouillet origin such as the Columbia. This has the effect of improving the quality and value of the lamb carcass while still maintaining the better wool qualities of the fine-wooled breeding herd.

Introduction of Exotic Animals

The biological advantage of common use (see Chapter 8) suggests that exotic animals, with diets different from livestock, could make more efficient utilization of the forage resource, provide a tool for range improvement, and add a new range product.

It should be emphasized, however, that new animals cannot be added to those already present on the range. Room must be made for

Fig. 10.4 Mixed herd of crossbred calves in Uganda resulting from crossing Boran, Angus, and Red Polled bulls to zebu, Ankola, and Boran mothers. *(Courtesy F.A.O.)*

them and care must be exercised not to exceed the total carrying capacity of the range (see Chapter 11). Diets of different kinds of animals almost always overlap; hence, numbers of resident animals must be reduced to maintain the plants used in common. Competition from an exotic for only 6 weeks during a critical period of conception or birth could cause severe damage to the native population. Thus, it is not only necessary to balance numbers of all animals combined with the carrying capacity of the range, but seasonal food preferences must be matched with seasonal availability of forage.

Certain conditions should exist before exotics are introduced: (1) There should be vegetation in excess of that required for soil protection at the end of the grazing season which cannot be efficiently utilized by native animals and is palatable to exotic animals. (2) Unpalatable plants should be increasing in abundance and be palatable to the proposed species. (3) The introduced exotic should not increase pressure on plants essential to native wildlife or livestock. (4) The introduced animals must be compatible with native animals and livestock. (5) The numbers of the introduced animal must be susceptible to control in order to obtain the proper balance of grazing pressure on forage plants. (6) The introduced animals must be susceptible to confinement to a single property without undue expense. (7) There must be no legal obstacles to harvest and sale of the animals. There are few of these conditions that can reasonably be met with the present state of our knowledge, and much more must be known before the introduction of exotics should be attempted.

MANAGEMENT FOR INCREASED REPRODUCTION

The percentage yield of calves and lambs is one of the most important factors affecting the success of a ranch, for these are the principal salable products, except for wool. Calf and lamb crops can be varied more than almost any other through skillful management, since they are intimately dependent upon quality of the range and the breeding and feeding practices.

The turnoff, the salable livestock, from a given range is directly related to the reproduction rate achieved, which depends upon the health and nutrition of the breeding herd. If the nutritive requirements described in Chapter 7 are met and the animals protected from disease, the chances for a high level of reproduction are good.

Several management practices, other than providing adequate nutritive requirements, affect turnoff rates. These include (1) controlling the breeding season, (2) removing nonbreeders, (3) extending the breeding life of females, (4) encouraging multiple births, and (5) protecting young animals.

Control of Breeding Season

The first step in intensified livestock production on rangeland is control of the breeding season. If animals are allowed to breed throughout the year, offspring may be dropped during unfavorable periods when survival is difficult. Cows in monsoonal Australia breed soon after the beginning of a 6-month wet period and drop their calves about midway into the dry season. Calf mortality rates of 40 to 50 percent are common (Smith and Alexander, 1966). Supplemental feeding and controlled breeding to ensure that cows drop their calves in the period of ample forage could almost double calf survival.

Breeding pastures where the herd can be more closely confined during the breeding season is one means of controlling breeding. Increased calf crops of as much as 30 percent over that on open mountain range have been attributed to breeding pastures. Where pastures are not possible, range *riding* to keep bulls well distributed can be effective. Vincke and Arnett (1927) reported increased calf crops of 15 percent in Montana due to this practice.

Controlled oestrus, by giving hormones, either as implants or orally, offer possibilities on moderate-sized, self-contained ranches. In addition to increased calf crops, these techniques insure production of even-aged, uniformly sized calves which command better prices at marketing.

Artificial insemination is another means of improving herd productivity. Running a vasectomized bull in the herd facilitates locating cows in heat which can then be artificially inseminated. This not only greatly enhances the chance of conception, for low-virility bulls may dominate virile ones when they are allowed to range freely, but sperm from superior bulls can be spread among more animals, thus eliminating the need for maintaining large numbers of bulls (Fig. 10.5). These promising techniques are not well suited to conditions on the open range, for they require much greater supervision and control of livestock than is possible under present practices. Many of the low-quality ranges now being used may prove unsuited for cattle production as these methods, production testing, and carcass-quality control are more widely employed in range-livestock production.

Removal of Nonbreeders

It is important to remove nonbreeding animals from the herd as soon as they can be identified. Early removal lessens demands for forage allowing more forage for the females with offspring, and reduces pressures on the range. If the nonbreeding animals are identified early, the rancher has a choice of when he sells them and may take advantage of favorable markets.

Fig. 10.5 Artificial insemination makes it possible to spread more widely the better genetic qualities of superior bulls like this Hereford.

Pregnancy checking, by rectal palpation, shortly after breeding is a technique available to most ranchers for early identification of non-breeders.

Extending the Breeding Life of Females

The number of years a female bears young has a direct effect on the productivity of a ranch. The smaller the number of prebreeding females and the greater the number of active breeders, the greater the potential turnoff. Under most range conditions, cattle are bred to calve at 3 years of age. Cows will breed much younger, and there are definite advantages to breeding yearling heifers. However, precautions should be taken and extra labor made available at calving time. Large, healthy yearlings (250 kg or more in weight) can be bred safely, especially if they are bred to genetically small-bodied bulls. They should be on a high plane of nutrition and calves should be weaned early; otherwise many will not breed as 2-year-olds. If these precautions are followed, an extra year can be removed from the prebreeding period of most of the cow herd.

Multiple Births

Cattle normally have only one calf. Twins are more common in sheep and goats, and some breeds of sheep normally have more than one offspring. Recent work with hormones indicates that there is a potential for inducing multiple ovulation in both sheep and cattle (Laster, 1973). The use of genetic lines with this trait offers a possible means of increasing range-livestock production.

Supplementation for Increased Reproduction

The condition of the female is important among the factors affecting calf or lamb crops. Conditioning animals by supplemental feeding or *flushing* during the breeding season sometimes is recommended. This is of special value during drought years or when weather is unusually severe.

Feed Concentrates   Research on feeding concentrates to range ewes in Montana showed that feeding before breeding time and during breeding time, early pregnancy, and late pregnancy resulted in increases of 10, 9, 5, and 4 percent, respectively, in number of lambs born. A lamb crop average of 145 percent was obtained by feeding throughout all periods, contrasted with 110 percent when no feeding was practiced (Darroch et al., 1950).

Flushing ewes with alfalfa, cottonseed meal, and barley on California ranges increased lamb crops (due primarily to a greater number of multiple births) from 32 to 64 percent in separate trials (Torrell et al., 1972). The added energy in the ration accounted for nearly half the increase; protein accounted for only 14 percent. No gains were attributed to vitamins.

Supplements of soybean pellets and monosodium phosphate both resulted in higher lamb crops, as much as 21 and 53 percent, respectively, on winter ranges in the Great Basin. Barley supplements did not prove beneficial (Harris, 1968).

Studies conducted on annualgrass ranges in the California foothills showed good responses from supplementing cows with concentrates from August 1 to February 1. Depending on intensity of grazing, supplemented cows produced calf crops of 78.4 to 86.5 percent, whereas unsupplemented herds produced only 60.0 to 77.8 percent (Bentley and Talbot, 1951).

Urea and maize supplements fed to Africander cows during the dry season in Africa led to 17 percent greater conception rates (Topps, 1972). Supplementing range cows with protein on a West Texas range did not increase calf crop in good years, but did increase the number of cows bred during dry years (Kothmann and Mathis, 1970). Under West Texas conditions supplementation could be expected to be economical on moderately grazed ranges only once in each 4 or 5 years.

Mineral Supplementation Supplementing with phosphorus or fertilizing pastures with phosphorus in South Texas increased average calf crop born from 76.4 percent to well over 90 percent (Reynolds et al., 1953).

In tropical areas where forage is deficient in phosphorus, applications of superphosphates increased the fertility of cows. Edye et al. (1971) reported increased conception rates in Droughtmaster cows in Australia of 7 to 18 percent and 8 to 20 percent increases in calving rates due to fertilization of pasture with superphosphates.

Reproductive Performance of Range Livestock

The percentage calf or lamb crop is an important measure of breeding-herd efficiency. This measure varies greatly throughout the world. In general it is low in open range areas and somewhat better under fenced conditions.

Calf Crop Cattle ranchers, almost universally, have been content with too low a calf yield. Calving rates of 30 to 40 percent are common in

Australia (Jenkins and Hirst, 1966) and are that low or lower over much of Africa. Percentages of 90 or better are possible in intensive range operations in North America. Remarkable increases result from improved handling methods and improved range conditions. On the Santa Rita experimental range in Arizona, a calf crop of 82.7 percent was obtained over an 11-year period, the highest for an individual year being 88.8 percent. Operators on the surrounding ranges for the same period secured a calf crop of 55 percent (Culley, 1938).

Nine-year studies in South Dakota involved grazing mixed prairie vegetation to obtain utilizations of 63, 46, and 37 percent. Calf crops weaned were 55, 60, and 85 percent, respectively. Weaning weights per cow were 208.3, 226.2, and 361.8 kg and per hectare 20.2, 14.6, and 15.7 kg, respectively (Johnson et al., 1951).

Unsupplemented cows on California annualgrass ranges produced calf crops of 60.0, 77.8, and 72.5 percent under heavy, medium, and light grazing intensities, respectively (Bentley and Talbot, 1951).

On Montana short-grass ranges, calf crops of 79, 85, and 85 percent were secured under heavy, medium, and light grazing, respectively (Hurtt, 1946).

Results of experimental grazing in Oklahoma, however, did not show significant differences among heavily, moderately, and lightly grazed pastures in percent of calf crop which varied from 92 to 96. Weaning weights were much greater on lightly grazed range (McIlvain et al., 1953).

Lamb Crop   Percentage lamb crop is even more variable throughout the world. Percentages of 150 or more are possible. On fenced range areas where lambs are not protected, percentages may drop to 50 or less. Where ewes are shed lambed (North America) or have a high degree of individual attention on open range (nomadic pastoralism) percentages are often in excess of 100.

MANAGEMENT FOR INCREASED PRODUCTION

Great variations occur in the amount of forage produced from year to year. Studies in central New Mexico showed range-capacity variations from 63 percent above to 61 percent below normal (Lantow and Flory, 1940). Over a 20-year period, forage production in Texas in some years was 67 percent above or below median yields (Skeete, 1966). Even when forage supplies are ample in amount they have nutritional deficiencies in certain seasons. These conditions can be met by (1) stocking conservatively, (2) varying numbers of livestock, (3) augmenting forage supplies, or (4) supplying additional nutrient by supplementation.

Variation in Range-Forage Production

There is a close relationship between forage production and climate, particularly precipitation. Precipitation on rangelands is unusually low and undependable, and periods of low precipitation often are protracted (see Chapter 1). Subnormal rainfall usually is accompanied by a decrease in forage production and a loss in many of the more desirable forage species. The greatest loss in plants occurs in periods with several growing seasons below normal. During these drought periods, severe range deterioration may take place (Fig. 10.6).

The extent to which forage production is decreased by weather varies with site, climate, vegetation type, and grazing management. For instance, on the Snake River Plains of southern Idaho, between 1932 and 1935, total vegetation on ungrazed areas decreased to 84 percent of the 1932 ground cover, grasses alone decreasing to only 48 percent of the 1932 cover. In the Southwest, over 13 years during which two

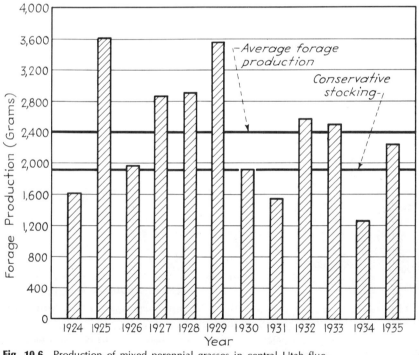

Fig. 10.6 Production of mixed perennial grasses in central Utah fluctuates greatly from year to year, and conservative stocking must be 20 percent or more below average production to furnish adequate forage in all but the lowest years. *(From Campbell, 1936.)*

drought periods were experienced, one lasting for 3 years and the other for 5 years, the cover of *Bouteloua eriopoda* decreased to a low point of 10.9 percent of the original, even though ungrazed (Nelson, 1934). This marked reduction in forage yield resulting from poor growing conditions alone makes it evident that no plans for grazing should be made without allowing for variation in production attributable to weather conditions.

Studies on salt-desert range in Utah show remarkable correlation between forage production and precipitation (Fig. 10.7). Expected air-dry herbage yield in pounds per acre can be calculated with considerable accuracy in this area by multiplying 12-month precipitation in inches by 45.8 and subtracting 92.46 (Hutchings and Stewart, 1953). Sneva and Hyder (1962) and Currie and Peterson (1966) have developed precipitation-production relationships for Oregon and the Great Plains, respectively. Other studies on the cold-desert ranges of Oregon and on the central Great Plains indicate that forage yields can be predicted from stored soil moisture and rainfall.

Drought damage to plants is affected materially by grazing use.

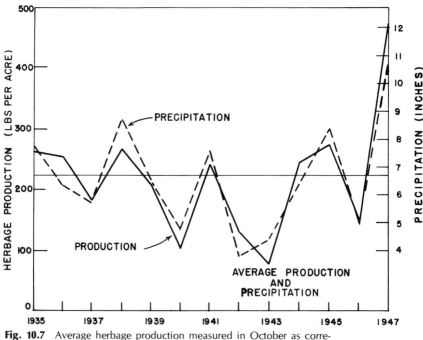

Fig. 10.7 Average herbage production measured in October as corre-lated with precipitation during the preceding 12-month period at the U.S. Forest Service Desert Experimental Range, a salt-desert winter sheep range located in southwestern Utah. *(After Hutchings and Stewart, 1953.)*

Studies indicate that conservative grazing is little or no more harmful than is total protection (Nelson, 1934). Under heavy grazing, however, plants suffer great damage and are more susceptible to drought injury. On the Snake River Plains, depletion of spring-fall ranges during drought periods was found to be proportional to the degree of grazing during the spring (Craddock and Forsling, 1938). Similarly, decreases in production of the principal grasses of the Great Plains under drought conditions were found to be directly proportional to the intensity of grazing (Savage, 1937). The reason is to be found in the reduction of root growth under heavy use.

Stocking to Prevent Damage to Plants during Drought

Variable production of forage must be anticipated in the management plan. Even under conservative stocking based upon average production, certain subnormal years may be expected during which forage shortages will occur. When livestock numbers cannot be varied, the stocking level should not exceed average forage production. The plant resource and grazing animals must be in balance at all times. If there is to be a temporary unbalance, it should be in favor of the plant, rather than the animals, if long-term production is the manager's goal. Usually 65 to 80 percent of average forage production is a safe base for calculating grazing capacity. Under no plan of stocking at a constant level will it be possible to prevent excess use in poor years and underuse in good years. Forage remaining in years of high production is not wasted, for it adds to the vigor of the plants, permitting them to recover better from close use during poor years.

Adjusting Livestock Numbers to Varying Forage Production

Adjusting a livestock program to adapt it to varying forage production is of great importance, although many ranchers find it difficult to vary numbers to balance stocking against the whims of the weather. In years of abundant forage, it is difficult to purchase livestock because of the demand from other producers. In poor years, the flood of animals on the market depresses prices. An attempt to regulate numbers according to range production by a buying-and-selling program is hazardous. The producer is tempted to hold over high numbers during poor years in a play for better markets. Severe overuse of the range, already weakened by drought, is likely to be the result. Nevertheless, adjustments are possible. Lantow and Flory (1940) suggested the following means of adjusting livestock numbers.

To decrease the herd during below-normal periods: (1) sell early enough to ensure sufficient feed to carry the breeding herd; (2) sell cattle whose meat value is near maximum, such as fat, dry cows and calves of acceptable market age; (3) sell weaned calves and old or thin cows which would be expensive to carry over; (4) sell heifers that would not contribute to building up the herd; and (5) sell inferior or nonbreeding cows.

To increase the herd during above-normal periods: (1) keep steer calves another year; and (2) increase the cow herd by retaining heifers or buying.

Many ranchers in arid areas find that keeping a basic, conservative number of breeding females that can be carried on the range in most years offers them needed flexibility. They can shift from a cow-calf operation to a cow-calf-yearling program during good years. In economies where grass-fat stock are sold from the range (e.g., Australia, East Africa) only a small part of the herd or flock consists of females, and castrates are used to obtain the desired flexibility.

A variable livestock-stocking system using a basic herd of cattle which is supplemented by a short-term, 9-month ewe herd proved practical and profitable on an Edwards Plateau range (Skeete, 1966). Each August a ewe band is purchased, the number depending upon the available forage supply and soil moisture. In February ewes and lambs both are marketed and the range allowed to rest for 3 months before purchasing another ewe band.

For ultimate success, a variable stocking program depends on a reasonable measure of predictability of precipitation, and consequently forage production. To the extent that this is possible, statistical theory may provide guides to variable stocking programs (Rogers and Peacock, 1968).

Augmenting Fluctuating Forage Supplies

Compensating for variability of range feed by the use of supplemental forage or adjusting length of grazing season offers a more practical method of management in most areas than does adjusting livestock numbers. On many range areas, forage in short supply is nutritionally deficient in certain seasons, and it is often advisable to augment or supplement range forage from other sources. There are several methods that can be used to balance these deficiencies: (1) providing additional pasturage, (2) feeding harvested forage, and (3) supplementing existing forage.

Permanent Pastures   Early spring is a critical period for Western ranges. During this season of scarce forage, farm feeding is common, but harvested feed is costly compared with pasturage. As the pressure on rangelands increases, the need for permanent pasturage to supplement range forage will increase. Unfortunately, the possibilities of producing additional forage from this source have too frequently been overlooked. Farm pastures too often are the poorer lands, much overused, and inputs in the form of seeding, fertilizing, draining, and irrigating too often have not been applied to them.

Equally good opportunities exist in many areas for range seedings to increase forage supplies and extend the grazing season (see Chapter 14). The possibilities for increased production are shown by forage-production studies from native prairie and crested wheatgrass in Canada (Table 10.5). Two things are accomplished: the yearly forage supply is balanced, and overuse of range is prevented.

Temporary Pastures   Much can be done to increase feed for livestock during emergency periods by planting such crops as annualgrasses which are well liked by livestock, are easily grown, and produce usable forage within a few weeks of planting. The great advantage of such crops is their flexibility. They can be produced as the need is foreseen and can be omitted during years of good range production. Although perennial grasses generally are more satisfactory, annual grains, especially winter rye, volunteer readily and, even though closely grazed, may persist for many years without reseeding. This procedure can be used

Table 10.5 Possible Rotations of Crested Wheatgrass and Mixed-Grass Prairie Based on Clipped Plot Yields 1948 to 1951 Inclusive

Data from Campbell (1952)

	Possible rotations grazing dates		Summer yield, # lbs per acre			
No.	Crested wheatgrass	Mixed-grass prairie	Forage	Total	Crude Protein	Total
1	Apr. 15-May 16		270		44	
		May 16-Sept. 1	240	510	58	102
2	Apr. 15-June 5		395		64	
		June 5-Sept. 1	385	780	64	128
3	Apr. 15-June 20		430		69	
		June 20-Sept. 1	410	840	73	142
4	Apr. 15-June 5		455		71	
		June 5-Sept. 1	440	895	55	126
5	Apr. 15-July 20		460		70	
		July 20-Sept. 1	360	800	36	136

only if cropland is available on the ranch. Plants which can be sown for this purpose are Sudan grass, winter rye, wheat, and oats; these are widely used in the North American plains. Townsville lucerne grown for dry-season feed in the Australian tropics, and forage sorghums grown for summer feed in Mexico accomplish similar purposes.

Range Feeding   Feeding harvested forages is one means of improving range production and conserving the range. *Range feeding* to augment forage supplies should be distinguished from *range supplementation* for the purpose of correcting nutritional imbalances (Cardon, 1966). Several factors which make necessary an adjustable feeding program as part of a range-livestock-production program are (1) great annual variation in forage production from year to year, (2) lack of balance between seasonal classes of range, (3) frequency of heavy snows in some regions which prohibits winter grazing, or (4) unexpected drought periods.

The possibilities for profit on ranges that can be grazed yearlong have been estimated to be about twice as great as on ranges that require a long period of winter maintenance on hay, although much of this difference is absorbed by higher land prices on yearlong ranges (Saunderson, 1940). Schuster and Albin (1966) could find no advantage in production from wintering cows in the feedlot over pasturing on range or stubble fields.

The decision to feed must be made on the basis of whether range forage supplies are adequate. Frequently, ranchers maintain animals on a range long after suitable forage is gone. Not only would the provision of additional feed prevent serious damage to the range, but monetary returns might be greatly increased as well. It is obviously poor management and poor economy to misuse range forage in order to avoid expense. It is also poor management to feed in such a way as to encourage overuse of the range. If feeding increases rather than lessens pressure on range forage, it cannot be justified. Where feeding is required, animals should be removed from the range while harvested forages are being supplied. Correcting nutritional deficiencies is the only valid objective of the range manager except in emergencies.

Supplementing Nutritionally Deficient Forage

Even when amounts of forage are adequate, nutritional deficiencies may exist. This is especially true of grass ranges which decline in quality with maturity and subsequent leaching. Although range livestock may not suffer from these deficiencies, production may be lowered.

Feeding Concentrates on the Range   Concentrated feeds often are used during periods of stress such as hard winters or drought. Concentrates may be grains, cottonseed, or soybean products such as meals or cake. Because of their high-protein content, cottonseed and soybeans most nearly balance the material furnished by most range forages in the dry season (Fig. 10.8).

Although energy may be inadequate on browse ranges, such as in the Great Basin, supplying carbohydrate concentrates should be done with caution since there is evidence that providing easily digested carbohydrates depresses the utilization of fiber in the range forage (Topps, 1972). A useful rule is to supply only enough energy supplement to increase consumption of forage, but not replace it. Harris (1968) reports a small increase in lamb crops when ewes were supplemented in winter with 182 gm of barley per day. Doubling the barley supplement resulted in lower lamb crops, 29 percent less than the control herd.

Supplementing sheep which are attended by a herder is a relatively easy thing, but adding to the diet of cattle often involves considerable extra labor. Feeding concentrates to cattle by mixing them with salt appears to be a practicable procedure. Regulating the percentage of salt regulates the consumption of supplement. The percentage varies with (1) supplement consumption desired, (2) distance from water, (3) amount and quality of natural forage available, (4) season of the year,

Fig. 10.8 Cattle being fed cottonseed cake on forested grazing lands in Louisiana, a standard practice in the Southeast in winter when forage is dry and deficient in protein. Several Eastern and Southern states surpass Western states in beef-cow population. *(See Edwards, 1973.)*

and (5) the degree of salt tolerance of the cattle, which seems to increase as feeding continues.

Nitrogen usually is supplied by a natural high-protein feed, but it can be given in the form of a nonprotein nitrogen such as urea or biuret, also. Positive responses have been observed through the addition of urea to the diet of range livestock in Africa. There was a small increase in the digestibility of the forage and, more importantly, an increase of 50 percent in forage intake (Topps, 1972). The choice between natural and nonplant sources of nitrogen should be made on the basis of convenience and cost, since there is evidence that nonplant nitrogen is not superior to high-protein concentrates (Clanton, 1973). Moreover, substances such as urea may require additional carbohydrate supplements for efficient utilization of the nitrogen. Crampton and Lloyd (1959) suggest an energy-to-protein ratio of 4:1.

Although the nongrowing season is most critical, concentrates may be effective in other periods. Launchbaugh (1957) increased gains of steers by feeding cottonseed meal during the last 3 months of the summer grazing season in Kansas. Gains continued into October at greater rates than under light grazing. Harris (1968), however, found no greater summer gains in calves and yearlings on crested wheatgrass when fed a supplement of soybean meal and dicalcium phosphate, although cows did maintain their weights better from August to December.

Raleigh (1970) increased daily gains from supplementing yearling steers with nitrogen and energy during the summer in Oregon (Table 10.6).

The biological necessity and economic advisability of supplementing must be decided on an individual basis. In general, some weight loss is expected during winter, and healthy animals soon regain their weight when green forage again is available. The need for supplementation can

Table 10.6 Average Daily Gains of Yearling Cattle on Supplements Compared to Those without Supplements at Squaw Butte, Oregon

Data from Raleigh (1970)

| Period | Daily nutrient intake | | Daily gain, kg | |
	Nitrogen, gm	Energy, kcal	No supplement	Supplemented
5/10-6/11	4.9	922	1.04	1.24
6/12-7/12	20.8	1295	0.88	1.00
7/13-8/12	41.0	3052	0.47	0.87
Mean gain			0.80	1.04

be determined by evaluating the nutritional quality of range forage by digestion techniques, or by field experiments in which suspected deficiencies are supplied and animal responses measured. Allden and Jennings (1969) suggest that protein deficiencies in range forage can be detected by analysis for fecal nitrogen and set 1.6 to 1.7 percent as the threshold. When nitrogen in the feces falls below this level, supplements may be needed.

Mineral Supplements The more common mineral supplements are calcium, phosphorus, and salt. Calcium and phosphorus usually are provided in the form of dicalcium phosphate. Iodine often is provided in the form of iodized salt. Other minerals are not likely to be required under range conditions, but local conditions may require them. Copper, for example, has been found to correct the ill effects of grazing pastures of high molybdenum content in New Zealand (Hogan et al., 1971), and in Oregon (Kubota et al., 1967). Cobalt additives have proved effective in Australia (Lee and Marston, 1969) and in some areas in the United States but not in others (Clanton and Rowden, 1963).

Experiments in Texas (Reynolds et al., 1953) showed that phosphorus deficiency could be prevented by feeding bone meal in self-feeders, by adding disodium phosphate to the water supply by automatic regulator devices, or by fertilizing the range with phosphorus. Unsupplemented cows produced 104.2 kg of calf per hectare; cows fed bone meal produced 130 kg; cows receiving phosphated water, 160.2 kg; and cows on fertilized range, 197.1 kg. Except during drought, fertilizing range solved the deficiency problem and at the same time increased the grazing capacity of the range as much as 84 percent and the phosphorus content of the grasses by 150 to almost 300 percent.

Hormone Treatments

Male animals normally are castrated at an early age on the range. Removal of the testes results in behavioral changes that make the animals more desirable, but it decreases the output of testosterone and other growth hormones. Supplementing the animals with a synthetic hormone, diethylstilbesterol, can increase animal gain, often as much as 10 percent. The use of synthetic hormones now is prohibited by federal regulation. However, male modification techniques that have the beneficial effects of castration but allow growth hormones to continue to be produced are being tried. In one technique, developed in New Mexico, the testes are pushed into the body cavity where body heat prevents sperm formation, but growth hormones continue to be produced.

POISONOUS PLANTS ON RANGELANDS

Poisonous plants cause untold losses each year through death of animals, physical malformations, abortions, and lowered gains. Some of these losses can be avoided by good management; others occur with such irregularity due to unpredictable conditions that they constitute an ever-present hazard.

Poison plants are normal components of range ecosystems, and there is little if any evidence that native herbivores were affected by them. Two factors have made them more serious on intensively used rangelands. Many poisonous plants are increasers, *Actinea odorata,* and have become more abundant under range misuse. Others are exotic species such as *Halogeton glomeratus* which, since its appearance about 1935, has spread widely through the intermountain region. That losses are not greater is attributable to wider recognition of poisonous plants and the hazards they present (Fig. 10.9).

Little progress has been made toward elimination of poisonous plants. Most are too widely distributed to permit economically feasible eradication programs. Some like *Delphinium* and *Veratrum* grow in dense patches and can be treated with herbicides, although to date control efforts are not promising (see Chapter 13). The solution to this problem lies in (1) recognizing poisonous plants, (2) knowing at what times they are most dangerous and for which animals, and (3) maintaining a good cover of suitable forage plants. Learning to live with poisonous plants is not a new idea for stockmen have been doing so for many years. Poison-

Fig. 10.9 Tall larkspur *(Delphinium)* is the most serious cattle-poisoning plant on high mountain ranges in the Western United States. It frequently grows in dense patches inviting control efforts with herbicides although to date only partial success has been possible. *(Photograph courtesy Kleon Kotter.)*

ous plants are not abundant in climax or near climax vegetations (Table 10.7), and good range management will minimize the threat they pose.

Factors that Affect Poisoning

Most poisonous species kill animals only if eaten in large amounts —often almost a straight diet of the poisonous plants. Where there is plenty of forage, animals naturally vary their diet, hence eating large amounts of any one species is unusual. With few exceptions, a normal range is safe for domestic livestock under proper handling.

Plants do not fall readily into a poisonous and a nonpoisonous group. Probably thousands of plants would be poisonous if eaten in sufficiently large quantity. Some of these are excellent forage (*Prunus* for example) when not eaten in too large quantities. Many plants definitely classed as poisonous are eaten daily by animals with no ill effect because they are taken in small amounts and the poison is eliminated by the animal as rapidly as it is consumed. Although a few species, such as water hemlock *(Cicuta),* are violently poisonous at certain times of the year, even in small amounts, *most poisonous plants are not dangerous to animals, except in large amounts.*

Poisonous plants are not equally poisonous at all seasons of the year. Species in which hydrocyanic acid is formed vary considerably in their virulence throughout the season; they are more dangerous in early spring or when growth is arrested by drought. *Lupines* are most dangerous after formation of fruits which have higher concentrations of the poison than the foliage.

Seasonal differences occur with respect to animal susceptibility. Lupines have been shown to cause malformations in calves, "crooked calf disease," where cows ate them during the fortieth to seventieth day of gestation (Shupe et al., 1967). Similarly, *Veratrum californicum* when eaten by sheep at the fourteenth day of pregnancy caused malformed lambs, especially malformation of the head (Binns et al., 1965). Many fetuses died also.

Table 10.7 General Ecological Status of Some Important Poisonous Plants

Locally abundant climax constituent	Present in climax but not abundant	Early regression stages	Late regression with severe misuse
Tall larkspur	Death camas	Low larkspur	Milkweed
Arrowgrass	Loco	Rubberweed	Tobacco
Chokecherry	Horsebrush	Saint Johnswort	Nightshade
Lupine	Sneezeweed	Burroweed	Halogeton
Black laurel	Water hemlock		Bitterweed
Oak			
Greasewood			

Different kinds of animals vary with respect to their susceptibility to poisoning from a particular plant. Losses of sheep from *Halogeton glomeratus* are sometimes very great, although cattle use halogeton-infested ranges, and eat it without ill effects. Conversely, *Delphinium* is not considered a problem on sheep ranges, but it is a major threat to cattle (Fig. 10.10). Some of these differences in susceptibility may be due to the amounts taken, but in the case of *Delphinium,* sheep are able to consume two to five times, in relation to size, the toxic dose for cattle without ill effect (Marsh et al., 1916).

There is a relationship between degree of gut fill and watering of animals and poisoning. Sheep on halogeton-infested ranges are especially susceptible when after having been off water they are brought to water and subsequently consume large quantities of forage. If halogeton is present, losses can then be great (James et al., 1970). If other forage high in calcium content is present in the stomach, sheep can tolerate greater amounts of halogeton than if they are gaunt (Cook and Stoddart, 1953).

Some plants such as loco weeds *(Astragalus* and *Oxytropis)* are habit forming and once animals develop a taste for them they may eat little else.

Fig. 10.10   Dead sheep poisoned by halogeton in southwestern Utah as a result of allowing them to fill up on water in the vicinity of a dense stand of this highly poisonous weed. Plants were dry in midwinter. *(Photograph by D. M. Beale.)*

General Management Rules to Prevent Poisoning

Prevention of poisoning is much more easily accomplished than is curing an animal, which has little chance of success anytime and almost none on the range. Although management is not a complete solution to the poisonous-plant problem, in most instances it is all that is economically feasible. A few rules of good range management and livestock husbandry, if carefully followed, will prevent the majority of losses.

1. Do not misuse the range so as to bring about the invasion of new species or the spreading of poisonous species which already exist in amounts not dangerous to animals.

2. Avoid areas where poisonous plants are abundant. This may require fencing of certain areas on cattle ranges, but such areas can be avoided through proper herding of sheep.

3. Do not move animals hastily through an area where poisonous plants are present. Unhurried animals select a variety of forage and are less likely to consume poisonous plants in toxic quantities.

4. Do not force animals to remain on the range after they have utilized the good forage species, or ultimately they will turn to the less desirable and, often, poisonous ones.

5. Do not allow animals on the spring range until the good forage species have made sufficient growth to support them; otherwise, they may be forced to consume the early-growing poisonous species. Some species are more poisonous in fall, others while in fruit, still others in spring. These factors should be considered in grazing plans.

6. When animals have been on dry feed, or after they have been deprived of forage, as during shipping, trailing, or corralling, they should not be put on ranges containing poisonous species until they are well fed.

7. Provide ample water so that animals will not be induced to eat increased amounts of forage following water deprivation and subsequent watering.

8. Use plenty of salt; shortage of salt may cause animals to eat plants not normally eaten. Shortage of other minerals, especially phosphorus, induces abnormal appetite, usually evidenced by bone chewing. Animals so affected are sure to eat abnormally of low-value vegetation such as poisonous plants. Feeding bone-meal supplement has been shown to reduce poisoning losses.

9. Graze with the kind of stock not poisoned by the plant in question. Many plants seriously poisonous to one kind of animal are not poisonous to another, or, at least under practical range conditions, are not dangerous.

10. If poisonous species are localized, spraying may prove economical.

A summary of the poisonous plants on North American ranges is included in Table 10.8.

Table 10.8 Poisonous Plants in Summary

Name	Location and habitat	Dangerous season	Grazing animal endangered	Poisoning conditions and general information
1. Arrowgrass (*Triglochin maritima*)	Wet and alkaline bottomlands and seacoasts	All	All	Hydrocyanic acid. Dangerous when frozen or in drought.
2. Azalea (*Rhododendron occidentale*)	California mountains, wet places, acid soils	All, especially spring	Sheep	Few ounces of leaf poisonous. Unpalatable.
3. Bitterweed (*Actinea odorata*)	West Texas and west to California, misused land	Winter to summer	Sheep	Cumulative, unpalatable annual. Overgrazed range. Dry years.
4. Black laurel (*Leucothoe davisiae*)	California mountains, acid soil, moise sites	Winter, spring	Sheep	0.2 lb per day fatal, evergreen, shrub.
5. Bracken fern (*Pteridium aquilinum*)	Woodlands throughout U.S., burned land, rich soil	Fall, even hay	All, especially cattle and horses	Cumulative poison, large amounts, Unpalatable.
6. Chokecherry (*Prunus spp.*)	Roadsides and valley bottoms, mountains	All	All, especially sheep	Large quantities, dangerous when frozen or wilted. Hydrocyanic acid.
7. Cocklebur (*Xanthium spp.*)	Fields and waste areas, wet places throughout U.S.	Spring	All, especially cattle and pigs	0.75 percent of body weight, not cumulative. First leaves or cotyledons, old plants safe.
8. Copperweed (*Oxytenia acerosa*)	Colorado River drainage, dry washes, and alkali flats	All, especially fall	All, especially cattle	Dangerous on fall trail. Compositae, 3 ft high, in clumps.
9. Coyotilio (*Karwinskia humboldtiana*)	Southwestern Texas, dry lands, gravelly hills	All, especially in fruit	All	Large shrub. Black fruit very toxic. Continued paralysis. Slow-acting poison.

Table 10.8 (Continued).

Name	Location and habitat	Dangerous season	Grazing animal endangered	Poisoning conditions and general information
10. Death camass (*Zygadenus spp.*)	Foothills and wetter desert lands of Western U.S.	Early spring	All, especially sheep	Dry by early summer. White flower, odorless bulb, 0.5 percent of weight.
11. Drymary (*Drymaria spp.*)	Misused, dry range, West Texas, Arizona, and New Mexico	All	All, especially cattle	Unpalatable, 0.5 percent of weight. Annual 3 in. high, increasing from overgrazing.
12. Dutchman's-breeches (*Dicentra cucullaria*)	Woodlands of North-eastern U.S.	Spring	Cattle	Trembling. Blue flower. Unpalatable, shortage of feed.
13. Greasewood (*Sarcobatus vermiculatus*)	Western U.S. alkaline bottomlands and washes	Spring	All, but mostly sheep	Oxalic acid. Large quantity on spring trails, eaten alone.
14. Halogeton (*Halogeton glomeratus*)	Salt deserts, northern intermountain region	Fall, winter	All, but mostly sheep	Oxalic acid. Very unpalatable annual. Misused ranges.
15. Horsebrush (*Tetradymia spp.*)	Intermountain region, mostly dry semideserts	Spring	Sheep	Spring trail, bighead, photosensitivity. Early yellow flower.
16. Horsetail (*Equisetum spp.*)	Wet meadows and mountains, Western U.S.	Hay	All, especially cattle and horses	Slow, habit-forming. Leafless, harsh texture, cone-bearing, herb.
17. Larkspur (low) (*Delphinium spp.*)	Foothills, deserts and plains	Early spring	Cattle	6- to 12-in. ephemeral, tuberlike root. Graze after June 1.
18. Larkspur (tall) (*Delphinium spp.*)	Mountain ranges	Early summer	Cattle	4-ft clumps, hollow stem. Dangerous all summer.

Table 10.8 (Continued).

Name	Location and habitat	Dangerous season	Grazing animal endangered	Poisoning conditions and general information
19. Laurel (*Kalmia spp.*)	Moist woods, swamps, mountains throughout U.S.	All, especially winter and spring	All, especially sheep	0.2 to 0.4 percent of weight. Unpalatable, eaten on overgrazed pastures. Ornamental.
20. Loco (*Astragalus and Oxytropis*)	Mountains to deserts and plains, Western U.S.	All, especially spring	All	Cumulative, habit-forming, some acute poisons, crazed action.
21. Lupine (*Lupinus spp.*)	Mountain, foothill, and semidesert areas of U.S.	Most when in fruit	Sheep	Pods and seeds of most species dangerous. Palmately compound leaf.
22. Milkweed (*Asclepias spp.*)	Sandy soils, moist bottoms, waste areas	All, especially spring	All, mostly sheep	May be poisonous when dry. Narrow and whorled-leaf species dangerous.
23. Nightshade (*Solanum nigrum*)	Waste areas throughout U.S.	Summer	All	Unpalatable annual. Small white flower, black berry.
24. Oak (*Quercus spp.*)	Foothills, sandy soils, Colorado, Utah, and Southwest	Spring only	All, especially cattle	Many species dangerous as straight diet, emaciation. Buds and new leaves.
25. Oleander (*Nerium oleander*)	Roadsides and woods, Southern U.S.	All, even dry	All	Small quantity. An ornamental shrub.
26. Paperflower (*Psilostrophe spp.*)	Arizona, Southern Utah, West Texas, and New Mexico	All, especially spring	Sheep	Emaciation. Unpalatable, overgrazing, slow poisoning.

348

Table 10.8 (Continued).

Name	Location and habitat	Dangerous season	Grazing animal endangered	Poisoning conditions and general information
27. Peganum *(Peganum harmala)*	West Texas and New Mexico, introduced	All	All	Unpalatable, 3-ft perennial large white to yellow flowers
28. Poison bean *(Daubentonia spp.)*	Gulf coast, Florida to Texas, sandy soils	Mostly winter	All	Seeds poisonous, leaves slightly. Small shrub, legume.
29. Poison hemlock *(Conium maculatum)*	Waste areas, roadsides, moist ground, all U.S.	All, mostly spring	All, especially cattle	Unpalatable, purple spotted Umbelliferae. Introduced. Rank odor.
30. Rayless goldenrod *(Aplopappus heterophyllus)*	Texas, Colorado, New Mexico, Arizona, along springs and waterways	Green and dry	All	Perennial. Causing trembling. Cumulative poisoning.
31. Rubberweed *(Actinea richardsonii)*	Dry mountains, Colorado, Utah, Arizona, and New Mexico	Spring and fall	All, especially sheep	Cumulative. Unpalatable perennial, small yellow-rayed head. Overgrazing.
32. St. Johnswort *(Hypericum perforatum)*	Waste places, Northwestern U.S., invading ranges	All	All, especially white	Photosensitivity, dermatitis. Rhizomes, aggressive competer.
33. Senecio *(Senecio spp.)*	Species worldwide, various habitats	Spring and summer	All	Some species poisonous in large amount, many cumulative.
34. Snakeroot *(Eupatorium rugosum)*	Woodlands, rich soils, Eastern U.S.	All, especially fall	All	Trembling. Several days' feeding. Transmitted to man by milk.
35. Sneezeweed *(Helenium hoopesii)*	Mountains, Montana to Arizona, open areas	Summer	Sheep	Cumulative. Spewing sickness. Large yellow composite.

Table 10.8 (Continued)

Name	Location and habitat	Dangerous season	Grazing animal endangered	Poisoning conditions and general information
36. Sorghum (*Sorghum spp.*)	Cultivated fields, moist areas, Southern U.S., escaped	All, especially when young	All	Hydrocyanic acid, wilted or frozen. Good forage otherwise.
37. Water hemlock (*Cicuta spp.*)	Wet meadows, ditch and stream banks, all U.S.	All, especially spring	All, especially cattle	Violent poison, especially tubers. Tops poisonous only in early spring.
38. Tobacco (*Nicotiana spp.*)	Western and Southern U.S.	All	All, especially cattle	Rapid. Unpalatable. On overgrazed range. Annual. Sticky leaf.

BIBLIOGRAPHY

Allden, W. G., and A. C. Jennings (1969): The summer nutrition of immature sheep: The nitrogen excretion of grazing sheep in relation to supplements of available energy and protein in a Mediterranean environment, *Austral. Jour. Agr. Res.* **20**:125–140.

Allen, Clive (1973): Breeding experiments at Kiboka, Final Report, FAO Range Management Project, Kenya, Rome (mimeo.).

Anonymous (1926): Water requirements for range sheep. *Nat. Woolgrower* **16**(6):33.

Baudelaire, J. P. (1972): Water for livestock in semiarid zones, *World Animal Rev.* **3**:1–9.

Bentley, J. P., and M. W. Talbot (1951): Efficient use of annual plants on cattle ranges in the California foothills, *U.S. Dept. Agr. Circ.* **870.**

Binns, Wayne, James L. Shupe, Richard F. Keeler, and Lynn F. James (1965): Chronologic evaluation of teratogenicity in sheep fed *Veratrum californicum, Jour. Am. Vet. Med. Assoc.* **147**:839–842.

Bowns, James E. (1971): Sheep behavior under unherded conditions on mountain summer ranges, *Jour. Range Mgt.* **24**:105–109.

Campbell, J. B. (1952): Farming range pastures, *Jour. Range Mgt.* **5**:252–258.

Campbell, R. S. (1936): Climatic fluctuations, in U.S. Dept. of Agr., Forest Service, *The Western Range*, U.S. Congress 74th, 2d Session, Senate Doc. 199, pp. 135–150

Cardon, B. P. (1966): Range supplements . . . when? where? *Western Livestock Jour.* **44**(33):18–21.

Clanton, D. C. (1973): KEDLOR Feed Supplement and dehydrated alfalfa in range supplement, *The Practicing Nutritionist* **7**(1):1–4.

———, and W. W. Rowden (1963): Cobalt supplementation on Nebraska ranges, *Jour. Range Mgt.* **16**:16–17.

Cook, C. Wayne, and L. A. Stoddart (1953): The halogeton problem in Utah, *Utah State Agr. Expt. Sta. Bull.* **364.**

Craddock, G. W. and C. L. Forsling (1938): The influence of climate and grazing on spring-fall sheep range in southern Idaho, *U.S. Dept. Agr. Tech. Bull.* **600.**

Crampton, E. W., and L. E. Lloyd (1959): *Fundamentals of Nutrition,* W. H. Freeman and Co., San Francisco.

Culley, Matt. J. (1938): An economic study of cattle business on a southwestern semidesert range, *U.S. Dept. Agr. Circ.* **448.**

Cundiff, Larry V. (1973): Effects of heterosis in Hereford, Angus, and Shorthorn cattle, U.S. Animal Research Center 1973, Beef Cattle Progress Report, pp. 11–22.

Currie, Pat O., and Geraldine Peterson (1966): Using growing-season precipitation to predict crested wheatgrass yields, *Jour. Range Mgt.* **19**:284–288.

Darroch, J. G., A. W. Nordskog, and J. L. Van Horn (1950): The effect of feeding concentrates to range ewes on lamb and wool productivity, *Jour. Aninal Sci.* **9**:431–445.

Edwards, Joseph (1973): Trends in beef cattle production in the United States, *World Animal Rev.*, no. **5**:11–15.

Edye, L. A., J. B. Ritson, K. P. Haydock, and J. Griffiths Davies (1971): Fertility and seasonal changes in live weight of Droughtmaster cows grazing a Townsville stylo-spear grass pasture, *Austral. Jour. Agr. Res.* **22**:963–967.

Furr, Dale (1968): Texas Tech University Kilgore Beef Cattle Research Center, Field Day Report, Amarillo (mimeo.).

Gerlaugh, Paul, L. E. Kimple, and D. C. Rife (1951): Crossbreeding beef cattle; a comparison of Hereford and Aberdeen Angus breeds and their reciprocal crosses, *Ohio Agr. Expt. Sta. Bull.* **703**.

Gregory, K. E., et al. (1965): Heterosis in preweaning traits of beef cattle, *Jour. Animal Sci.* **24**:21–28.

———, et al. (1966a): Heterosis effects on growth rate of beef heifers. *Jour. Animal Sci.* **25**:290–298.

———, et al., (1966b): Heterosis effects on growth rate and feed efficiency of beef steers, *Jour. Animal Sci.* **25**:299–310.

Harris, Lorin E. (1968): Range nutrition in an arid region, Utah State University, 36th Honor Lecture, Logan.

Henry, W. A., and F. B. Morrison (1928): *Feeds and Feeding*, 19th ed., The Henry-Morrison Co., Madison, Wis.

Herbel, C. H., and A. B. Nelson (1966): Activities of Hereford and Santa Gertrudis cattle on a southern New Mexico range, *Jour. Range Mgt.* **19**:173–176.

———, F. N. Ares, and A. B. Nelson (1967): Grazing distribution patterns of Hereford and Santa Gertrudis cattle on a southern New Mexico range, *Jour. Range Mgt.* **20**:296–298.

Hogan, K. G., D. F. L. Money, D. A. White, and R. Walker (1971): Weight responses of young sheep to copper and connective tissue lesions associated with the grazing of pastures of high molybdenum content, *New Zealand Jour. Agr. Res.* **14**:687–701.

Hurtt, Leon C. (1946): Penalties of heavy range use, *Am. Hereford, Jour.*, July.

Hutchings, Selar S. (1946): Drive the water to the sheep, *Nat. Wool Grower*, **36**:10–11.

———, and George Stewart (1953): Increasing forage yields and sheep production on intermountain winter ranges, *U.S. Dept. Agr. Circ.* **925**.

Ingram, D. C. (1930): Ranges are made usable by hauling water for livestock, *U.S. Department of Agriculture Yearbook*, Washington, pp. 446–449.

James, Lynn F., John E. Butcher, and Kent R. Van Kampen (1970): Relationship between *Halogeton glomeratus* consumption and water intake by sheep, *Jour. Range Mgt.* **23**:123–127.

Jardine, J. T. (1915): Grazing sheep on range without water, *Nat. Wool Grower* **5**(9):7–10.

Jenkins, E. L., and G. G. Hirst (1966): Mortality of beef cattle in northwest Queensland, *Quart. Rev. Agr. Econ.* **19**:134–151.

Johnson, Leslie, Leslie R. Albee, R. O. Smith, and Alvin L. Moxon (1951): Cows, calves, and grass; effects of grazing intensities on beef cow and calf production and on mixed prairie vegetation on western South Dakota ranges, *South Dakota Agr. Expt. Sta. Bull.* **412**.

Knapp, Bradford, Jr., A. L. Baker, and R. T. Clark (1949): Crossbred beef cattle for the Northern Great Plains, *U.S. Dept. Agr. Circ.* **810.**

Kothmann, M. M., and G. W. Mathis (1970): Cow-calf production from different grazing systems, stocking rates, and levels of supplement, *Tex. Agr. Expt. Sta. Prog. Rep.* **2776.**

Kubota, Joe, Victor A. Lazar, G. H. Simonsen, and W. W. Hill (1967): The relationship of soils to molybdenum toxicity in grazing animals in Oregon, *Soil Sci. Soc. Am. Proc.* **31:**667–671.

Lantow, J. L., and E. L. Flory (1940): Fluctuating forage production, *Soil Conservation* **6:**137–144.

Laster, D. B. (1973): Studies on induced twinning in beef cattle, U.S. Meat Animal Research Center, *Beef Cattle Progress Report* **11:**55–60.

Launchbaugh, J. L. (1957): The effect of stocking rate on cattle gains and on native shortgrass vegetation in west-central Kansas, *Kan. Agr. Expt. Sta. Bull.* **394.**

Lee, H. J., and H. R. Marston (1969): The requirement for cobalt of sheep grazed on cobalt-deficient pastures, *Austral. Jour. Agr. Res.* **20:**905–918.

Macfarlane, W. V. (1964): Terrestrial animals in dry environments: Ungulates, in *Handbook of Physiology* D. B. Diel, E. F. Adolph, and C. G. Wilbur (eds.), section 4, American Physiological Society, Washington, D. C., pp. 509–539 **4:**509.

———— (1965): *Studies in Physiology,* Springer Verlag, Heidelberg, p. 191.

———— (1972): Prospects for new animal industries: Functions of mammals in the arid zone, *Proceedings of the South Australian Water Research Foundation,* Adelaide, S. Australia.

Marsh, C. Dwight, A. B. Clawson, and Hadleigh Marsh (1916): Larkspur poisoning of livestock, *U.S. Dept. Agr. Bull.* **365.**

McIlvain, E. H., et al. (1953): Seventeen-year summary of range improvement studies at the U.S. Southern Great Plains Field Station, Woodward, Okla., 1937–1953 (mimeo.).

Nelson, E. W. (1934): The influence of precipitation and grazing upon black grama range, *U.S. Dept. Agr. Tech. Bull.* **409.**

Nichol, A. A. (1938): Experimental feeding of deer, *Ariz. Agr. Expt. Sta. Tech. Bull.* **75.**

Norman, M. T. J. (1966): Katherine Research Station 1956–64: a review of published work, Commonwealth Scientific and Industrial Research Organization. Division of Land Research Technical Paper, no. 28.

Ragsdale, A. C., H. J. Thompson, D. M. Worstell, and Samuel Brody (1950): Environmental Physiology: IX. Milk production and feed and water consumption responses of Brahman, Jersey, and Holstein cows to changes in temperature, 50° to 105°F and 50° to 8°F, *Mo. Agr. Expt. Sta. Res. Bull.* **460.**

Raleigh, R. J. (1970): Symposium on pasture methods for maximum production in beef cattle: Manipulation of both livestock and forage management to give optimum production, *Jour. Animal Sci.* **30:**108–114.

Reynolds, E. B., et al. (1953): Methods of supplying phosphorus to range cattle in south Texas, *Tex. Agr. Expt. Sta. Bull.* **773.**

Rogers, LeRoy F., and Walter S. Peacock, III (1968): Adjusting cattle numbers to fluctuating forage production with statistical decision theory, *Jour. Range Mgt.* **21:**255–258.

Rouse, John E. (1970): *World Cattle,* 2 vols., University of Oklahoma Press, Norman.

Saunderson, Mont H. (1940): Some economic aspects of western range-land conservation, *Jour. Land Pub. Utility Economics* **16:**222–226.

Savage, D. A. (1937): Grass survival of native grass species in the central and southern Great Plains, *U.S. Dept. Agr. Tech. Bull.* **549.**

Schuster, Joseph L., and Robert C. Albin (1966): Drylot wintering of range cows—adaption to the ranching operation, *Jour. Range Mgt.* **19:**263–268.

Shupe, James L., Wayne Binns, Lynn F. James, and Richard F. Keeler (1967): Lupine, a cause of crooked calf disease, *Jour. Am. Vet. Med. Assoc.* **151:**198–203.

Siebert, B. D., and W. V. Macfarlane (1969): Body water content and water turnover of tropical *Bos taurus, Bos indicus, Bibos banteng,* and *Bos bubalus bubalis, Austral. Jour. Agr. Res.* **20:**613–622.

Skeete, George M. (1966): Can ranchers adjust to fluctuating forage production?, *Jour. Range Mgt.* **19:**238–262.

Smith, Arthur D. (1954): How much water does a deer drink?, *Utah Fish and Game Bull.* **10**(9):1, 8.

Smith, I. D., and G. Alexander (1966): Prenatal mortality in Shorthorn cattle in northern Queensland, *Proc. Austral. Soc. Animal Prod.* **6:**63–65.

Sneva, Forrest A., and D. N. Hyder (1962): Estimating herbage production on semi-arid ranges in the intermountain region, *Jour. Range Mgt.* **15:**88–93.

Squires, V. R. (1970): Growth of lambs in a semi-arid region as influenced by distance walked to water, *Comm. Sci. Ind. Res. Organ.* **8:**219–225.

——, A. D. Wilson, and G. T. Daws (1972): Comparisons of some walking activities of Australian sheep, *Proc. Austral. Soc. Animal Prod.* **9:**376–380.

Stanley, E. B. (1938): Nutritional studies with cattle on a grassland-type range in Arizona, *Ariz. Agr. Expt. Sta. Tech. Bull.* **79.**

Stoddart, L. A. (1935): Osmotic pressure and water content of prairie plants, *Plant Physiology* **10:**661–680.

Talbot, M. W. (1926): Range watering places in the Southwest, *U.S. Dept. Agr. Dept. Bull.* **1358.**

Thompson, H. J., D. M. Worstell, and Samuel Brody (1949): Environmental physiology with special references to domestic animals: V. Influence of temperature, 50° to 105°F, on water consumption in dairy cattle, *Mo. Agr. Expt. Sta. Bull.* **436.**

Topps, J. H. (1972): Urea or biuret supplements to low protein grazing in Africa, *World Animal Rev.,* no. 3:14–18.

Torrell, D. T., I. D. Hume, and W. C. Weir (1972): Effect of level of protein and energy during flushing on lambing performance of range ewes, *Jour. Animal Sci.* **34:**479–482.

Treus, V., and D. Kravchenko (1968): Methods of rearing and economic utilization of eland in the Askaniya-Nova Zoological Park, in *Comparative Nutrition of Wild Animals,* M. A. Crawford (ed.), Symposia of the Zoological Society of London, no. 21, Academic Press, London, pp. 395–416.

Vinke, L., and C. N. Arnett (1927): Beef cattle in Montana, *Mont. Agr. Expt. Sta. Circ.* **133.**

Wilson, A. D. (1970): Water economy and food intake of sheep when watered intermittently, *Austral. Jour. Agric. Res.* **21:**273–281.

Wiltbank, J. N., et al. (1966): Effects of heterosis on age and weight at puberty in beef heifers, *Jour. Animal Sci.* **25:**744–751.

―――― et al. (1967): Fertility in beef cows to produce straightbred and crossbred calves, *Jour. Animal Sci.* **26:**1005–1010.

Chapter 11

Wildlife and Rangelands

Wild animals, more than any other range resource, are of intimate concern to range management. These animals evolved on the land and developed a dynamic equilibrium with plants and their environment. When domestic livestock, or exotic wildlife, are introduced, the natural balance between producers (plants) and consumers (wildlife) is changed and the structure of the ecosystem is altered. Domestic livestock were introduced and, in many areas, the major wildlife species were removed thus creating entirely new relationships between producers and consumers (See Chapter 6). The range manager's job is to determine how domestic animals and wildlife affect the range resource and how they interact with each other.

Some of these interactions may be complementary to both wildlife and livestock; others will be competitive to the point of being destructive. The extent to which the various interactions are understood relates directly to the value society has perceived for wildlife. Big-game animals have economic value and their role has been extensively studied. Rodents and small mammals are less clearly of beneficial value and their role is less well known. Insects and predators have usually been considered pests and their ecological place in the range ecosystem not fully appreciated.

INTERACTIONS BETWEEN WILDLIFE AND LIVESTOCK

When domestic animals and wildlife use the same range, many interactions can be expected to occur. For the most part, these interactions are competitive, but some mutualistic or even cooperative interactions are possible. The major categories of interactions are (1) direct competition, (2) indirect effects on the environment, (3) environmental effects resulting from man's activities, and (4) interactions among species other than livestock and big game.

Direct Competition

Competition among herbivores may be for any life requirement, such as space, forage, or water. Of the many possible areas of competition, only that of forage has been extensively studied. Observation that white-tailed deer in Texas (Merrill et al., 1957) and pronghorn in Montana (Pyrah, 1971) were more abundant or were more frequently seen in livestock-free areas indicate there may be social antagonism between big game and domestic livestock, although this has not been proved. In these instances, competition for forage rather than spacial factors may be the cause. Klein and Strandgaard (1972) theorized that herd density may have depressed body size among roe deer, but conceded that availability of forage was certainly a factor. Judging from the distribution of wild ungulates in an area in Africa (Talbot and Stewart, 1964), one can question the existence of spacial intolerance among native species, for there they appear to comingle readily. This view also is supported by Bell (1970).

Water requirements of most domestic animals and some wild animals are fairly well known. Competition for water, however, exists only during drought periods when water is limited and is not a major problem under most range conditions.

The problem of forage competition, however, may be a major factor in both livestock and game production. The degree of competition between species or classes of animals depends upon (1) the similarity in diets, (2) the kind and amount of forage present, (3) the relative size and numbers of each, (4) the intensity of grazing, and (5) the degree to which they use the same part of the range.

Large herbivores place great pressure on the world's rangelands.

Table 11.1 Animal-Unit Month (AUM) Equivalents of Big-Game Populations in 12 Western States, 1966

Data from Public Land Law Review Commission (1969)

State	AUMs	State	AUMs
Arizona	487,744	New Mexico	790,801
California	1,641,727	Oregon	2,775,463
Colorado	1,048,705	South Dakota	1,237,819
Idaho	1,414,487	Utah	650,825
Montana	1,556,809	Washington	1,362,068
Nevada	397,224	Wyoming	1,294,196
	Total 14,657,868		

Almost 15 million animal-unit months of forage are consumed by big-game animals in the Western United States (Table 11.1). In Africa, even greater and more varied populations of native herbivores exist, many times surpassing the numbers of domestic livestock.

Although a good deal is known about forage preferences of wild ungulates, much less is known about their forage requirements. Accurate assessments of forage intake can only be made with captive animals,

Table 11.2 Daily Consumption of Some North American Big-Game Species

Species	Kind of feed	Daily consumption, gm/kg	Source
White-tailed deer	Mixed commercial	36.0	French et al. (1955)
	Mixed commercial—winter	23.0	Dahlberg and Guettinger (1956)
	Native and commercial feeds	31.7	Dahlberg and Guettinger (1956)
	Native browse—winter	36.2	Dahlberg and Guettinger (1956)
	Mixed commercial	25.0	Ullrey et al. (1971)
Mule deer	Mixed browse—winter	26.5	Smith (1950b)
	Sagebrush—winter	24.5	Smith (1950a)
	Mixed browse and forbs—summer	34.9	Smith (1953)
	Sagebrush, oak, juniper—winter	20.8	Smith (1959)
	Oak—winter	22.5	Smith (1959)
	Mixed browse—winter	22.5	Smith (1959)
Pronghorn	Mixed browse—winter	16.7	Smith et al. (1965)

and in most instances in which this has been done with wild herbivores, the animals were undergoing digestion trials and were confined to small cages. In such circumstances, forage intakes undoubtedly are depressed by confinement, and data thus collected are poorly representative of actual intakes. Consumption of sagebrush by two male mule deer when they were free in pens, was over half again as much as when they were subsequently confined in digestion cages (Smith, 1950b). A few data are available on North America big-game species where intakes could be accurately determined, and animals, though confined to pens, were free to move about freely (Table 11.2).

Determining Forage Preferences of Wild Ungulates   Forage preferences of wild ungulates may be determined by (1) measuring utilization of range plants, (2) observing grazing animals, (3) examining rumen contents, (4) examining fecal material, and (5) feeding captive animals. Each of these has certain advantages and disadvantages.

Determining utilization is a difficult task (See Chapter 6) even with livestock which commonly quite fully use the major species. Utilization by big game, except where they congregate in herds or are concentrated on winter ranges, is often much lighter and thus more difficult to detect. Moreover, size of individual bites and the number of bites differ making detection commonly more difficult with game animals. Further, once the forage has been removed, one can only infer how the plant looked prior to grazing.

Field observations of the time spent grazing upon specific plants are liable to error, first, because of the difficulty of determining which plant of two or more in close proximity is being eaten. This difficulty can be minimized by using tamed animals (McMahan, 1964), but an animal raised in captivity may have developed different taste responses than one that has grown in the wild. Second, the amount of forage removed from a plant in a given time differs with different plants, and presumably, with the condition of a plant. Quite different preference ratings resulted using time spent as compared to the amounts of forage actually eaten by mule deer as the basis for rating (Smith and Hubbard, 1954). Consumption per minute differed as much as 2.9 times among the species fed; the differences would probably have been even greater if browse and herbaceous species had been fed.

Rumen analyses, also, are subject to certain limitations. Usually, it is impractical to secure enough rumen samples adequately to sample the animal population. Further, some plants are broken down more rapidly than others when taken into the digestive tract. Succulent green material passes through the digestive tract more rapidly than dry or woody material. Deer rumens taken when both green and dry forage are being eaten may show a disproportionately high amount of the dry feed. Rumen analyses yield useful qualitative data, but considerable variations can be expected due to method of analyses, i.e., whether volumetric, gravimetric, or point-frame (Table 11.3).

Determination of diets of wild herbivores by identification of fecal fragments shows promise, especially for those animals difficult to study and observe closely. The method depends upon the fact that epidermal fragments of plants are altered little in the digestive tract so that cell structure, stomata, and epidermal appendages (trichomes) are identifiable. The method presents certain difficulties. (1) Laboratory facilities are required. (2) Processing and preparation of material may greatly affect identification. (3) Closely related species may not be distinguishable. (4) There is no certainty that the same volume-intake to fecal-fragment ratio exists among different species. Griffiths and Barker (1966) found that analyses of ground rumen material from kangaroo and sheep treated in the same manner as fecal matter underestimated or overestimated individual species by as much as 7 percent:

	Percent identified	Percent in sample
Amphigon caricinus	18	25
Triodia mitchelli	21	25
Goodenia glabra	24	18

Table 11.3 Percent Composition of Forage Classes in Rumens of White-Tailed Deer by Volume, Weight, and Point-Frame Techniques

Data from Coblentz (1970)

Number of rumens	Woody browse			Grass and grasslike			Forbs		
	Vol	Wt	Pt-fr	Vol	Wt	Pt-fr	Vol	Wt	Pt-fr
10	26	43	39	52	31	50	22	26	11
6	70	83	89	17	7	9	13	10	2
5	100	100	100	0	0	0	0	0	0
13	87	94	86	6	3	12	7	3	2
5	100	100	100	0	0	0	0	0	0
8	100	100	100	0	0	0	0	0	0
2	100	100	100	0	0	0	0	0	0
16	58	63	66	18	13	20	24	24	14
10	14	14	58	46	42	16	40	44	26
12	88	96	92	8	3	8	4	1	0
Mean	68.3	74.1	77.9	17.7	12.2	14.8	14.0	13.7	7.3

Until volume-fragment ratios are established for each species, the method can provide only qualitative data, and even this may not be possible. Slater and Jones (1971) found even greater disparities than those shown above; no white clover was found in the feces of sheep, though it made up 37 percent of the diet (see also Free et al., 1970).

Feeding captive animals has proved fairly successful, but all species do not adapt to confinement. Results obtained in this way may be of limited application, since they apply only to areas possessing vegetation in similar proportions to the forage offered to the animal, and different forage species may not remain equally acceptable after harvesting. When related to weight changes of the animals, controlled feeding offers an accurate means of determining forage requirements of game animals.

Forage Preferences of Big Game   Herbivores show distinct preferences for certain forage species, and this extends to classes of forage (grass, browse, or forbs) as well. For general comparisons, preference for forage classes is used widely to characterize different herbivores as grazers (those preferring grass or, often, any herbaceous vegetation) and browsers (those preferring shrubby plants). Although these forage-class preferences exist, different mixtures of plants from area to area and different intensities of use among areas make generalizations difficult. Preferences go to individual species and grazers, such as horses, readily accept a forb such as alfalfa. As utilization becomes heavier, there is less chance for animals to express preference and dissimilarities in animal diets decrease.

Difference in classifications of plants by different investigators obscure preference differences. Thus, African workers recognize browsers and grazers, not clearly indicating where forbs (broad-leaved herbs) fall into this dichotomy. In America, distinctions are made between three forage classes—grass, browse, and forbs. The half shrub *Artemisia frigida* is regarded as a forb in some areas and a shrub in others. Since, where it occurs, it makes up a high part of the diet of pronghorn (*Antilocapra americana*), how it is classified makes a great difference when characterizing the forage preference of pronghorn. Nevertheless, it is possible to characterize broadly the class preferences of some major game species (Table 11.4).

Clearly defined species preferences are evident. Elk *(Cervus canadensis)* prefer grass and sedges over other classes of forage even in winter. They do, however, utilize a great many forbs in summer. Similarly, bighorn depend most heavily on grasses or grasslike plants but use forbs in spring and summer.

Both mule deer *(Odocoileus hemionus)* and white-tailed deer *(O. virginianus)* make little use of grass generally, but young green foliage of

Table 11.4 Percentage of Forage Classes in the Diets of North American Big-Game Species (selected for yearlong coverage)

Species and area	Season	Percent of diet					Source	Other references
		Grass & sedge	Forb	Browse	Other			
Bighorn, Idaho	Spring	77*		22	1		Smith (1954)	Constan (1972)
	Summer	86*		14	–			
	Fall	66*		25	9			
	Winter	56*		39	5			
	Yearlong	70*		27	3			
Elk, California	Spring	61	4	34	1		Harper et al. (1967)	DeNio (1938)
	Summer	58	20	22	–			Knight (1970)
	Fall	57	22	21	–			Mackie (1970)
	Winter	76	1	22	1			Constan (1972)
Moose, Montana	Summer	1	71	29			Knowlton (1960)	
	Fall	1	7	91	–			
	Winter	1	1	99	–			
Mule deer, New Mexico	Spring	2	30	58	–		Boeker et al. (1972)	Smith (1953)
	Summer	2	42	56	–			Anderson et al. (1965)
	Fall	6	8	86	–			Klebenow (1965)
	Winter	2	4	94	–			McCulloch (1969)
	Yearlong	2	16	75	7			Mackie (1970)
Pronghorn, Wyoming	Spring	15	20	42	–		Taylor (1972)	Buechner (1950a)
	Summer	2	32	55	–			Ferrel and Leach (1952)
	Fall	4	15	75	–			Dirschl (1963)
	Winter	7	0	92	–			Smith et al. (1965)
	Yearlong	3	5	92	–		Severson et al. (1968)	Beale and Smith (1970)
								Mitchell and Smoliak (1971)
White-tailed deer, Montana	Spring	38	18	43	1		Allen (1968)	Buechner (1950a)
	Summer	1	54	45	–			McMahan (1964)
	Fall	2	17	81	–			Chamrad and Box (1968)
	Winter	6	29	65	–			Coblentz (1970)

*Forbs included with grasses.

certain species is taken in spring and when other kinds of forage are in
short supply. Browse is the most important food item yearlong, although
during the growing season forbs may form more than 50 percent of the
diet of mule deer (Fig. 11.1).

Pronghorn, despite being a plains animal, is not a grass eater.
Browse is the most important source of forage except in the height of
the growing season when forbs are abundant (Beale and Smith, 1970) or
when few shrubs are available. Pronghorn show special preference for
the genus *Artemisia,* eating whatever species are present. In the intermountain area, black sagebrush *(A. nova)* constituted from 7 to 40 percent of pronghorn diet even in summer. On the Red Desert of Wyoming, Taylor (1972) found two species of *Artemisia* contributing 21 percent
of antelope forage in June and 74 percent in April. In winter, sagebrush
is preferred over other browse species (Smith et al., 1965).

Similar species may differ considerably in their forage preferences;
e.g., red kangaroos have almost identical preferences to sheep, while
grey kangaroos utilize more grass (Table 11.5).

Factors Affecting Competition for Forage   Obviously, forage preference is
the most important factor in competition. Commonly, it is assumed that
if one kind of animal prefers browse and another grass there is little

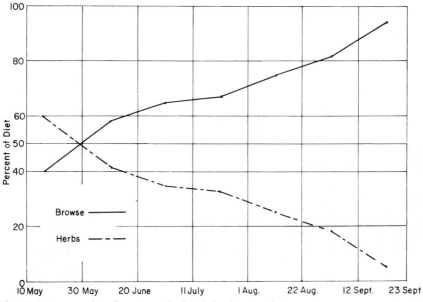

Fig. 11.1 Summer diet of captive mule deer fed cafeteria style in percent
showing relative importance of browse and herbaceous species throughout the summer months. *(From Smith, 1953.)*

Table 11.5   Forage Preferences of Sheep and Kangaroos
Grazing Together in Paddocks

Data from Griffiths and Barker (1966)

	Diet, percent			
	Jan.		**Sept.–Nov.**	
	Grass	Dicots	Grass	Dicots
Red kangaroos and sheep	46	54	68	32
Grey kangaroos	64	35	79	21

competition involved (Fig. 11.2). This ignores the fact that forage selection is made from individual plant species. Some species irrespective of
their form class may be readily eaten by several kinds of animal, while
others are eaten by only one animal. For example, in northern Utah
mule deer prefer geranium *(Geranium fremontii)* above other species;
sheep and cattle eat it very lightly. *Senecio serra* is eaten avidly by sheep,
only lightly by cattle, and not at all by mule deer. Bitterbrush *(Purshia
tridentata)* is highly palatable to all three, and it may be overused despite
the fact that most of the forage of each of these animals is supplied by
other and different plants. If the amount of forage provided by such a

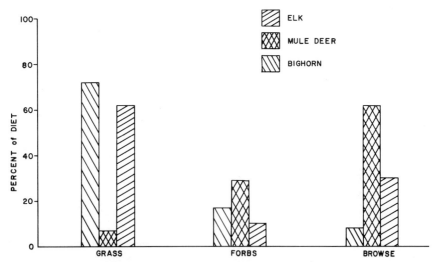

Fig. 11.2   Winter diets of bighorn, elk, and mule deer, by forage class in
percent, Montana *(Data from Constan, 1972.)*

key species is inadequate to supply the demands of whatever kinds of animals eat it, competition exists.

This is well illustrated in the intermountain valleys of the Western United States which are grazed in winter months by livestock and by pronghorn yearlong. Sheep consume a great deal of grass but they also use black sagebrush heavily. Pronghorn in winter are largely dependent on sagebrush and make almost no use of grass. Thus, despite a great dissimilarity between the forage preferences of sheep and pronghorn, when characterized by forage class, intense competition can arise for this one species. Competition is not subject to generalization; it must be assessed under each set of circumstances. The time of year, vegetative mix, the species of animals, and the intensity of use of the range all affect the degree of competition.

Competition may be lessened if the range areas used by wild and domestic animals differ. This is more likely in areas of rough topography and steep slopes where wild animals accept, if not actually prefer, the more inaccessible areas (Fig. 11.3). In one instance in Utah, deer were found to frequent 92 percent of a range area, while cattle used only 52 percent. Similarly, deer use appeared to be about 25 times as heavy in oak types as compared to sagebrush types. Conversely, cattle use was 7 times as heavy in sagebrush types as in oak (Julander and Robinette, 1950). Similar instances of spatial separation of deer and sheep have been noted (Smith and Julander, 1953). Under Western mountain-

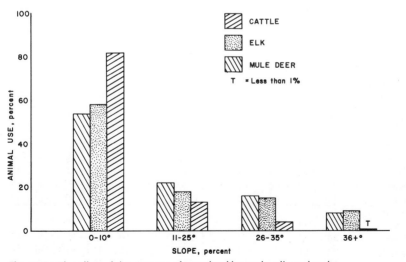

Fig. 11.3 The effect of slope on use of rangeland by cattle, elk, and mule deer. Cattle are much more responsive to topography than either elk or deer. *(Adapted from Mackie, 1970.)*

range conditions, it has been estimated that deer obtain only about 50 percent of their forage from areas that are regularly used by cattle and about 75 percent from areas regularly used by sheep (Stoddart and Rasmussen, 1945).

Condition of the range is an important factor in competition also. The more diversity in plant species, the greater opportunity there is for animals to express their preferences and the less need for accepting other less desired forage. If range deterioration results in a limited number of plant species, as it often does, competition is increased.

Intensity of grazing affects competition. In Texas tame animals were introduced into pastures which had different prior-grazing histories, from heavy to none. Some marked differences in forage selection were found among deer and domestic livestock presumably due to the reduction in choice resulting from prior-grazing use (Table 11.6). In winter and spring the amount of grass consumed by cows in the unused pasture was much less than in the grazed pastures.

The zone of foraging may operate to minimize or to intensify competition. Because of differences in size or build, different animals are able to harvest forage from widely different vertical zones (Table 11.7). The giraffe, for example, feeds from the tops of trees and competes

Table 11.6 Forage Classes Selected by White-tailed Deer and Domestic Livestock by Season in Pastures with Two Prior Grazing Histories

Data from McMahan (1964)

Animals and grazing intensity	Spring B*	Spring F†	Spring G‡	Summer B	Summer F	Summer G	Fall B	Fall F	Fall G	Winter B	Winter F	Winter G
	(Percent)											
Heavy use												
Deer	41	44	15	91	4	5	85	2	13	89	5	6
Cow	0	2	98	0	0	100	21	0	79	10	0	90
Sheep	8	27	66	13	4	83	18	1	81	15	1	84
Goat	50	7	43	82	2	16	59	1	40	70	0	30
No prior use												
Deer	37	56	7	32	68	1	57	37	6	89	8	3
Cow	30	22	49	5	10	85	17	10	73	73	15	12
Sheep	27	61	12	14	65	22	28	31	41	51	9	40
Goat	59	36	5	55	38	7	50	41	10	95	2	3

B* = browse; F† = forbs; G‡ = grass.

Table 11.7   Estimated Upper Limits of the Foraging Zone
for Some Wild and Domestic Herbivores

Species	Height, m	Species	Height, m
Giraffe	5.49	Mule deer	1.63
Elephant	5.49	White-tailed deer	1.52
Camel	2.44	Sheep	1.07
Elk	1.83	Bison	1.00
Cattle	1.52		

little with other animals; only the elephant is able to match it in reach
(Fig. 11.4). Few animals have such a clear-cut advantage, but smaller
animals are disadvantaged by larger ones. Thus, elk have driven deer
from their accustomed winter ranges when the numbers of elk become
large. Small animals may also be benefited by larger ones, as in the case
of Thomson's gazelle which occupies areas where vegetation is kept
short by larger herbivores (Bell, 1970).

Quantifying Competition among Herbivores

It is possible to arrive at the degree of competition only when different
animals use the same range either concurrently or sequentially, for unless the same forage choices are available to the animals being compared
there is no way in which competition can be assessed. When foraging
habits are determined under similar conditions and related to species
used, a rough guide to competition can be developed by determining the
percentage of the diet that each animal enjoys alone, compared to the

Fig. 11.4   Zebras and giraffes in South Africa, an example of two species
which feed in different vertical zones and from different kinds of forage.
Zebras eat the ground cover of grass; giraffes the browse from the tops of
small trees. *(Courtesy F.A.O.)*

percentage that comes from species which are used in common. Despite the fact that mule deer largely depend on browse and forbs and cattle eat more grass, in northern Utah, three-fourths of the diet of deer came from species which also were used by sheep and cattle (Fig. 11.5). Both sheep and cattle obtained greater percentages of their diets from plants unused by deer.

Grazing Capacities under Common Use

Common use (see Chapter 8) is generally thought of as comingling of animals on a range, but they need not be present at the same time, as when livestock graze a range in spring and fall and big game use the same area over winter.

It usually is assumed that common use results in maximum production from the range. This is often true, though in some instances increased production may be achieved only at the expense of some forage species which are preferred by more than one animal. The forage species that are used in common become the *key species* upon which management is based. If this key species is utilized fully by one animal under proper stocking, the addition of another animal on the range can be made only if numbers of the first animal are reduced or the key

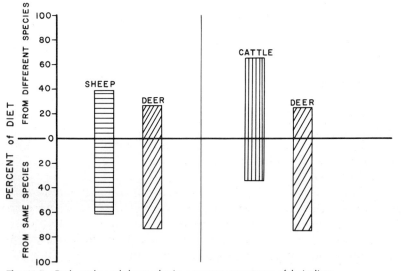

Fig. 11.5 Both cattle and sheep obtain a greater percentage of their diets from plants unused by mule deer (upright bars) than vice versa; the greater part of a deer's diet comes from plants also used by sheep and cattle (pendulous bars). *(Data from Smith, 1961.)*

species is sacrificed, a possibly unacceptable alternative. If no plants are used in common, no competition exists and the capacity of the range is equal to the sum of capacities for each separately. This situation occurs among certain African big-game animals, some being strictly grazers and others browsers (Denney, 1972). Between these extremes other competitive relations exist, the precise nature of each depending upon the relative numbers and kinds of animal using the range.

Calculating Common-Use Grazing Capacities Carrying capacity of a range under common use can be determined using the key-species concept (Smith, 1965). It rests on the assumption that the key species makes up approximately the same part of the animal's diet throughout the grazing season. The method involves (1) identification of the key species which both animals use, (2) finding the maximum level of use that the key species can tolerate, (3) determining the proportion of the diet that the key species makes up for each animal, (4) determining the grazing capacity for each animal alone, and (5) determining the attained utilization of other species when the key species is used fully by either animal.

Combined grazing capacities are determined by means of budget calculations in which the utilization of the key species is successively reduced by increments for one animal and increased for the other (accomplished in practice by changing animal numbers) so that the combined utilization of the key species is kept at its highest permissible use. For example, if 70 percent is proper use of the key species, one animal is allotted 70, 65, 60, etc., and the other is allotted 0, 5, 10, etc., percent of the key species. Utilization of each plant is likewise proportionately decreased for one animal and increased for the other. If the use factor of another species is 55 percent, incremental changes would be 3.9 percent for each 5 percent change in the key species. Multiplying these arbitrarily assigned utilization figures by the amount of each plant species present and summing the products for each animal separately, results in a grazing-capacity index for any combination of animal numbers (Table 11.8). When these indexes are plotted, a number of distinct common-use situations can be identified depending upon the character of the vegetation, the dissimilarity in forage preference of the animals, and the key-species relationships that exist. Two situations identified in northern Utah, one involving mule deer and sheep and the other involving mule deer and cattle are shown in Fig. 11.6. These were developed from different areas and cannot be related to each other.

On the range shared by deer and cattle, only one key species was involved. The carrying capacity for deer was greater than it was for cattle, and greatest productivity in the short run would result from deer

Table 11.8 Calculated Grazing Capacities of a Range for Selected Combinations of Deer and Sheep with Two Key Species

Data from Smith (1965)

Stock	Amelanchier alnifolia Util.*	Amelanchier alnifolia F.F.†	Purshia tridentata Util.	Purshia tridentata F.F.	Prunus melanocarpa Util.	Prunus melanocarpa F.F.	Rosa sp. Util.	Rosa sp. F.F.	Aster chilensis Util.	Aster chilensis F.F.	Total combined forage factor
(Amelanchier key species)											
Deer	60	0.0240	42	0.0210	40	0.0160	58	0.0058	14	0.0042	0.0710
Deer	55	0.0220	39	0.0195	36	0.0144	53	0.0053	13	0.0039	0.0775
Sheep	5	0.0020	4	0.0019	8	0.0031	12	0.0012	14	0.0042	
Deer	50	0.0200	35	0.0175	33	0.0132	48	0.0048	12	0.0036	0.0838
Sheep	10	0.0040	8	0.0039	15	0.0062	24	0.0024	27	0.0082	
Deer	45	0.0180	32	0.0158	30	0.0119	43	0.0043	11	0.0032	0.0901
Sheep	15	0.0060	12	0.0058	23	0.0092	37	0.0036	41	0.0123	
Deer	40	0.0160	28	0.0140	26	0.0104	38	0.0038	10	0.0029	0.1021
Sheep	20	0.0080	16	0.0080	31	0.0124	49	0.0049	54	0.0162	
(Aster key species)											
Sheep	22	0.0088	17	0.0085	34	0.0136	54	0.0054	60	0.0180	0.0543
Sheep	20	0.0080	16	0.0080	31	0.0124	49	0.0049	55	0.0165	0.0748
Deer	21	0.0084	15	0.0075	14	0.0056	20	0.0020	5	0.0015	
Sheep	18	0.0072	14	0.0070	29	0.0116	45	0.0045	50	0.0150	0.0931
Deer	40	0.0160	28	0.0140	27	0.0108	40	0.0040	10	0.0030	

Util.* = utilization; F.F.† = forage-factor product of utilization and amount of species present, which is not shown.

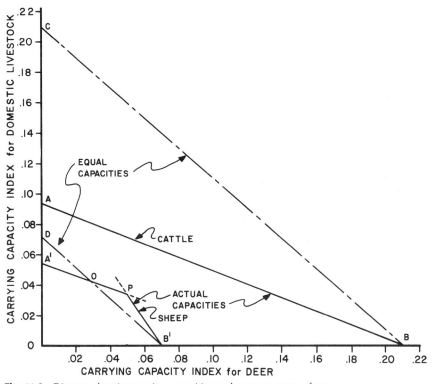

Fig. 11.6 Diagram showing grazing capacities under common use of two ranges, one used by deer and sheep and one by cattle and deer. In the former case all combinations of deer and sheep represented by the lines *OB'P* give greater capacities under common use, maximum production being attained at point *P*. Reducing deer numbers below point *O* and increasing the number of sheep results in lower grazing capacities than if deer only were present. With cattle, no mix of animal numbers increases grazing capacity over deer alone in the case pictured here. *(From Smith, 1965.)*

use alone. Adding cattle and reducing deer numbers result in lowered grazing capacities.

The situation involving sheep was more complicated. Here there were two key species, one for deer *(Amelanchier)* and one for sheep *(Aster)*, but each was eaten by both animals. When sheep were most abundant, aster was the plant most heavily used; when deer were in greater numbers, serviceberry *(Amelanchier)* was the plant in short supply. Maximum grazing capacity is attained at point *P* where both key species are fully used. Common use in this instance would increase grazing capacity at any combination of sheep and deer that falls above the line of equal capacities. This potential increase in grazing capacities is represented by the triangle *OPB'*.

Other situations than these two are possible (Smith, 1965), most of which show maximum grazing capacities at a specific animal mix. Any other combination than this results in lowered grazing capacities, some combinations yielding less grazing capacity than a single animal. Common use does not necessarily assure increased grazing capacity; and maximum production is attained at one particular mix of animal numbers.

Relative Advantage between Big Game and Livestock

Where wild ungulates and domestic livestock compete for forage, the latter have certain advantages: (1) in times of feed shortage in winter, or during drought, range feeding of livestock can be practiced, a course that meets with little success with wild animals; (2) domestic livestock are less selective in their foraging habits than are wild animals and accept alternative forages more readily; (3) many big-game species become habituated to certain range areas and do not readily move to other areas in response to shortages of forage. Consequently, competition exerted on wild animals by livestock is generally more severe than is the reverse. This is illustrated in data showing the overlap of dietary preferences of mule deer and cattle (Fig. 11.5). A very high percentage of the mule deer's diet is from species also used by cattle, while only a small part of cattle's diet comes from mutually shared species.

Data by McMahan and Ramsey (1965) dramatically demonstrated the advantage of livestock over white-tailed deer in Texas where deer confined to pastures with livestock suffered high death losses among adult deer and poor fawn survival. In 8 years, no fawns survived to yearling age under heavy livestock use. Livestock, under the same conditions, survived and reproduced.

EFFECTS OF SMALL MAMMALS ON RANGELANDS

Rangeland is almost all occupied by small mammals. Their impact on vegetation is probably greater than is commonly believed, especially in the case of rodents. Many of these small animals consume much the same vegetation as domestic animals, and they remove amazing amounts of forage when large populations occur. A classical example of damage is that of the European rabbit *(Orctolagus)* which was introduced into Australia and New Zealand. These countries have had no range problem more serious, and only after costly control campaigns have rabbits been brought under a measure of control (Gibb, 1967; Chapline, 1971). The problems were aggravated by the fact that rabbits had been introduced without normal biological controls; numbers, therefore, became astronomical. Australia's rabbit problem was made more serious because of the immensity and sparse population of the country. Thousands of miles of fences were built to halt their spread. A myxomytosis-causing virus also was employed (Herman, 1953). Though much diminished, the rabbit problem remains a serious one (Ingersoll, 1964).

Native animal populations are more stable, although they too build to peak numbers cyclically. Throughout western America, the jackrabbit (*Lepus* spp.) and the ground squirrel (*Spermophilus* spp.) are common residents of the range. In the Southwest the numbers of kangaroo rats (*Dipodomys* spp.) sometimes become large, and prairie dogs (*Cynomys* spp.) reach considerable numbers locally. Small burrowing rodents such as the pocket gophers and voles (*Microtus*) are distributed widely on rangelands throughout western North America. That these may sometimes exert significant impacts on the range is shown by studies in the Southwest where consumption of forage by rodents amounted to 28.7 percent of all vegetation and 38.8 percent of the most valuable grasses (Vorhies and Taylor, 1933).

It is likely that the effects of animals which occupy the same niches in the ecosystems of other regions are comparable. For example, Formozov (1966) lists more than 50 species of small mammals found in the Eurasian steppes, most of which have counterparts in the Great Plains, many being of the same genera as those in America. It is doubtful that small mammals other than jackrabbits, are as important on Western rangelands as they once were. As the original sod was broken, subsequently poorly farmed, and often abandoned, smaller rodents, ground squirrels especially, greatly increased. When abandoned lands were revegetated, farming methods improved, and poisoning campaigns waged, rodent numbers declined. Locally they may still have considerable effects, but they are nowhere as widespread and damaging as in the past.

Forage Consumption by Small Mammals

Small mammals have received most attention because of their foraging habits, although it is by no means certain that they are most important from this standpoint. When their environmental role is better understood, other considerations may be of more importance than the amounts of forage consumed.

Jackrabbits   Because of their size, their wide distribution, and their cyclical population patterns, jackrabbits have substantial impacts on range forage. Food eaten by black-tailed jackrabbits in Colorado in spring and summer was made up of herbaceous species, grasses being most important. In fall and winter, grasses and forbs were about equally represented in the diet, with shrubs contributing about one-fifth (Sparks, 1968). Many of the grasses preferred were not the same as those most utilized by cattle which lessened competition. By contrast, in Nevada browse species were more important than were herbaceous plants; grasses and forbs were important in spring and early summer while some shrubs, especially *Larrea,* were eaten most months of the year (Hayden, 1966). Elsewhere, species of *Chrysothamnus* and *Artemisia* are especially important to jackrabbits in winter.

The white-tailed jackrabbit, which is most common on the plains, showed a preference for forbs in summer (30 percent), grasses in the fall (43 percent), and shrubs in winter (76 percent), mostly *Chrysothamnus parryi* (Bear and Hansen, 1966). (See also Hansen and Flinders, 1969).

Black-tailed jackrabbits frequently attain high populations in the Western United States. They have become so numerous that control campaigns, in which they are driven into temporary corrals and killed, were once common when they became concentrated in winter. Several thousand have been removed in this manner in a single day. Stoddart (1972) reported populations of 0.12 to 1.0 per hectare in northwestern Utah, and Goodwin (1960) reported populations of more than 9 rabbits per hectare when they were concentrated in winter. In southern New Mexico, populations of 1.1 black-tailed jackrabbits and 1.2 cottontails per hectare were recorded (Norris, 1950). Based on their rate of consumption and forage preference, 62 black-tailed and 48 antelope jackrabbits in the Southwest were believed to be the equivalent of a cow (Arnold, 1942). Currie and Goodwin (1966) considered the forage consumed by 5.8 black-tailed jackrabbits to equal that of a sheep on winter range in northwest Utah, winter being the time when forage preferences are likely to be most similar.

Prairie Dogs Prairie dogs *(Cynomys* spp.) assemble in areas called "towns," where populations become very high; when this happens, the removal of vegetation in its entirety from the vicinity is common. Moreover, through selective feeding they alter vegetation, favoring the grasses that reproduce vegetatively over those depending upon seed (Koford, 1958).

Prairie dog stomachs, mostly from Montana, Wyoming, Colorado, and Arizona, showed that the black-tailed prairie dog *(Cynomys ludovicianus)* and the Gunnison prairie dog *(Cynomys gunnisoni)* consumed largely grasses, though the white-tailed prairie dog *(Cynomys leucurus)* resorted mostly to the goosefoot family (Table 11.9). Valuable forage plants constituted over 78 percent of the food of these three species, with low-value plants adding only 19 percent. Extensive poisoning campaigns, which are notably successful against prairie dogs, have all but eliminated them over much of the Western United States. In fact, some are included among rare and endangered species.

Kangaroo Rats Throughout the Southwestern United States, the kangaroo rat *(Dipodomys* spp.) occurs in great numbers. These animals feed largely upon seeds which they collect in cheek pouches and store in burrows (Table 11.10). When kangaroo rats congregate, plants in the vicinity of a community are denuded of seeds. Examination of 24 burrows showed an average of 1.68 kg of material stored in each. With a population estimated at 49 rats per hectare, this would amount to 8.3 kg per hectare (Vorhies and Taylor, 1922).

Ground Squirrels Ground squirrels *(Citellus* spp.) are distributed widely, one species or another occurring throughout the Western United States. At times their numbers are considerable.

Shaw (1920) estimated Columbian ground squirrels' *(Citellus columbianus)* daily consumption was 17.2 percent of their weight. He calculated that 385 Columbian ground squirrels were equivalent to 1 cow, and 96 were equivalent to 1 sheep. Grinnell and Dixon (1918) estimated that the forage eaten on a densely populated section of land by the Oregon ground squirrel was sufficient to feed 90 steers throughout the growing season, 750 squirrels consuming as much as 1 steer.

Consumption may not fully express the loss in forage due to ground squirrels for much is cut and left on the ground. Ground squirrels confined in pens at densities of 4.86 per hectare were observed to reduce the herbage crop by 35 percent, 10 times more than their forage consumption would indicate (Fitch and Bentley, 1949). Their impacts have

Table 11.9 Composition of the Diet of Various Species of Prairie Dog

Data from Kelso (1939)

Animal species	Grass family, percent	Goosefoot family, percent	Miscellaneous plant groups, percent	Animal material, percent
Black-tailed prairie dog	61.55	12.73	24.32	1.40
White-tailed prairie dog	28.09	50.63	20.42	0.86
Gunnison prairie dog	47.26	13.80	33.61	5.33

Table 11.10   Major Plant Types Recovered in
Merriam Kangaroo Rat Pouches

Data from Reynolds (1950)

Type of plant	Percentage of occurrence in pouches	Percentage of total vegetation
Grasses:		
Perennial	33	5
Annual	69	35
Forbs:		
Perennial	34	15
Annual	34	39
Woody shrubs	13	5
Cacti	13	0.5

measurable effects on animal production. Howard et al. (1959) compared winter weight gains of heifers on California ranges and found
cattle gains where ground squirrels had been controlled were increased
by 0.47 kg per day. Based on the number of squirrels destroyed, this
increase was equivalent to 2 kg of heifer weight for each ground squirrel
killed.

Dissemination of Seeds by Native Fauna

Native fauna undoubtedly are a factor in dissemination of range plants,
particularly those with hard seed coats. In the Eastern United States, the
location of once-existing fence lines can be traced in abandoned fields by
the rows of *Juniperus virginianus,* presumably due to seeds from bird
droppings. Viable seeds of herbaceous plants have been found in the
feces of both deer and sheep, though they were more numerous in
sheep feces (Heady, 1954). Although germination percentages of seed
recovered from sheep and rabbits are low (Lehrer and Tisdale, 1956),
the fact that any seeds are passed at all makes animals a factor in the
spread of plants (Fig. 11.7).

Seed-eating rodents with cheek pouches undoubtedly play a great
part in dissemination of seed. In Nevada, seeds of annuals were found
four times more frequently in the pouches of kangaroo rats than were
native and introduced perennials. Introduced wheatgrasses occurred
more frequently in pocket mouse pouches (La Tourrette et al., 1971).
Undoubtedly, some of the seed is stored in burrows and later germinates. Kangaroo rats have been blamed for the increase of mesquite and
cactus by spreading their seed (Reynolds, 1950). Seed-collecting rodents
are an important factor in establishment of bitterbrush (West, 1968).

Fig. 11.7 Coyote scat full of mesquite beans near Florence, Arizona. Seed-eating animals, domestic and wild, are important in the spread of plants on rangeland. *(Courtesy C. R. Madsen.)*

Destruction of Range Plants by Rodents

Periodically, populations of nonhibernating rodents cause great damage to established perennial plants. Shrubs are girdled and the roots and stem bases of perennial herbs are consumed. When heavy stands of sagebrush are thinned and grasses increase, greater livestock-grazing capacity results. There may be a loss of grazing capacity for deer, however, on winter deer range. When desirable species are attacked, the effects are detrimental. In 1964, canopy reductions of 30 and 50 percent were observed on sagebrush and bitterbrush, respectively in northern Utah (Smith and Doell, 1968). Equally great impacts have been observed on perennial grasses (Foster, 1965).

Pocket gophers *(Thomomys* spp.) are common on open ranges at high altitudes. Their food is largely made up of the fleshy roots of perennial forbs, but they also utilize herbage (Aldous, 1951). To a considerable extent, the major items of food are plants which are not the more desirable forage, being low in palatability to livestock, or plants which increase under heavy grazing. Among these are *Achillea lanulosa, Penstemon, Taraxacum officinale* and species of lupine (Hansen and Ward,

1966). On the short-grass plains prickly pear made up half their diet (Turner et al., 1973).

The effects of gophers on range is not firmly established by existing data. Howard and Childs (1959) considered them more serious competitors for forage than were ground squirrels in California, but could find no evidence that they influenced the composition of annual forage types. One study indicated the possibility that gophers hold depleted mountain meadows in poor condition (Moore and Reid, 1951). Both beneficial and detrimental effects were attributed to gophers in Montana, since both poor forage plants and good forage plants were reduced (Branson and Payne, 1958). Their preference for certain planted grasses over others, 80 and 90 percent of *Arrhenatherum elatius* plants being destroyed, makes species selection important in seedings (Garrison and Moore, 1956).

Ellison (1946) found no evidence that they contributed to erosion or reduced forage production, and thought it possible their preference for forbs over grass would be beneficial.

Meadow voles *(Microtus* spp.) periodically become abundant and during winter they forage by removing bark from range shrubs such as sagebrush, bitterbrush, and snowberry *(Symphoricarpos* spp.). Usually only part of the crowns of individual plants are killed but occasionally complete kills result. Mueggler (1967) recorded crown kills of sagebrush of 35 to 97 percent on extensive areas of sage-grass range during an irruption of voles in Montana. Young plants were affected less than old.

EFFECTS OF INVERTEBRATE ANIMALS ON RANGE FORAGE

Frequently, invertebrates become so numerous that serious grazing-capacity loss is entailed. Losses from this source are often more important than those from larger animals. Insects commonly attack range plants in the greatest numbers when drought conditions prevail, and hence their effects are intensified.

Damage to range forage from insects comes from (1) consumption of forage, (2) weakening or killing of plants, (3) making forage unattractive to animals, and (4) destroying seed. These effects may sometimes be combined.

Forage Consumption by Saltatorial Insects

Grasshoppers (called locusts in Africa) and crickets are probably the most important and widespread of forage-consuming invertebrates. There are biblical references to locust plagues (migratory grasshoppers), and references to them are found in early writings of Africa, Asia, and Australia. High populations develop on denuded rangelands of remote desert areas following rains which provide moisture for hatching of the eggs. From there they spread to humid areas appearing as a dark cloud and consuming all vegetation in their path. Locust-scouting programs to combat this menace are widespread throughout the world.

In America, grasshoppers are regarded by many as second only to drought in destructiveness to range forage. Many instances are known in which these insects have denuded land of virtually all forage. Because of the grasshopper's small size, the extent of their damage is not always appreciated. Forage consumption by grasshoppers may remove as much as 89 percent of all vegetation and 43 percent of the forage (Morton, 1936). Not all grasshoppers are equally destructive, some species preferring plants of low-forage value (Ball, 1936), while others consume the better-forage plants (Anderson and Wright, 1952). Estimates of the value of range forage lost in Montana during the years 1934 to 1936 amounted to $1,750,000 (Strand, 1937).

Mormon crickets *(Anabrus simplex)* are less destructive in the Western United States than are grasshoppers. Their occurrence is less frequent and their distribution more localized. Moreover, grasses make up a small part of their diet. Forbs, many of which are of low-forage value and some even poisonous to livestock, are the most preferred food item (Ueckert and Hansen, 1970).

Forage Loss to Other Invertebrates

Considerable loss of forage may result from infestations by insects. Data from northern Utah showed that consumption of snowberry *(Symphoricarpos)* by tent caterpillars in the summer of 1938 reduced the forage available to cattle by 10 percent and that available to sheep by 24 percent (Arthur D. Smith, 1940). Similar damage has been observed on bitterbrush in California (Clark, 1955). Ferguson et al. (1963) found four kinds of insects that attacked the flowers and seeds of bitterbrush in southern Idaho. Estimates from pastures in New York led to the conclusion that, under certain conditions, more grasses and other forage were consumed by insects than by cattle (Wolcott, 1937). Furniss (1972) has listed no fewer than 205 species of insects and mites that adversely affect browse plants alone.

Red harvester ants *(Pogonomyrmex* spp.) denude parts of many range areas. It is estimated that as much as 6 percent of a range may be left bare, and that seed is collected from as much as 30 percent of the range (Killough and LeSueur, 1953). This seed-collecting habit may be an important source of loss on seeded areas.

Ranges in southern Idaho have been decimated of shadscale *(Atriplex confertifolia)* presumably by a species of snout moth (Hutchings and Farmer, 1951). In this case the plants are killed by the insect, hence the damage is more serious than that inflicted by foliage consumption. Doubtless many other insects of importance have escaped notice or have been thought to have only minor influence.

Of recent concern in the intermountain region is the black grass bug *(Labops hesperius)*. Eggs are laid in the stems of grasses where they hatch and where the larvae feed, killing the grass stems in the process (Haws et al., 1973). The effects are intensified when monocultures such as seedings of crested wheatgrass are infected. Whole seedings may be rendered unfit for forage (Fig. 11.8).

Walking sticks *(Diapheromera velii)* were found not to affect range grasses, for they subsist almost entirely on scurf pea *(Psoralea tenuiflora)*, a plant of little importance as forage (Ueckert and Hansen, 1972).

INDIRECT ENVIRONMENTAL INTERACTIONS

Not all forage relationships among herbivores are direct, nor are they all adverse. Where overgrazing alters the climax vegetation, often secondary species highly acceptable to game animals replace the preferred

Fig. 11.8 Black-grass bugs *(Labops* spp.) lay their eggs on the stems of grasses where the young hatch arresting development (left). The wheatgrass plant on the right was unaffected and developed normally. *(Photograph by William P. Nye.)*

livestock forages. Buechner (1950a) has observed this in Texas on cattle-pronghorn range, and Leopold (1950) believes it to have been a factor in the irruption of deer herds throughout the Western United States. England and DeVos (1969) concluded that spot overgrazing by bison was favorable to the success of the pronghorn on the northern plains. The elephant in East Africa is a major factor in maintaining a grassland savanna and preventing the encroachment of trees (Spence and Angus, 1970), a situation favorable to wild and domestic animals which feed on herbaceous forage.

Some changes, though benefiting other fauna, are undesirable. Observations indicate that both insects and rodents prefer overgrazed ranges to those which are used properly. The recurrent locust plagues of Africa and Australia which affect vast areas originate on the barren "scalds" of overgrazed rangeland. Grasshoppers in Oklahoma (Charles C. Smith, 1940) increased enormously in numbers upon overgrazing, though most insects were more numerous under light grazing. Similar correlation of range conditions and outbreaks of grasshoppers have been observed in Canada (Treherne and Buckell, 1924), and, in Arizona, grasshoppers prefer rangelands dominated by weedy herbs over well-grassed ranges (Nerney, 1958). (See also Table 11.11.) This increase of invertebrate forms of animal life also accompanies abandonment of cropland as in the case of the beet-leaf hopper (*Eutettix tenellus*), abandoned lands providing more favorable conditions for the multiplication of this crop-disease-carrying pest than virgin lands (Piemeisel and Lawton, 1937). Kirkham and Fisser (1972), however, could find no relationship between harvester-ant populations and range use.

Rabbits show similar responses to land use and grazing histories, especially heavy grazing that reduces the grass cover and increases weedy species. Thus, these animals may be the result of, as well as the cause of, overuse of ranges. There is evidence that improvement in the grass cover of New Zealand rangelands acts to reduce rabbit populations (Howard, 1958). In New Mexico jackrabbits frequented heavily grazed areas, being more numerous than on well-grassed range (Norris, 1950).

Table 11.11 Population of Invertebrate Animals, Thousands per Acre, Classified into Orders under Two Range Conditions

Data from Taylor et al. (1935)

Range condition	Cole- optera	Dip- tera	Hemip- tera	Homop- tera	Hymen- optera	Orthop- tera	Total
Overgrazed	118	30	100	214	140	180	782
Normal	50	28	8	52	28	20	186

In Oklahoma, cottontail rabbits, deer mice, and pocket mice were most numerous on eroded overgrazed prairie (Charles C. Smith, 1940). Abandoned farm lands that have reverted to weeds attract large numbers of jackrabbits, the greatest populations being found on the most sparsely vegetated lands (Piemeisel, 1938). Similar observations have been made concerning prairie dogs (Osborn and Allan, 1949). In Virginia, rabbits at first increased for about 4 years after croplands were abandoned and then declined as broomsedge *(Andropogon virginicus)* replaced weeds (Byrd, 1956).

Many believe that pocket gophers increase on deteriorated high mountain meadows, although others attribute poor range conditions to the gophers. The former view is supported by the fact that gophers prefer the tap-rooted forb species that increase as range conditions worsen. Herbicidal sprays which reduce these weedy species have been shown to reduce gopher populations as well as the forbs (Keith et al., 1959; Johnson and Hansen, 1969). Thereafter, grass species were much more important in the gopher's diet, which give some support to the view that they contribute to poor range conditions. Uncontrolled gopher populations have been credited with failure of range seedings in Utah (Julander et al., 1969).

Effect of Rangeland Use on Game Birds

There is substantial evidence that heavy grazing of rangelands has been a factor in the decline of grouse native to western North America, and on the success of introduced species. Trampling of nests was once thought to be the chief factor in affecting production. Successional changes in vegetation now seems the most important causative factor. Hoffman (1963) attributed the decline in the lesser prairie chicken *(Tympanuchus pallidicinctus)* in eastern Colorado to destruction of native prairie vegetation. As improved range management practices were adopted, grouse numbers increased.

Native vegetation is important at some stage in the productive cycle of several grouse species. The sharptail grouse in northern Utah are dependent upon relict areas of native grass-sagebrush vegetation during the spring mating activities, although they also accept certain cultivated crops for nesting and brooding (Hart et al., 1950). In eastern Washington, the native palouse prairie provided effective nesting cover for Hungarian partridge *(Perdix perdix);* overgrazed areas were not used for nesting (Yocom, 1943). Even the forest-dwelling blue grouse *(Dendrogapus obscurus)* used the openings among the trees covered with native grasses and forbs for brooding cover. They preferred vegetation from 13 to 25 cm in height, a condition not met under grazing (Mussehl, 1963). Zwickel (1972) could not conclusively establish higher production of blue grouse in ungrazed over grazed forest openings, though he found some indication of it.

Total exclusion of grazing may not be required, however. Campbell et al. (1973) believed moderate grazing was beneficial to the scaled quail in New Mexico, providing more food choices of broad-leaved vegetation than did ungrazed grasslands. Bennett (1938) found evidence that moderate grazing of prairie potholes opened up the vegetation and aided in survival of blue-winged teal ducklings, although heavy grazing was detrimental.

ENVIRONMENTAL EFFECTS OF MAN'S ACTIVITIES

Man, by manipulating vegetation by fire, spraying, or mechanical means markedly affects wildlife populations. These effects may be beneficial or detrimental, and although we do not well understand them at present some have been documented.

Effects of Spraying Sagebrush on Sage Grouse

As its name suggests, sage grouse *(Centrocercus urophasianus)* depend heavily on various species of *Artemisia* for both food and cover (Klebenow, 1969). Although sagebrush is valuable as winter forage, it is little used by livestock in other seasons. This has led to widespread efforts to control it. In Montana herbicidal sprays were applied in strips alternating with unsprayed strips on sage-grouse ranges. Subsequent inventories of nests revealed that practically none were found in the sprayed areas where an almost complete kill of sagebrush resulted (Martin, 1970). Based upon observations in Idaho, Klebenow (1970) judged that 5 to 10 years would elapse before sage grouse would find sprayed areas suitable for extensive use.

In addition to the effects of spraying on sagebrush, other food items may also be affected. Juvenile grouse are largely dependent on succulent forbs and secondarily on insects; shrubs are unimportant. Grasses are unimportant either to adults or young (Peterson, 1970). Spraying with agents that reduce vegetation to pure grass stands would be especially serious to the survival of young grouse. Smaller birds may also be affected by sagebrush spraying, especially those such as Brewer's sparrow *(Spizella breweri)* that nests in shrubs (Best, 1972).

Fences

Fencing rangelands has a marked impact on the movement of large migrating animals, which includes many of the large herbivores in the more northerly portions of the Western United States. The effects are most devastating on the pronghorn, which by nature does not jump obstacles (Spillett et al., 1967). In severe winter weather when they are forced to migrate, sheep-tight fencing prevents migration and considerable death losses have occurred, particularly on the Red Desert in Wyoming.

An adverse effect of quite a different nature occurs in Africa. Because standard fences do not exclude many of the large ungulates (elephants, rhino, giraffe, buffalo) and because fences that would exclude them are prohibitively expensive, the animals are removed. In consequence of their removal shrubs, which make up a major part of the diet of these large ungulates, have increased necessitating brush-control programs with, as yet, unknown consequences.

Control of Woody Vegetation

One of the more drastic alterations of habitat occurs when woodland types such as piñon-juniper stands are "chained." The results are both beneficial and detrimental for big-game animals. Where browse plants are released from competition by the trees, increased forage results if the browse species present are those not greatly harmed by the treatment. Bitterbrush is only temporarily affected, and old decadent plants resprout becoming more productive. By contrast, cliffrose *(Cowania mexicana)* does not sprout readily; only the smaller plants survive the chaining treatment. Further, when excessive cultivation and seeding of grasses follows tree removal, the range has limited value for overwintering animals such as mule deer which depend heavily upon browse in winter.

Throughout the year deer made more use of juniper areas where trees were killed with herbicides and left standing than they did untreated areas. No greater use was found on a cabled area (130 ha), and it was not used at all during inclement weather (Neff, 1972).

Almost nothing is known about the effects of chaining piñon-juniper on other wildlife species. Deer mice *(Peromyscus)* showed somewhat increased numbers immediately following juniper removal, although these were not long sustained (Baker and Frischknecht, 1973). Kundaeli and Reynolds (1972) found increased cottontail rabbit *(Sylvilagus audoboni)* populations following juniper removal if the uprooted trees were left.

Oak-brush control in Texas destroys squirrel habitat, and removal of roost trees causes population reduction of wild turkeys. Spraying sagebrush in Wyoming increased elk use, presumably in response to greatly increased production of herbaceous vegetation (Wilbert, 1963).

Water Development

Although there is little documentation of the water requirements of wildlife species, development of water for livestock undoubtedly affects distribution of wildlife species and may promote their production and survival as well. Skovlin (1971) reports the influx of game animals onto a ranch in East Africa during a drought, presumably because of the availability of water. Pronghorn depend heavily on stock-watering facilities in the arid ranges of the Great Basin, although just how dependent they are on water is not known. Stock-watering ponds were used by waterfowl broods in the northern Great Plains (Lokemoen, 1973), and no doubt elsewhere. The size and type of pond were important determinants of use and of the success of broods using them. Ponds larger than 0.4 ha received greatest use, and grassy shorelines were preferred.

Bue et al. (1952) found a direct relationship between use of prairie ponds by waterfowl and the character of the vegetation, which was determined by the intensity of grazing. Out of 20 waterfowl nests, 19 were found where stocking rates were no greater than 36 cattle days per hectare. No nests were found where stocking intensity exceeded 74 cattle days per hectare. Bank slopes with horizontal-to-vertical ratio of 5:1, areas greater than 0.8 ha, vegetation cover 17 to 30 cm high, and a fenced zone at least 8 m wide were recommended to make stock ponds useful to waterfowl (Hamor et el., 1968).

Both turkeys *(Meleagris)* and sage grouse water daily and undoubtedly profit by water sources developed for livestock whenever they are present. Water developments should be so constructed as to serve both livestock and wildlife. (See Wright, 1959.)

OTHER INTERACTIONS AMONG RANGE FAUNA

Disease and predation are the most obvious of nonforage interactions between native and domestic herbivores, and neither are well understood. For diseases of big game see Neiland and Dukeminier (1972).

Disease

Wild animals are susceptible to some of the same diseases as domestic livestock, and they may be reservoirs of infection. During an outbreak of hoof-and-mouth disease from 1914 to 1925 in California, thousands of deer were killed in the resulting control program (Leopold, 1933).

Brucellosis organisms, which cause infectious abortion in cattle, have been reported in deer, elk, and in bison, but the small numbers of the latter make them of little practical significance as a reservoir of disease. The incidence of reactors among white-tailed deer was very low, less than 1 percent (Fay, 1961).

Leptospira organisms have been found among white-tailed deer in Illinois, though infrequently, and deer were not believed to be prime natural reservoirs of the disease (Ferris and Verts, 1964). Many smaller vertebrates also are affected by some species of *Leptospira* (Roth et al., 1961).

Bighorn sheep are widely believed to have been adversely affected by lungworms transmitted by domestic sheep. Klemme (1940) reported that domestic sheep were believed to have transmitted a disease to native "big horns" in Mongolia which virtually eliminated them. Buechner (1960) cited several instances where bighorn herds had been decimated by disease, but seems to suggest that competition for forage from other game animals and livestock was a major factor in the overall decline in North American bighorn.

Beneficial interactions may also occur. Sklovin (1971) credits wild ungulates in East Africa—specifically the giraffe and eland—with minimizing the threat to livestock from the tsetse fly by killing the scrub and making the habitat unsuited to it.

Predation

No problem is more controversial than is that of predation. It is a significant factor in East Africa where several large carnivores prey on cattle (Skovlin, 1971). In the United States the coyote (*Canis* spp.) is by far the most important predator upon sheep and, occasionally, young calves.

Although considerable losses of livestock are caused by the coyote, domestic animals do not constitute its major source of food (Fig. 11.9). Coyote stomachs taken from nearly all states of the West have shown that other items, rabbits especially, make up a much greater percentage of their food than do domestic stock (Table 11.12). Careful monitoring of lamb losses among 10 bands of sheep in Utah revealed known predator losses ranging from less than 1 percent to 4.1 percent, the average for the 10 herds being 1.5 percent (Davenport et al., 1973). It is little consolation to a rancher who experiences losses, however, to know that those losses are small percentagewise. Predators were responsible for 48 percent of the dead lambs that actually were examined.

Bear, both black and grizzly, and golden eagles are known to prey on livestock, particularly sheep and goats. Grizzlies are effective predators of young elk (Cole, 1972). Of these, only the eagles are widely distributed on Western rangelands. Mollhagen et al. (1972) studied animal remains in golden eagle nests on Texas and New Mexico sheep and

Fig. 11.9 The coyote is a symbol of the Western range, although not one universally admired. *(Taken at Kim Colorado by E. R. Kalmbach.)*

goat ranges. Rabbits accounted for over half of the remains, sheep and goats for only 7.0 percent.

Unquestionably, predator losses among big-game species occur, though to what extent is little known. Knowlton (1968) found white-tailed deer making up 50 percent of the diet of coyotes in Texas when rodents were scarce, and even 75 percent during fawning. Coyotes are widely believed to be a factor in keeping down production among pronghorn (Udy, 1953), although there is little concrete evidence to support this view. Bobcats effectively preyed on pronghorn fawns in southwestern Utah, but the situation there may not have been typical of other areas (Beale and Smith, 1973) (Fig. 11.10). Wright (1960) identified 10 predators as being responsible for 211 kills among 15 East African big-game species. The lion was responsible for the most kills and wildebeests were most frequently taken by them. Thomson's gazelle was preyed upon by all 10 predators.

The cougar, puma, or mountain lion (*Felis* spp.) is one of the most successful North American predators, though its activities are concen-

Table 11.12   Percentages of Various Classes of Food Found in
Coyote Stomachs

Data from Sperry (1941); Ferrel et al. (1953); and Grier (1957)

	Volume, percent		Frequency, percent
Class of food	Sperry	Ferrel	Grier
Animal food			
Mammals			
Rabbits	32.0	22.3	57.1
Livestock (carrion)*	26.0	11.1	38.1
Rodents	17.5	26.5	36.4
Domestic livestock:			
Sheep and goat	13.0	7.2	
Calf-colt-pig	1.0		
Deer	3.5	13.8	
Miscellaneous	1.0	10.5	2.2
Birds		4.6	
Poultry	1.0		15.7
Game birds	1.0		3.7
Nongame birds	1.0		5.7
Other	1.0		7.6[†]
Vegetable food	2.0	4.0	

*Flesh of mature cows and horses as well as old weathered animal mate-
rial was believed not to have been killed by coyotes. See also Korschgen
(1957).
†Animal and vegetable.

trated upon wild animals, especially deer and elk. Deer kills occur about
a week to 10 days apart in winter; elk kills are less frequent (Hornocker,
1970). Horses (particularly colts), cattle, and sheep all are occasionally its
prey (Robinette et al., 1959); but, though individuals do considerable
damage under certain conditions, the cougar is not considered by most
stockmen as a great danger.

The threat of predators may have serious consequences for range
resource and range production. In Africa, most livestock are herded and
confined to pens (bomas) at night, thus shortening the grazing day and
greatly depressing livestock production. In North America, sheep her-
ders operating in areas of heavy coyote populations may "close herd"
sheep rather than allow them to spread widely over the range, thus
putting more pressure on range forage.

Fig. 11.10 Carcass of pronghorn antelope fawn located by means of the radio transmitter which was affixed soon after birth. Determining cause of mortality in wild or domestic animals is difficult; radio telemetry facilitates this task.

CONTROL OF UNDESIRABLE RANGE WILDLIFE

Much effort has been expended in Australia, New Zealand, and the Western United States to control populations either of forage-eating mammals or predators. Shooting, trapping, and poisoning are the principal methods that have been employed. These have met with uncertain success.

Although there is evidence that control programs using 1080 (sodium monofluoroacetate) reduced coyote populations, and possibly sheep losses, other methods used prior to the development of 1080 were not effective (Advisory Committee on Predator Control, 1972). In some cases, predator reductions were accomplished at a cost greater than the value of sheep lost (Advisory Board on Wildlife Management, 1964). There is no means of knowing the value of the sheep saved.

An important consideration is the effect of coyote control upon other predators. There is evidence that destruction of a larger predator results in an increase of a smaller one. According to this view, coyotes became a menace when larger predators were reduced or eliminated (Presnall, 1948). Robinson (1961) observed increased populations of bobcats and other smaller carnivores when coyote numbers were reduced. Apparently foxes, raccoons, and other small carnivores increased throughout the West when 1080 reduced coyote numbers (Wagner, 1972).

The effectiveness of rodent-control efforts also is largely unknown. Although demonstrably large numbers of rodents have been killed by these efforts, there is little evidence that, except for the prairie dog, populations were greatly affected. There is more evidence that proper land use—good cultural practices on farmland and proper grazing on rangelands—is more important in determining rodent numbers than are control programs. Thus, when range ecosystems are better understood in all their ramifications, proper land use may make extensive control programs unnecessary.

We do not yet know the complex ecological relationships that exist among livestock, predators, small mammals, and insects. In America, reductions in the numbers of large predators undoubtedly contributed to the buildup of deer and elk, which then competed with livestock for forage (Rasmussen, 1941). Reductions in coyote numbers may contribute to the buildup in jackrabbit populations, since there is evidence that coyotes are the major factor in rabbit mortality (Wagner and Stoddart, 1972). Thus, reduction in predators may be accompanied by increased consumption of range forage by rabbits. Conversely, poisoning of rodents and hunting of big game may increase the losses of livestock from predators. It seems likely that even destruction of insects such as crickets and grasshoppers may affect predation upon livestock.

Indiscriminate poisoning and shooting of wildlife is capable of seriously upsetting the balance that nature attempts to establish between plants and animals and between the hunted and the hunter. The complexity of these interrelationships is such as to make very unwise any widespread control activity that has not been carefully considered (Shelford, 1942).

BIBLIOGRAPHY

Advisory Board on Wildlife Management (1964): Wildlife management in the national parks, Report to Secretary of the Interior Udall, Washington.

Advisory Committee on Predator Control (1972): Predator control—1971; report to the Council on Environmental Quality and the Department of the Interior.

Aldous, C. M. (1951): The feeding habits of pocket gophers (*Thomomys talpoides moorei*) in the high mountain ranges of central Utah, *Jour. Mammal.* **32**:84–87.

Allen, Eugene O. (1968): Range use, foods, condition, and productivity of white-tailed deer in Montana, *Jour. Wildlife Mgt.* **32**:130–141.

Anderson, Allen E., Walter A. Snyder, and George W. Brown (1965): Stomach content analyses related to condition in mule deer, Guadalupe Mountains, New Mexico, *Jour. Wildlife Mgt.* **29**:352–366.

Anderson, Norman L., and John C. Wright (1952): Grasshopper investigations on Montana range lands, *Mont. Agr. Expt. Sta. Bull.* **486.**

Arnold, Joseph F. (1942): Forage consumption and preferences of experimentally fed Arizona and antelope jack rabbits, *Ariz. Expt. Sta. Tech. Bull.* **98.**

Baker, Maurice F., and Neil C. Frischknecht (1973): Small mammals increase on recently cleared and seeded juniper rangeland, *Jour. Range Mgt.* **26**:101–103.

Ball, E.D. (1936): Food plants of some Arizona grasshoppers, *Jour. Econ. Etomology* **29**:679–684.

Beale, Donald M., and Arthur D. Smith (1970): Forage use, water consumption, and productivity of pronghorn antelope in western Utah, *Jour. Wildlife Mgt.* **34**:570–582.

——— and ——— (1973): Mortality of pronghorn antelope fawns, *Jour. Wildlife Mgt.* **37**:343–352.

Bear, G. D., and R. M. Hansen (1966): Food habits, growth and reproduction of white-tailed jackrabbits in southern Colorado, *Colo. Agr. Expt. Sta. Tech. Bull.* **90.**

Bell, R. H. V. (1970): The use of herb layer by grazing ungulates in the Serengeti, in Adam Watson (ed.), *Animal Populations in Relation to Their Food Resources,* Blackwell Scientific Publications, Oxford.

Bennett, Logan J. (1938): *The Blue-winged Teal,* Collegiate Press, Ames, Iowa.

Best, Louis B. (1972): First-year effects of sagebrush control on two sparrows, *Jour. Wildlife Mgt.* **36**:534–544.

Boeker, E.L., et al. (1972): Seasonal food habits of mule deer in southwestern New Mexico, *Jour. Wildlife Mgt.* **36**:56–63.

Branson, F. A., and G. F. Payne (1958): Effects of sheep and gophers on meadows of the Bridger Mountains of Montana, *Jour. Range Mgt.* **11**:165–169.

Bue, I. G., Lytle Blankenship, and William H. Marshall (1952): The relationship of grazing practices to waterfowl populations and production on stock ponds in western South Dakota, *Trans. North Am. Wildlife Conference* **17**:396–414.

Buechner, Helmut K. (1950a): Life history, ecology, and range use of the pronghorn antelope in Trans-Pecos, Texas, *Am. Midland Naturalist* **43**:257–354.

——— (1950b): Range ecology of the pronghorn on the Wichita Mountains Wildlife Refuge, *Trans. North Am. Wildlife Conference* **15**:627–644.

——— (1960): The bighorn sheep in the United States, its past, present, and future, *Wildlife Monogr.* no. 4.

Byrd, M. A. (1956): Relation of ecological succession to farm game in Cumberland County in the Virginia Piedmont, *Jour. Wildlife Mgt.* **20:**188–195.

Campbell, Howard, Donald K. Martin, Paul E. Ferkovich, and Bruce K. Harris (1973): Effects of hunting and some other environmental factors on scaled quail in New Mexico, *Wildlife Monogr.* no. 34.

Chamrad, Albert D., and Thadis W. Box (1968): Food habits of white-tailed deer in South Texas, *Jour. Range Mgt.* **21:**158–164.

Chapline, W. R. (1971): Range management and improvement in New Zealand, *Jour. Range Mgt.* **24:**329–333.

Clark, Edwin C. (1955): The Great Basin tent caterpillar in relation to bitterbrush in California, *Calif. Fish and Game* **42:**131–142.

Coblentz, Bruce E. (1970): Food habits of George Reserve deer, *Jour. Wildlife Mgt.* **34:**535–540.

Cole, Glen F. (1972): Grizzly bear-elk relationships in Yellowstone National Park, *Jour. Wildlife Mgt.* **36:**556–561.

Constan, Kerry J. (1972): Winter foods and range use of three species of ungulates, *Jour. Wildlife Mgt.* **36:**1068–1076.

Currie, Pat O., and D. L. Goodwin (1966): Consumption of forage by black-tailed jackrabbits on salt-desert ranges of Utah, *Jour. Wildlife Mgt.* **30:**304–311.

Dahlberg, Burton L., and Ralph C. Guettinger (1956): The white-tailed deer in Wisconsin, *Wis. Conserv. Dept. Tech. Wildlife Bull.* **14.**

Davenport, John W., James E. Bowns, and John P. Workman (1973): Assessment of sheep losses to coyotes—A problem to Utah sheepmen—A concern of Utah researchers, *Utah Agr. Expt. Sta. Res. Rep.* **7.**

DeNio, R. M. (1938): Elk and deer foods and feeding habits, *Trans. North Am. Wildlife Conference,* **3:**421–427.

Denney, Richard N. (1972): Relationships of wildlife to livestock on some developed ranches on the Laikipia Plateau, Kenya, *Jour. Range Mgt.* **25:**415–425.

Dirschl, Herman J. (1963): Food habits of the pronghorn in Saskatchewan, *Jour. Wildlife Mgt.* **27:**81–93.

Ellison, Lincoln (1946): The pocket gopher in relation to soil erosion on mountain range, *Ecology* **27:**101–114.

England, Raymond E., and Antoon DeVos (1969): Influence of animals on pristine conditions on the Canadian grasslands, *Jour. Range Mgt.* **22:**87–94.

Fay, L. D. (1961): The current status of brucellosis in white-tailed and mule deer in the United States, *Trans. North Am. Wildlife and Nat. Res. Conference* **26:**203–210.

Ferguson, Robert B., Malcolm M. Furniss, and Joseph V. Basile (1963): Insects destructive to bitterbrush flowers and seeds in southwestern Idaho, *Jour. Econ. Entomology* **56:**459–462.

Ferrel, Carol M., and Howard R. Leach (1952): The prong-horn antelope of California with special reference to food habits, *Calif. Fish and Game* **38:**285–293.

————, ————, and Daniel F. Tillotson (1953): Food habits of the coyote in California, *Calif. Fish and Game* **39:**301–341.

Ferris, D. H., and B. J. Verts (1964): Leptospiral reactor rates among white-tailed deer and livestock in Carroll County, Illinois, *Jour. Wildlife Mgt.* **28:**35–41.

Field, C. R. (1972): The food habits of wild ungulates in Uganda by analyses of stomach contents, *E. African Wildlife Jour.* **10:**12–14.

Fitch, Henry S., and J. R. Bentley (1949): Use of California annual-plant forage by rodents, *Ecology* **30:**306–321.

Formozov, A. N. (1966): Adaptive modifications of behavior in mammals of the Eurasian steppes, *Jour. Mammal.* **47:**208–223.

Foster, Ronald B. (1965): Effect of heavy winter rodent infestation on perennial forage plants, *Jour. Range Mgt.* **18:**286–287.

Free, James C., Richard M. Hansen, and Phillip L. Sims (1970): Estimating dryweights of food plants in feces of herbivores, *Jour. Range Mgt.* **23:**300–302.

French, C. E., et al. (1955): Nutritional requirements of white-tailed deer for growth and antler development, *Pa. Agr. Expt. Sta. Bull.* **600.**

Furniss, M. M. (1972): A preliminary list of insects and mites that infest some important browse plants of western big game, U.S. Department of Agriculture, Forest Service, Research Note INT-155.

Garrison, George A., and A. W. Moore (1956): Relation of the Dallas pocket gopher to establishment and maintenance of range grass plantings, *Jour. Range Mgt.* **9:**181–184.

Gibb, J. A. (1967): What is efficient rabbit destruction?, *Tussock Grasslands and Mountain Lands Inst. Review* no. 12:9–14.

Goodwin, D. L. (1960) Seven jackrabbits equal one ewe, *Farm and Home Science* **21:**38–39, 51.

Grier, H. T. (1957): Coyotes in Kansas, *Kan. Agr. Expt. Sta. Bull.* **393.**

Griffiths, M., and R. Barker (1966): The plants eaten by sheep and by kangaroos grazing together in a paddock in southwestern Queensland, *Comm. Sci. Ind. Res. Org. Wildlife Res.* **11:**145–167.

Grinnell, Joseph, and Joseph S. Dixon (1918): Natural history of the ground squirrels of California, *Calif. State Commission Hort. Monthly Bull.* **7:**597–708.

Gwynne, M. D., and R. H. V. Bell (1968): Selection of vegetation components by grazing ungulates in the Serengeti National Park, *Nature* **220:**390–393.

Hamor, Wade H., Hans G. Uhlig, and Lawrence V. Compton (1968): Ponds and marshes for wild ducks on farms and ranches in the northern plains, *U.S. Dept. Agr. Farmers Bull.* **2234.**

Hansen, R. M., and J. T. Flinders (1969): Food habits of North American hares, Range Science Department, Science Series No. 1, Colorado State University, Fort Collins.

———— and A. L. Ward (1966): Some relations of pocket gophers to rangelands on Grand Mesa, Colorado, *Colo. Agr. Expt. Sta. Tech. Bull.* **88.**

Harper, James A., Joseph H. Harn, Wallace W. Bentley, and Charles F. Yocom

(1967): The status and ecology of the Roosevelt elk in California. *Wildlife Monogr.* no. 16.

Hart, Chester M., Orville S. Lee, and Jessop B. Low (1950): The sharp-tailed grouse in Utah; its life history, status, and management, *Utah State Dept. of Fish and Game Pub.* no. 3.

Haws, B. Austin, Don D. Dwyer, and Max G. Anderson (1973): Problems with grasses? Look for black grass bugs!, *Utah Sci.* **34:**3–9.

Hayden, Page (1966): Food habits of black-tailed jack rabbits in southern Nevada, *Jour. Mammal.* **47:**42–46.

Heady, Harold F. (1954): Viable seed recovered from fecal pellets of sheep and deer, *Jour. Range Mgt.* **7:**259–261.

Herman, Carlton M. (1953): A review of experiments in biological control of rabbits in Australia, *Jour. Wildlife Mgt.* **17:**482–486.

Hoffman, Donald M. (1963): The lesser prairie chicken in Colorado, *Jour. Wildlife Mgt.* **27:**726–732.

Hornocker, Maurice G. (1970): An analysis of mountain lion predation upon mule deer and elk in the Idaho primitive area, *Wildlife Monogr.* no. 21.

Howard, Walter E. (1958): The rabbit problem in New Zealand, *New Zealand Dept. Sci. Indust. Res., Information Series* no. 16.

———, and Henry E. Childs, Jr. (1959): Ecology of pocket gophers with emphasis on *Thomomys bottae* Mewa, *Hilgardia* **29:**277–358.

———, K. A. Wagnon, and J.R. Bentley (1959): Competition between ground squirrels and cattle for range forage, *Jour. Range Mgt.* **12:**110–115.

Hutchings, Selar, and Lowell Farmer (1951): Biological factors, moth infestation of shadscale in Raft River, Idaho, Office Memorandum, Intermountain Forest and Range Experiment Station.

Ingersoll, Jean M. (1964): The Australian rabbit, *Am. Scientist* **52:**265–273.

Johnson, Donald R., and Richard M. Hansen (1969): Effects of range treatment with 2,4-D on rodent populations, *Jour. Wildlife Mgt.* **33:**125–132.

Julander, Odell, Jessop B. Low, and Owen W. Morris (1969): Pocket gophers on seeded Utah mountain range, *Jour. Range Mgt.* **22:**325–329.

———, and W. Leslie Robinette (1950): Deer and cattle range relationships on Oak Creek Range in Utah, *Jour. Forestry* **48:**410–415.

Keith, James O., Richard M. Hansen, and A. Lorin Ward (1959): Effect of 2,4-D on abundance and foods of pocket gophers, *Jour. Wildlife Mgt.* **23:**137–145.

Kelso, L. H. (1939): Food habits of prairie dogs, *U.S. Dept. Agr. Circ.* **529.**

Killough, John R., and Harold LeSueur (1953): The red harvester ant, *Our Public Lands* **3:**4, 14.

Kirkham, Dale R., and Herbert G. Fisser (1972): Rangeland relations and harvester ants in northcentral Wyoming, *Jour. Range Mgt.* **25:**55–60.

Klebenow, Donald A. (1965): A Montana forest winter deer habitat in western Montana, *Jour. Wildlife Mgt.* **29:**27–33.

——— (1969): Sage grouse nesting and brood habitat in Idaho, *Jour. Wildlife Mgt.* **33:**649–662.

——— (1970): Sage grouse versus sagebrush control in Idaho, *Jour. Range Mgt.* **23:**396–400.

Klein, David R., and Helmut Strandgaard (1972): Factors affecting growth and body size of roe deer, *Jour. Wildlife Mgt.* **36:**64–79.

Klemme, Marvin (1940): *An American Grazier Goes Abroad,* The Deseret News Press, Salt Lake City, Utah.

Knight, Richard R. (1970): The Sun River elk herd, *Wildlife Monogr.* no. 23.

Knowlton, Frederick F. (1960): Food habits, movements and populations of moose in the Gravelly Mountains, Montana, *Jour. Wildlife Mgt.* **24:**162–170.

——— (1968): Coyote predation as a factor in management of antelope in fenced pastures, *Proceedings of the Third Biennial Antelope States Workshop, Casper, Wyo.,* pp. 65–74.

Koford, Carl B. (1958): Prairie dogs, whitefaces, and blue grama, *Wildlife Monogr.* no. 3.

Korschgen, Leroy J. (1957): Food habits of the coyote in Missouri, *Jour. Wildlife Mgt.* **21:**424–435.

Kundaeli, John N., and Hudson G. Reynolds (1972): Desert cottontail use of natural and modified pinyon-juniper woodland, *Jour. Range Mgt.* **25:**116–118.

La Tourrette, Joseph H., James A. Young, and Raymond A. Evans (1971): Seed dispersal in relation to rodent activities in seral big sagebrush communities, *Jour. Range Mgt.* **24:**118–120.

Lehrer, W. P., Jr., and E. W. Tisdale (1956: Effect of sheep and rabbit digestion on the viability of some range plant seeds, *Jour. Range Mgt.* **9:**118–122.

Leopold, A. Starker (1950): Deer in relation to plant succession, *Trans. North Am. Wildlife Conference* **15:**571–580.

Leopold, Aldo (1933): *Game Management,* Scribner, New York and London.

Lokemoen, John T. (1973): Waterfowl production on stock-watering ponds in the northern plains, *Jour. Range Mgt.* **26:**179–184.

Mackie, Richard J. (1970): Range ecology and relations of mule deer, elk, and cattle in the Missouri River breaks, Montana, *Wildlife Monogr.* No. 20.

Martin, Neil S. (1970): Sagebrush control related to habitat and sage grouse occurrence, *Jour. Wildlife Mgt.* **34:**313–320.

McMahan, Craig A. (1964): Comparative food habits of deer and three classes of livestock, *Jour. Wildlife Mgt.* **28:**798–808.

———, and Charles W. Ramsey (1965): Response of deer and livestock to controlled grazing in central Texas, *Jour. Range Mgt.* **18:**1–7.

McMulloch, Clay Y. (1969): Some effects of wildfire on deer habitat in pinyon-juniper woodland, *Jour. Wildlife Mgt.* **33:**778–784.

Merrill, Leo B., James G. Teer, and O. C. Wallmo (1957): Reactions of deer populations to grazing, *Texas Agr. Progress* **3:**10–12.

Mitchell, George J., and Sylvester Smoliak (1971): Pronghorn antelope range characteristics and food habits in Alberta, *Jour. Wildlife Mgt.* **35:**238–250.

Mollhagen, Tony R., Robert W. Wiley, and Robert L. Packard (1972): Prey remains in golden eagle nests: Texas and New Mexico, *Jour. Wildlife Mgt.* **36:**784–792.

Moore, A. W., and Elbert H. Reid (1951): The Dalles pocket gopher and its influence on forage production of Oregon mountain meadows, *U.S. Dept. Agr. Circ.* **884.**

Morris, Melvin S., and Roger Hungerford (1952): Food consumption and weight response of elk under winter conditions, *Proc. Western Assoc. State Game and Fish Comm.* **32:**185–187.

Morton, F. A. (1936): Summary of 1936 rangeland grasshopper studies, Bureau of Entomology and Plant Quarantine, Bozeman Mont. (Mimeo).

Mueggler, W. F. (1967): Vole damage to big sagebrush in southwestern Montana, *Jour. Range Mgt.* **20:**88–91.

Mussehl, Thomas W. (1963): Blue grouse brood cover selection and land-use implications, *Jour. Wildlife Mgt.* **27:**547–555.

Neff, D. J. (1972): Responses of deer and elk to Beaver Creek watershed treatments, *Proceedings of the Sixteenth Arizona Watershed Symposium*, Arizona Water Commission, Report no. 2, pp. 18–24.

Neiland, Kenneth A., and Clarice Dukeminier (1972): A bibliography of the parasites, diseases and disorders of several important wild ruminants of the northern hemisphere, *Alaska Dept. Fish and Game Wildlife Tech. Bull.* **3.**

Nellis, Carl H., and Robert L. Ross (1969): Changes in mule deer food habits associated with herd production, *Jour. Wildlife Mgt.* **33:**191–195.

Nerney, N. J. (1958): Grasshopper infestations in relation to range condition, *Jour Range Mgt.* **11:**247.

Norris, J. J. (1950): Effect of rodents, rabbits, and cattle on two vegetation types in semidesert range land, *N. Mex. Agr. Expt. Sta. Bull.* **353.**

Osborn, Ben, and Philip F. Allan (1949): Vegetation of an abandoned prairie-dog town in tall grass prairie, *Ecology*, **30:***322*–332.

Peterson, J. G. (1970): The food habits and summer distribution of juvenile sage grouse in central Montana, *Jour. Wildlife Mgt.* **34:**147–155.

Piemeisel, R. L. (1938): Changes in weedy plant cover on cleared sagebrush land and their probable causes, *U.S. Dept. Agr. Tech. Bull.* **654.**

———, and F. R. Lawton (1937): Types of vegetation in the San Joaquin Valley of California and their relation to the beet leaf hopper, *U.S. Dept. Agr. Tech. Bull.* **557.**

Presnall, Clifford C. (1948): Applied ecology of predation on livestock ranges, *Jour. Mammal.* **29:**155–161.

Public Land Law Review Commission (1969): Fish and Wildlife Resources on the Public Lands, Vol. I, U.S. Department of Commerce Clearing House, P. B. 187 246, Springfield, Va.

Pyrah, Duane B. (1971): Antelope range use, seasonal home range and herd units on and adjacent to sagebrush control plots, Montana, Pittman-Robertson Project W 105-R-5 Job Progress Report Job W 4.1, pp. 36–54.

Rasmussen, D. I. (1941): Biotic communities of Kaibab Plateau, Arizona, *Ecol. Monogr.* **11:**229–275.

Reynolds, Hudson G. (1950): Relation of Merriam kangaroo rats to range vegetation in southern Arizona, *Ecology* **31:**463.

Robinette, W. Leslie, Jay S. Gashwiler, and Owen W. Morris (1959): Food habits of the cougar in Utah and Nevada, *Jour. Wildlife Mgt.* **23:**261–272.

Robinson, Weldon B. (1961): Population changes of carnivores in some coyote-control areas, *Jour. Mammal.* **42:**510–515.

Roth, Earl. E., et al. (1961): Leptospirosis in wildlife and domestic animals in the United States, *Trans. North Am. Wildlife and Nat. Res. Conference* **26:**211–219.

Severson, Keith, Morton May, and William Hepworth (1968): Food preferences, carrying capacities, and forage competition between antelope and domestic sheep in Wyoming's Red Desert, *Wyo. Agr. Expt. Sta. Science Monogr.* **10.**

Shaw, W. T. (1920): The cost of a squirrel and squirrel control, *Wash. Agr. Expt. Sta. Popular Bull.* **118.**

Shelford, V. E. (1942): Biological control of rodents and predators, *Sci. Monthly* **55:**331–344.

Skovlin, Jon M. (1971): Ranching in East Africa: A case study, *Jour. Range Mgt.* **24:**263–270.

———, Paul J. Edgerton, and Robert W. Harris (1968): The influence of cattle management on deer and elk, *Trans. North Am. Wildlife and Nat. Res. Conference* **23:**169–181.

Slater, Joanna, and R. L. Jones (1971): Estimation of the diets selected by grazing animals from microscopic analysis of the faeces—a warning, *Jour. Austral. Inst. Agr. Res.* **37:**238–239.

Smith, Arthur D. (1940): Consumption of range forage by caterpillars, *Proc. Utah Acad. of Sci., Arts and Letters (Abst.)* **17:**3.

——— (1950a): Feeding deer on browse species during winter, *Jour. Range Mgt.* **3:**130–132.

——— (1950b): Sagebrush as a winter feed for deer, *Jour. Wildlife Mgt.* **14:**285–289.

——— (1953): Consumption of native forage species by captive mule deer during summer, *Jour. Range Mgt.* **6:**30–37.

——— (1959): Adequacy of some important browse species in overwintering of mule deer, *Jour. Range Mgt.* **12:**8–13.

——— (1961): Competition for forage by game and livestock, *Utah Agr. Expt. Sta. Farm and Home Sci.* **22:**8–10, 22.

——— (1965): Determining common use grazing capacities by application of the key species concept, *Jour. Range Mgt.* **18:**196–201.

———, Donald M. Beale, and Dean D. Doell (1965): Browse preferences of pronghorn antelope in southwestern Utah, *Trans. North Am. Wildlife and Nat. Res. Conference* **30:**136–141.

———, and Dean D. Doell (1968): Guides to allocating forage between cattle and big game on big game winter range, *Utah State Div. of Fish and Game Pub.* no. 68–11.

———, and Richard L. Hubbard (1954): Preference ratings for winter deer forages from northern Utah ranges based on browsing time and forage consumed, *Jour. Range Mgt.* **7:**262–265.

Smith, Charles Clinton (1940): The effect of overgrazing and erosion upon the biota of the mixed grass prairie of Oklahoma, *Ecology* **21:**381–397.

Smith, Dwight R. (1954): The bighorn sheep in Idaho; its status, life history and management, *Idaho Dept. Fish and Game, Wildlife Bull.* no. 1.

Smith, Justin G., and Odell Julander (1953): Deer and sheep competition in Utah, *Jour. Wildlife Mgt.* **17:**101–112.

Sparks, Donnie R. (1968): Diet of black-tailed jackrabbits on sandhill rangeland in Colorado, *Jour. Range Mgt.* **21**:203–308.

Spence, D. H. H., and A. Angus (1970): African grassland management—burning and grazing in Murchison Falls National Park, Uganda, in *The Scientific Management of Animal and Plant Communities for Conservation, The 11th Symposium of the British Ecological Society, University of East Anglia,* Blackwell Scientific Publications, Oxford, pp. 319–331.

Sperry, C. C. (1941): Food habits of the coyote, *U.S. Department of the Interior Wildlife Res. Bull.* **4.**

Spillett, J. Juan, Jessop B. Low, and David Sill (1967): Livestock fences—how they influence pronghorn antelope movements, *Utah Agr. Expt. Sta. Bull.* **470.**

Stoddart, L. A., and D. I. Rasmussen (1945): Deer management and range livestock production, *Utah Agr. Expt. Sta. Circ.* **121.**

Stoddart, L. Charles (1972): Population biology of the black-tailed jackrabbit *(Lepus californicus)* in northern Utah, PhD. diss., Utah State University, Logan.

Strand, A. L. (1937): Montana insect pests for 1935 and 1936, *Mont. Agr. Expt. Sta. Bull.* **333.**

Talbot, Lee M., and D. R. M. Stewart (1964): First wildlife census of the entire Serengeti-Mara region, East Africa, *Jour. Wildlife Mgt.* **28**:815–827.

Taylor, Elroy (1972): Food habits and feeding behavior of pronghorn antelope in the Red Desert of Wyoming, Proceedings of the Third Biennial Antelope States Workshop, Billings, Mont., pp. 211–221.

Taylor, W. P., C. T. Vorhies, and P. B. Lister (1935): The relation of jack rabbits to grazing in southern Arizona, *Jour. Forestry* **33**:490–498.

Treherne, R. C., and E. R. Buckell (1924): Grasshoppers of British Columbia with particular reference to the influence of injurious species on the range lands of the Province, *Canada Dept. Agr. Bull.* (n.s.), 39.

Turner, G. T., et al (1973): Pocket gophers and Colorado mountain rangeland, *Colo. Agr. Expt. Sta. Bull.* **554S.**

Udy, Jay R. (1953): Effects of predator control on antelope populations, *Utah State Dept. Fish and Game Publ.* no. 5.

Ueckert, Darrell N., and Richard M. Hansen (1970): Seasonal dryweight composition in diets of Mormon crickets, *Jour. Econ. Entomology* **63**:96–98.

———, and ——— (1972): Diet of walking sticks on sandhill rangeland in Colorado, *Jour. Range Mgt.* **25**:111–113.

———, ———, and C. Terwilleger, Jr. (1972): Influence of plant frequency and certain morphological variations on diets of rangeland grasshoppers, *Jour. Range Mgt.* **25**:61–65.

Ullrey, D. E., et al. (1971): A basal diet for deer nutrition research, *Jour. Wildlife Mgt.* **35**:57–62.

Vesey-FitzGerald, Desmond Foster (1960): Grazing succession among East African game animals, *Jour. Mammal.* **41**:161–172.

Vorhies, C. T., and W. P. Taylor (1922): Life history of the kangaroo rat, *Dipodomys spectabilis spectabilis* Merriam, *U.S. Dept. Agr. Bull.* **1091.**

————, and ———— (1933): The life histories and ecology of jack rabbits in relation to grazing in Arizona, *Ariz. Agr. Expt. Sta. Tech. Bull.* **49.**

Wagner, Frederic H. (1972): Coyotes and sheep, 44th Faculty Honor Lecture, Utah State University.

————, and L. Charles Stoddart (1972): Influence of coyote predation on black-tailed jackrabbit populations in Utah, *Jour. Wildlife Mgt.* **36:**329–342.

West, N. E. (1968): Rodent-influenced establishment of ponderosa pine and bitter-brush seedlings in central Oregon, *Ecology* **49:**1009–1011.

Wilbert, Don E. (1963): Some effects of chemical sagebrush control on elk distribution, *Jour. Range Mgt.* **16:**74–78.

Wolcott, G. N. (1937): An animal census of two pastures and a meadow in northern New York, *Ecol. Monogr.* **7:**1–90.

Wright, Bruce S. (1960): Predation on big game in East Africa, *Jour. Wildlife Mgt.* **24:**1–15.

Wright, J. T. (1959): Desert wildlife, *State Ariz. Game and Fish Dept. Wildlife Bull.* no. 6.

Yocom, Charles F. (1943): The Hungarian partridge *(Perdix perdix* Linn.) in the Palouse region, Washington, *Ecol. Monogr.* **13:**167–202.

———— (1952): Columbian sharp-tailed grouse *(Pedioecetes phasianellus columbianus)* in the state of Washington, *Am. Midland Nat.* **48:**185–192.

Zwickel, Fred C. (1972): Some effects of grazing on blue grouse during summer, *Jour. Wildlife Mgt.* **36:**631–634.

Chapter 12

Multiple-Use Relationships on Rangelands

Many rangelands are capable of producing more than one product, the importance of each depending upon the stage of development and goals of the particular country. In undeveloped countries the need for protein may outweigh other considerations and dictate livestock production. In more developed countries, concurrent demands arise for many range products. In the United States, public ranges are managed under a multiple-use philosophy in which an effort is made to accommodate all the possible land uses for which there is a need or demand. No one of these, however, can be optimized without positive or negative impacts upon the others. It is the task of the range manager to integrate all these uses in such a way that complementary relationships are enhanced and competitive relationships minimized.

No use of rangeland more intimately affects other possible uses than does livestock grazing. On many ranges, timber production is also of great concern. Because of the interrelationships between timber management and range management, the land manager must understand both.

Virtually all rangeland has some significance as a watershed. Although more arid ranges are not sources of live streams, they often are the sources of floodwater and sediment from surface flow following unusually heavy storms, thus affecting the quantity and the quality of water of the streams into which they discharge. Other rangelands are the headwaters of rivers and source of valuable water supplies. On such ranges, water production is of paramount importance, and grazing becomes strictly secondary.

Conservation of the soil is related to water production. Generally, good range management is good watershed management, and, if the watershed is grazed properly, problems of erosion and runoff are at a minimum. Floods cannot be eliminated by ordinary means of land management, but there can be no doubt that mismanagement of land accelerates flooding and creates problems even in years of relatively normal precipitation.

In the United States, recreational use of rangelands (for hunting, rock collecting, camping, hiking, and touring by vehicles) is increasingly important.

TIMBER PRODUCTION AND GRAZING

In many areas of the world, timber production and forage may be produced from the same areas, although many timber types have little or only transitory forage values (see Chapter 2). In the more open-growing stands, such as the ponderosa pine type in the Western United States, the pine forests of the Southwestern United States, and tropical savannas, considerable forage also is produced. In the dense coniferous forests of the Pacific Northwest United States, tropical rain forests, and the dense eucalyptus forests of Australia, minimal forage values are present, although during regenerative stages they may be productive for wildlife, especially big game. Since neither timber production nor range-forage production can be maximized on forested ranges without some impact upon the other use, understanding the interrelationships of the two is important.

Effect of Browsing upon Tree Establishment and Growth

Young trees, since they are within reach of browsing animals, often suffer from defoliation and removal of shoots and buds. The possible severity of injury from this cause is demonstrated in the use of goats to clear fields of brush and tree sprouts. While proper grazing from other kinds of livestock is less severe, injury to tree reproduction is sufficiently common to require consideration in range-management plans on timbered areas.

Defoliation is not the only manner in which injury to trees occurs. The disturbance of the soil through trampling has a marked effect upon seedling establishment. Once established, seedlings may be uprooted and mechanically injured by trampling. Retarded growth and even death from disease may result. These effects are especially severe where animals congregate, and may result in complete failure of regeneration in heavily frequented areas.

Damage from large herbivores may occur even after trees are well grown and after the leader is out of reach of browsing animals. Rubbing, horning, and similar activities result in broken branches or ruptures in the bark through which disease may enter. The American bison (*Bison bison*) is credited with preventing tree encroachment at the periphery of the prairies. Similarly, elephants, though grazing animals, uproot or push over trees keeping open grasslands where otherwise forests or savannas would develop (Wing and Buss, 1970).

The degree to which herbivores affect the establishment and growth of timber stands is dependent upon several factors: (1) the kind of animal and its forage preferences; (2) the density of the animal population; (3) the kind of timber, whether coniferous or deciduous; (4) the age of the stand; and (5) the kind and availability of alternative sources of forage. Given the proper combination of these factors damage may be slight; when they are adverse the result, for tree production, may be devastating. Under certain conditions grazing may be beneficial.

Browsing Injury to Conifers   In the United States, most problems from grazing by domestic livestock on coniferous trees occur in the pine forests of the Southeast, and the ponderosa pine and spruce fir forests of the West. These are sufficiently open so that considerable forage is present, making them attractive and suitable to livestock use. In other regions and timber types, damage to coniferous stands is of less general concern (Reid, 1947; Morris, 1947).

Foraging upon trees is of importance during reproductive stages when the terminal buds are within reach of animals. Browsing may cause death of seedlings, although with careful livestock management, losses due to this cause on ponderosa pine forests in the United States may be no more than 4 percent from sheep and less than 1 percent from cattle (Pearson, 1950). But even though they survive, repeated browsing may so deform the young trees as to reduce their value for timber and increase the incidence of disease. Rate of growth may also be slowed (Pearson, 1950).

Proper grazing can be expected to minimize damage where otherwise it might occur. Hedrick and Keniston (1966) observed no injurious browsing of Douglas fir seedlings in the Pacific Northwest by sheep if the utilization of the more palatable forbs did not exceed 50 percent, and after 10 years seedlings were 64 cm taller in grazed than in protected plots. Slash pine *(Pinus ellottii)* plantations in Louisiana were grazed without ill effects under moderate or light use, damage occurring only when grazing was heavy (Table 12.1).

Browsing Injury to Hardwoods   In general, deciduous trees are much more susceptible to browsing than are the conifers, although much of the concern arises out of damage to farm woodlots where grazing is uncommonly severe in many instances. Deciduous trees are more attractive as forage, and other forage may be minimal because of the dense canopy overhead.

Observations of grazed woodlots have shown them to be lacking in reproduction (Cheyney and Brown, 1927), and the growth rate of established trees is believed by some to be lessened (Deam, 1923). Furthermore, grazing often favors the less desirable or weed trees over better species. Continued grazing results in the formation of a heavy sod, which further discourages the reproduction of trees (Denuyl and Day, 1932; 1934). There appears to be little justification for grazing farm woodlots, owing to their low value as pastures and the harm done to tree reproduction.

Under open-range conditions, however, even hardwood forests can be grazed by livestock, as demonstrated in the Ozark region. Although there has been considerable damage from grazing during past years, the damage can be reduced to negligible proportions by proper livestock management. This requires closing the area to grazing during winter months and reducing numbers to achieve moderate grazing pressures (Hornkohl and Read, 1947).

In the Western United States the principal deciduous tree on rangeland is aspen. Because of the many palatable herbaceous species as-

Table 12.1   Slash Pine Regeneration under Regulated Grazing

Data from Pearson et al. (1971)

		Control	Light	Control	Moderate	Control	Heavy*
1st year	May	821	811	846	866	761	660
	October	741	714	784	776	711	615
2d year		725	653	777	764	691	565
5th year		697	618	776	745	683	559

*All significant (0.85).

sociated with it, aspen forms an important forage type. Heavy grazing by livestock can prevent aspen regeneration (Sampson, 1919), but, under proper stocking and management, regeneration of aspen is possible under either sheep or cattle use (Fig. 12.1).

The effect of grazing on forest stands is related directly to intensity of stocking and the kinds of animal used. If animals which prefer herbaceous plants and noncommercial trees are stocked at reasonable rates, forests can be grazed safely.

Mortality of Seedlings Due to Trampling by Livestock

Browsing is only one source of seedling injury. In coniferous forests, trampling may be of greater concern than is browsing, especially if livestock are allowed to congregate unduly as on driveways, bedgrounds, and shading areas (Ingram, 1931).

In northern Idaho, studies made on sheep ranges in western white pine forests revealed some seedling injuries due to trampling, but they were limited to seedlings under 5 years of age. Furthermore, the incidence of trampling was small except on driveways, and in no case was it

Fig. 12.1 Cutover aspen stand in southwestern Utah showing the effects of different animals. The area in the right foreground was grazed by sheep and no reproduction survived. The area beyond the pole fence in the background was grazed by cattle and deer, and deer only had access to the area to the left of the low fence in foreground. Neither cattle nor deer prevented reproduction.

such as to reduce the number of seedlings below that needed to establish an adequate timber stand (Young et al., 1942). Similarly, it has been observed that sheep trampling was of minor importance compared to the other causes of seedling mortality in ponderosa pine forests (Sparhawk, 1918).

Where sheep are properly herded, they are probably less destructive than cattle, which by reason of larger hooves and greater weight can be more destructive. Lull (1959) calculated cattle to exert more than twice the pressure sheep did. Documentation of sprout mortality in aspen *(Populus tremuloides)* showed no trampling losses from sheep but from 5 to 9 percent losses from cattle (Smith et al., 1972). Similarly, sheep trampling caused no damage to planted Douglas fir seedlings (Black and Vladimiroff, 1963). Aside from direct physical injury to seedlings, trampling may have an adverse effect upon seedling establishment through compaction of the soil, although this has not been established, and the effects may vary greatly with soil types and site (Table 12.2).

Kind of Animal and Injury to Reproduction

Overall damage to tree reproduction by sheep and goats is more severe than by cattle (Campbell, 1947). Since sheep show greater preference for browse forage than do cattle, the incidence of browsing on trees is somewhat higher. With proper consideration to availability of other forage and with moderate stocking rates, it is doubtful if material damage would occur in coniferous forests either from cattle or sheep. Goats may need to be excluded from timberlands (Williams, 1952), and certainly uncontrolled hogs are not desirable.

Effects of Wild Herbivores on Tree Reproduction

Wild herbivores may be more injurious than domestic livestock. Their predilection for browse, the fact that they use forest types more readily

Table 12.2   Bulk Densities of Forest Soil after
10 Years of Grazing under Different Intensities

Data from Linnartz et al. (1966)

Depth	Ungrazed	Moderate	Heavy
		(gm/cc)	
0–4 in.	1.32	1.39	1.41
6–10 in.	1.45	1.48	1.51
12–16 in.	1.48	1.50	1.51

than livestock, and their longer period of occupancy in many areas make their effects greater. On the other hand, livestock are likely to be in greater numbers and their use more intense. In southern Utah, it was found that aspen reproduction was only moderately affected by cattle when stocking was controlled in small paddocks, somewhat more affected by sheep when controlled, eliminated when sheep were uncontrolled, and only lightly affected by mule deer (Smith et al., 1972). The light impact made by deer in this instance was due to moderate grazing pressure. Under heavy stocking, deer would have had much greater impacts.

It is difficult to generalize the impact of wild herbivores. Injury to both coniferous and deciduous tree species has been observed in all sections of the United States and the effect of red deer *(Cervus elaphus)* in New Zealand is widely known (Wodzicki, 1950). Both Douglas fir and redwood have been browsed by deer in northern California, at times so severely that growth is slowed (Table 12.3). Of the two species, Douglas fir was more affected. Roy (1960) showed that slowing the height growth of Douglas fir was proportional to severity of browsing; unbrowsed reproduction reached heights of over 1 m while those browsed five or six times were only 43 cm tall. The severity of browsing varied inversely with the availability of other browse species (Fig. 12.2). In contrast, Black and Vladimiroff (1963) found only 9 of 3,400 Douglas fir seedlings browsed by black-tailed deer; much more browsing was done by mountain beaver *(Aplodontia rufa)* and rabbits *(Sylvilagus bachmani)*. Even so, the incidence of browsing from all causes was less than 4 percent.

Hardwood tree species are generally more susceptible to damage from deer than are conifers. Stoekler et al (1957) observed 20-year-old sugar maple on an area in Wisconsin only 1.2 m tall due to repeated deer browsing. Moose in Newfoundland have been observed to suppress growth and kill both balsam fir *(Abies balsamea)* and white birch *(Betula papyrifera)* regeneration (Bergerud and Manuel, 1968).

More commonly, the effects are less severe. Rayburn and Barkalow

Table 12.3 Effect of Deer upon Height Growth of Douglas Fir
and Redwood in California

Data from Browning and Lauppe (1964)

	Growth per year, in in.	Average height, in in.	Percent increase in 4 years
Unprotected	2.7	17.0	166
Protected	8.0	39.8	410

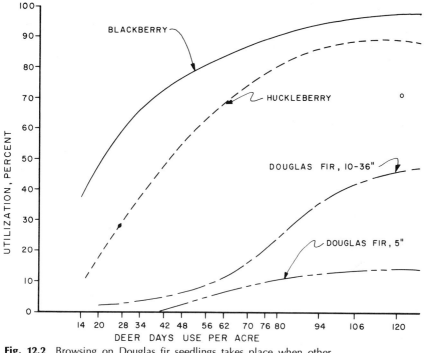

Fig. 12.2 Browsing on Douglas fir seedlings takes place when other browse is absent; not until blackberry and huckleberry were substantially gone did deer make use of Douglas fir. *(Data from Crouch, 1966.)*

(1973) report that neither tupelo *(Nyssa aquatica)* or green ash *(Fraxinus pennsylvanica)* regeneration, as indicated by height growth, was seriously affected by deer in North Carolina. Although reduction in height growth was demonstrated in both dogwood *(Cornus florida)* and tulip poplar *(Liriodendron tulipifera)* seedlings in a 5-year clipping study, the intensity of clipping was greater than would normally be expected from deer (Harlow and Halls, 1972). Removal of both terminals and laterals was required to cause mortality of seedlings.

Mature trees may also be affected. Krebill (1972) observed high mortality in aspen stands in Wyoming which he attributed to pathogens and insects following barking by elk. In California, the black bear *(Euarctos americana)* is known to strip the bark and cambium of 5- to 10-year-old second-growth redwood (Glover, 1955).

In most cases, damage to forest stands by wildlife occurs because of (1) increased numbers of wild animals over the naturally balanced ecosystem, or (2) animals introduced into the system. Elephants in Tsavo and Murchison Falls are examples of the former; red deer and other exotics in New Zealand are examples of the latter.

Effects of Forest Practices on Forage Production

On forested ranges, no management activity more affects forage reproduction than does timber harvesting. Once a timber stand reaches pole
size, understory vegetation is reduced greatly and, except for certain
types such as ponderosa pine which become open with age, grazing
capacities are thereafter low. Timber harvest or thinning results in
greatly increased production of herbaceous and woody growth, although thinning is not equally effective in all timber types. Thinning in
oak stands in the Ozarks may be so light, or sprouting by species that are
less desirable as forage may be so competitive, as to make thinning
ineffective in increasing forage for deer (Crawford, 1971). Thinning in
pine types either in the Southeast (Grelen et al., 1972) or in ponderosa
pine in the West are effective (McConnell and Smith, 1970). The response of herbaceous plants is related directly to the reduction of the
timber stands and opening the canopy (Fig. 12.3). Burning and thinning
have been shown to maintain satisfactory timber production requirements while increasing forage yield in pine or pine hardwood types in
East Texas (Halls and Alcaniz, 1971).

Game animals are especially benefited by forest cutting and fire,
since they make use of timber types more than livestock. Historically, in
all sections of the United States, deer numbers have risen after logging,
which was often followed by fires. There followed a period of high-
forage production and then a decline as second-growth timber stands
became established. Confronted with a shortage of forage, deer numbers declined also. Halls and Crawford (1960) describe such a cycle in
the Ozark region. Table 12.4 illustrates the increase in carrying capacity
for deer resulting from different intensities of timber removal in a
mixed oak-pine type in Virginia.

In some instances, timber harvesting practices may remove trees
that are most important to wildlife. Removal of oak trees in the Southwestern United States reduces the acorn crop, an important source of
energy for white-tailed deer in the fall when herbaceous forage is mature. In the African savanna, *Acacia tortulis* trees, which are cut for making charcoal, produce seedpods that are an important forage source in
the dry season for the flocks of nomadic tribes. Herdsmen further contribute to the problem by lopping branches from remaining trees to
provide green feed.

Beneficial Effects from Grazing

The relationships between tree reproduction and grazing animals are
not all harmful. Heavy utilization, even to the point of overgrazing,
actually may favor germination as has been shown in both ponderosa

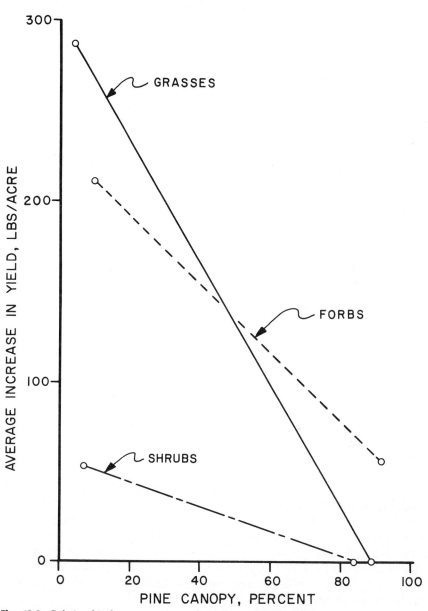

Fig. 12.3 Relationship between canopy of ponderosa pine and forage production. Similar relationships exist in other types and sites. *(From McConnell and Smith, 1970.)*

Table 12.4 Deer Carrying Capacities Resulting from Different Levels of Timber Removal in the Mixed Oak-Pine Type in Virginia

Data from Patton and McGinnes (1964)

Percent basal area removed	Number of years after cutting			
	1	2	3	4
	(Deer days/acre)			
30	10	12	15	17
40	15	16	19	23
50	20	23	27	31
60	27	32	37	43
70	37	44	51	59
80	51	60	70	82

pine and white pine types of Idaho where bedgrounds and driveways had more seedlings than lightly grazed areas (Sparhawk, 1918; Young et al., 1942). Trampling may aid germination by reducing heavy litter and forcing seed into moist mineral soil. After the seeds have germinated, moderate grazing may again serve to aid seedling survival by reducing competition from herbaceous plants. No gain may result, however, if the heavy use is prolonged after seedling growth is under way. Survival of Douglas fir in the Northwest has been observed to be greater on moderately grazed than on ungrazed areas (Ingram, 1931).

A study in the pond pine type in North Carolina revealed greater survival of pine seedlings as a result of grazing and also more rapid growth during the first 3 years (Shepherd et al., 1951). Although mortality of seedlings was greater under grazing, many more seedlings emerged on grazed areas. Close grazing for 5 to 6 years after a good seed crop, followed by lighter grazing, has been recommended as a means of ensuring greater survival of ponderosa pine (Pearson, 1931).

Grazing may serve a useful function in the early stages of reforestation by removing the heavy growth of herbaceous vegetation that develops in naturally or artificially regenerated forests (Hatton, 1920). Grazing reduces the amount of inflammable material rendering fires, either accidental or controlled, less damaging to timber reproduction (Wahlenberg, 1935). Vegetational changes induced by grazing, which is toward the lower-growing herbaceous species, reduce fire hazards also (Campbell and Rhodes, 1944). Westell (1954) ascribed beneficial effects to aspen production by deer in the Lake states through thinning the too-thick stands.

Timber Management for Wildlife and Livestock

It is entirely possible to combine the production of timber with the production of forage for livestock and big game, although the manner in which it is done and the results obtained differ with the region and the type of timber.

Both livestock and wildlife are attracted to cutover areas in response to increased amounts and quality of forage. Tame deer spent almost three times as much time foraging on logged areas as they did in uncut forests in Colorado (Wallmo et al., 1972). In Oregon, elk use on cutover areas increased more than deer use (Edgerton, 1972). Preference for logged areas reaches its maximum some years following cutting and then declines. Maximum use by elk in western Oregon was observed 6 to 8 years after cutting (Harper and Swanson, 1970).

The following procedures both protect timber and provide for a sustained wildlife population: (1) cutting timber heavily enough to give effective release to understory vegetation, (2) cutting sufficient acreages so that grazing pressures will be moderate or light on the cutover area, (3) adopting cutting cycles that coincide with the decline in forage yields in newly cut areas, (4) thinning in pole-sized stands to increase forage production, (5) maintaining proper stocking, especially during the seedling stage, and (6) in the case of domestic livestock, stocking with animals that least prefer the desired tree species. Where wildlife are involved and the tree species palatable as is aspen, the last course is not possible.

Indirect Effects from Grazing Timberlands

Considerable evidence exists to suggest that there may be indirect effects of livestock grazing in forests, which are fully as important as the direct ones. An apparent increase in the incidence of tree diseases in Western forest areas may be due in some cases to the increase of alternate host plants under heavy grazing. White pine blister rust is believed to have increased, owing to spread of *Ribes* by grazing animals (Young et al., 1942). Other similar relationships are suspected. It is possible that influences of this sort may be of equal or greater significance than are those associated with direct injury to trees.

WATER PRODUCTION AND GRAZING

All water which is available for man's use comes from precipitation. Without this means to recharge water supplies, there would be no streams and no ground water to be tapped by wells. Rangelands are predominantly arid or semiarid, and water is at a premium. In the United States where rangelands are a major source of water for irrigation and for industrial and domestic supplies for cities, often far removed from the range areas, water is doubly important as a rangeland product. What happens to the precipitation that falls is of extreme concern not only for the range manager but also to others. The investment in water storage facilities alone in the United States totals nearly $13 billion. The life of these is directly related to silting rates which are influenced by grazing practices.

Water requirements on and off site for producing animal products from the range are high. It has been estimated that under West Texas conditions it takes 240 tons of water to produce 1 kg of red meat at the supermarket (Thomas, 1969), and estimates as high as 350 tons have been made. Production of 1 kg of scoured wool may require 500 tons of water.

The Hydrological Cycle

The process of continual transfer of water from land and ocean surfaces to the atmosphere and back is diagramed in Fig. 12.4. Vegetation is seen to stand between the atmospheric moisture and water bodies between which the greater part of the atmospheric moisture circulates. It is here that land use with its modifications of vegetation and soil can alter the hydrological cycle. If infiltration is large, a high percentage of the pre-

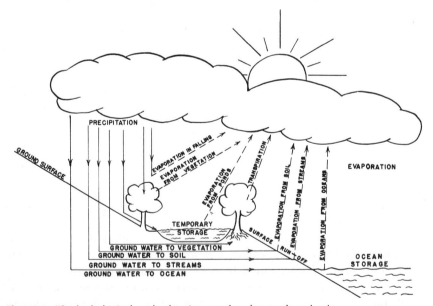

Fig. 12.4 The hydrological cycle showing transfer of water from land, plant, and ocean to the atmosphere and back.

cipitation penetrates to ground water, ultimately to appear as spring flow, or to be recovered through wells. In contrast, if surface runoff is high, peak flows occur immediately after storms accompanied by erosion and soil loss. Water yields and quality can be greatly affected by the way vegetation is altered under use, which in turn affect evaporation and transpiration losses, though there is, as yet, inadequate documentation of these relationships.

Transpiration and Evaporation Vegetation can be produced only through the expenditure of water. Vegetation reduces the amount of evaporation from the soil, but the aerial portions intercept part of the precipitation and return it to the atmosphere. That which penetrates the vegetal canopy and enters the ground can later be absorbed by the roots and returned to the atmosphere by transpiration. The magnitude of the amounts of water involved can be appreciated from contemplation of the *water requirement* or ratio between amount of water transpired and amount of herbage produced. For native vegetation this ratio is about 700:1, which means that growth of a kilogram of herbage would entail the transpiration of 700 kg of water, or more than 6 cm spread over a hectare. For a heavy forest stand in the Southeast, this transpiration loss has been determined to be from 43 to 56 cm (Hoover, 1944).

Lower values may be expected from less dense vegetation and under lesser precipitation. Losses from aspen stands in Utah due to evaporation and transpiration were 41 to 51 cm; losses from herbaceous vegetation were from 30 to 46 cm. The losses in excess of that from bare soil were about 20 and 10 cm for aspen and herbaceous cover, respectively (Croft and Monninger, 1953). These indicate the magnitude of the cost in terms of water loss for maintaining a vegetative cover to control runoff and erosion. Whether or not grazing to reduce foliage may measureably reduce loss of water through transpiration has not been adequately determined. Possibly, judicious grazing may increase water yield without impairing streamflow behavior (Buckhouse and Coltharp, 1968).

Infiltration Water which reaches the soil surface infiltrates the soil until the rate of application exceeds the rate of infiltration, after which the excess moves laterally as *surface runoff*, also called *overland flow*. At any instance and place the capacity of soil to absorb water has a definite limit. This limit is determined by many factors, among which are soil texture, soil structure, soil-moisture content, and length of time that water has been applied. Coarser-textured soils accept water more readily than do fine-textured soils. Fine-textured soils may be made more absorptive through improvement in their structure. Small particles grouped and cemented together to form aggregates act like larger particles. The ability of a clay soil to absorb water may, through alterations in its structure, become comparable to that of a sandy soil.

Characteristically, infiltration rates are highest at the beginning of water application, decreasing steadily to attain a nearly constant rate. While this constant rate may be limited by subsurface layers or by saturation of the soil mantle, very often it is determined by the condition of the soil surface. Fig. 12.5 shows characteristic infiltration curves.

Vegetation plays a major role in forming and determining the character of the soil. Packer (1963) found a direct relationship between vegetative cover and bulk density, an important determinant of infiltration. Moreover, vegetation has a pronounced effect upon the preservation of a permeable soil surface under the impact of precipitation. This is accomplished through the interception of the raindrops by the vegetal canopy and by the accumulated litter and plant residues, thus dissipating the energy of fall. Such mechanical intervention serves to prevent puddling of the soil surface which closes the soil channels preventing water penetration. Infiltration, therefore, is maintained at a high level under vegetation, with a consequent reduction in the amount of water appearing as overland flow.

Effects of Grazing upon Infiltration

Grazing may be expected to alter the natural infiltration-runoff relationships by reducing the protection afforded by herbage, by reducing and scattering the litter, and by compacting the soil through trampling. The magnitude of these changes is determined by the intensity of grazing as well as by soil type, climate, topography, livestock management, and vegetation type. There is ample evidence that heavy and indiscriminate use seriously affects the capacity of soil to absorb water and thus minimize runoff.

Soil structure can be altered materially by grazing (Table 12.5). The effects are related directly to grazing intensity as shown by comparisons of soils on undergrazed, overgrazed, and depleted ranges in New Mexico which had pore space of 68.1, 51.1, and 46.5 percent, respectively (Flory, 1936). The centimeters of water penetrating per hour were 10.5, 5.5, and 2.1, respectively.

Research on forest watersheds in North Carolina showed that 9 years of heavy grazing by cattle reduced large pore space in the first 5 cm of soil by 44 percent and in the 5- to 10-cm depth by 60 percent, and permeability was only about one-tenth that of comparable ungrazed forest (U.S. Department of Agriculture, 1948). Infiltration rates on un-

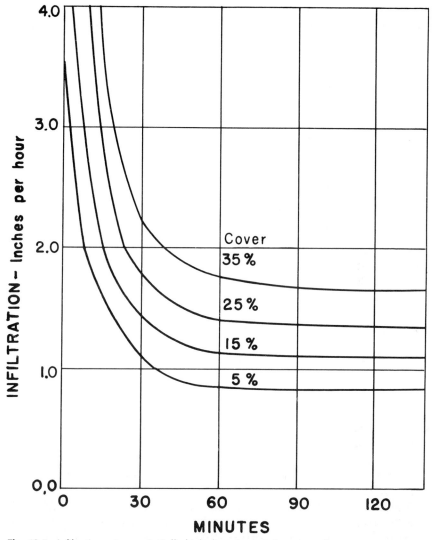

Fig. 12.5 Infiltration rates are initially high dropping to a lower, nearly constant rate. The curves show the effect of a browse crown cover on infiltration rates on a sandstone soil in Utah. *(Data from Woodward, 1943.)*

grazed range on the southern High Plains were nearly four times as great as on grazed ranges of like character (Brown and Schuster, 1969). On the Texas High Plains, the amount of water infiltrating silty clay loam decreased as the successional stage regressed from high to low (Dee et al., 1966).

Table 12.5   Average Porosity, Bulk Density, and Organic-Matter Content of Soil Cores from the Manti Canyon Plots 7 Years after Seeding*

Data from Meeuwig (1965)

Depth (in.)	Unseeded		Seeded	
	Ungrazed	Grazed	Ungrazed	Grazed
	Noncapillary porosity (percent)			
0-2	18.8	15.1	19.6	17.7
2-4	16.2	15.9	16.5	15.4
4-6	15.8	15.3	15.8	14.9
	Capillary porosity (percent)			
0-2	42.7	43.9	40.8	42.3
2-4	43.4	41.5	41.5	41.7
4-6	41.1	41.7	41.2	40.7
	Bulk density (gm per cc)			
0-2	0.96	0.99	1.02	1.04
2-4	1.06	1.05	1.05	1.04
4-6	1.10	1.09	1.08	1.10

*Only in the upper 2 in. were differences significant.

The amount of vegetal material greatly affects infiltration rates. Vegetation accounted for 73 percent of the variance among factors affecting water retention on runoff plots in Utah (Meeuwig, 1970). Infiltration rates on small watersheds in South Dakota having the same soil type were directly related to the quantity of plants and litter present under three intensities of grazing (Table 12.6). Interception by herbaceous plants and subsequent delivery to the soil may be a factor in penetration of moisture. Depth of water penetration under *Agropyron spicatum* was directly related to plant size; clipping reduced moisture penetration by 60 percent (Ndawula-Senyimba et al., 1971).

Relationship of Grazing to Runoff and Erosion

The consequence of reduced infiltration rates is increased runoff. This reduces the effectiveness of precipitation and makes ranges even more arid than they are normally, further reducing their ability to produce forage. Reductions in soil moisture available for plant production under heavy grazing amount to as much as 4.5 cm a year and averaged 2 cm or

Table 12.6   Air-Dry Herbage and Mulch and Rate of Water Intake on Small Native-Range Watersheds Grazed Heavily, Moderately, and Lightly, Cottonwood, South Dakota, 1964

Data from Rauzi and Hanson (1966)

Study area	Total herbage	Mulch	First 30-min	Second 30-min	Average for 1-hr period
	(lb/acre)		(In./hour)		
Heavily grazed	900	456	1.40	0.71	1.05
Moderately grazed	1,345	399	2.16	1.21	1.69
Lightly grazed	1,869	1,100	3.19	2.72	2.95

about 8 percent of annual rainfall, a substantial loss to plant production (Table 12.7).

Accompanying this loss of water is a greatly increased potential for soil erosion. Erosion varies with the velocity of the water in motion, and the natural erodibility of the soil. Velocity varies approximately as the square root of the slope; thus, increasing the slope by 4 times will double the velocity. Doubling the velocity will increase by 4 times the eroding power (to the second power), by 32 times the quantity of material that can be carried (to the fifth power), and by 64 times the size of particle

Table 12.7   May to August Precipitation and Runoff from the Differentially Grazed Watersheds

Data from Hanson et al. (1970)

	Heavy		Moderate		Light	
Year	Precipitation	Runoff	Precipitation	Runoff	Precipitation	Runoff
	(In.)					
1963	12.14	1.79	12.02	1.57	12.61	1.39
1964	8.59	0.66	8.58	0.28	7.74	0.05
1965	10.81	0.13	11.05	0.14	10.91	0.12
1966	9.40	0.16	9.18	0.02	9.45	0.00
1977	11.00	1.21	11.16	0.79	10.90	0.54
Mean	10.39	0.79	10.40	0.56	10.32	0.42

that can be moved (to the sixth power) (Ayres, 1936).

On rangelands having various rain intensities, soil conditions, and plant covers, runoff from 30 percent slopes average 41.2 percent of the precipitation, and soil losses averaged 614 kg per hectare. On 40 percent slopes, runoff averaged 49.5 percent, and soil losses were 2,100 kg per hectare (Craddock and Pearse, 1938). Soil losses were more than three times greater on fine-textured than on medium-textured soils (Hanson et al., 1973).

The factors of slope and soil character are not under ready control of the land administrator. However, management may influence the degree of accumulation of water, which in turn affects the velocity of flow. The velocity of water which collects in a channel 15 cm wide and 15 cm deep will be approximately 2½ times the velocity attained if the same volume of water is kept spread over a strip of land surface 91 cm in width. Similarly, in the absence of a channel, the increase of water depth on the ground surface from 6 to 13 mm would have the effect of increasing the velocity of flow 1.2 times. The range manager must consider the effect of use upon the disposition of precipitation and resultant erosion.

Ground-Cover Requirements to Prevent Overland Flow and Erosion

Little is known of minimum vegetation requirements to prevent critical water losses during storms. Under adverse conditions, as in cases of steep slopes, erodible soils, and intense storms, a considerable protection is required. Occasionally under such situations no impairment of vegetal growth through grazing may be permissible. Under less critical situations, completely satisfactory watershed conditions may exist under full use by grazing animals. Packer (1951) concluded that ground-cover densities, including living plants and litter, should be at least 70 percent to achieve adequate protection to the soil surface on erosive mountain lands in southern Idaho. A ground cover of 65 percent was considered necessary in the mountains of northern Utah (Fig. 12.6).

A major factor in soil erosion is the impact effect of raindrops which dislodges soil particles and seals the soil surface, reducing infiltration. Complete protection of soil requires about 550 kg of plant material per hectare (Osborn, 1956). The effectiveness of vegetation in preventing soil detachment by rainfall varies with the kind of vegetation; short grass is much more effective than taller vegetation (Table 12.8).

Effect of Grazing upon Water Yields

Theoretically, proper grazing might be expected to increase the amount of water going to ground water and increase watershed yields. The

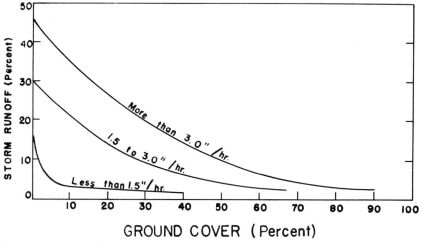

GROUND COVER (Percent)

Fig. 12.6 Relation of summer storm runoff to total ground cover under low-, moderate-, and high-intensity rainfall. *(Data from Marston, 1952.)*

complexities of the relationships involved are such, however, that significant effects have not been well documented.

Liacos (1962b) computed a water balance on plots with former grazing histories in the annual grass type of California and concluded that greatest water yields came from heavily grazed plots. One-fourth of the yield, however, came from surface runoff. Light grazing had a higher

Table 12.8 Amounts of Cover Required to Prevent Erosion by Raindrop Impact

Data from Osborn (1956)

Effectiveness, percent	Effective weight of cover	Weight of cover			
		Short-sod grasses	Mixed-range grasses	Ordinary crops and grasses	Tall, coarse crops and weeds
		(Lb/acre)			
98	3,000	4,000	5,000	6,000	
95	1,500	2,000	3,000	3,500	6,000
90	1,000	1,500	2,000	2,500	4,000
80	600	1,000	1,400	1,750	2,250
70	400	700	1,100	1,300	1,500
50	200	400	750	800	900
35	100	250	500	600	600
25	80	175	400	400	400

yield of water than protected plots and produced no runoff. Soil moisture content was higher under heavy grazing (Liacos, 1962a). Where
vegetation is altered more drastically than by grazing, benefits have been
observed. Conversion of oak forest to a mixture of grasses and legumes
resulted in increased water yields of 80 percent (Papazafiriou and
Burgy, 1970), and brushland converted to grass in Arizona resulted in
increased water yields of 3.8 to 35.5 cm per year (Hibbert, 1969).

Recognition of the fact that vegetation uses water has led to proposals to graze range watersheds heavily in order to reduce transpirational
losses. The contention has been that more water is obtainable from an
unvegetated slope than from a vegetated one. In one instance, prominent citizens and officials actually petitioned government land administrators for more intensive grazing of the forage on watersheds adjacent
to their communities under the premise that more water would be available for irrigation. This idea contains some truth. It fails, however, to
consider the possible deterioration of the watershed through erosion
and the threat of floods and damage to adjoining properties. Moreover,
greater water flow may mean less, not more, *usable* water (Fig. 12.7).

Effect of Grazing upon Time Distribution of Streamflow

Hypothetical water-flow curves show that the demand for irrigation
water comes late in the summer (Fig. 12.8). Water flow from an undisturbed drainage more nearly coincides with this demand period than

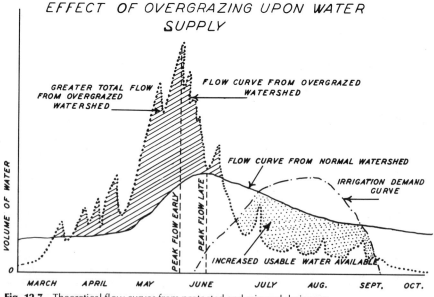

Fig. 12.7 Theoretical flow curves from protected and misused drainages,
together with irrigation-water demand curves. Water deficits are much
less under proper management.

Fig. 12.8 Nature designs streams to produce clear water and flow regularly through vegetated channels. This stream drains an area which had not been grazed for 50 years.

does the flow from misused watersheds. Not only does peak flow come too early for irrigation, but also it may contain quantities of silt so great as to interfere seriously with irrigation or other uses.

Effects of Range Treatment Practices on Water Regimes

Aside from the effects of grazing, there are a number of rangeland treatments that may influence water intake, water loss, and erosion. Since many of these are designed to increase forage production, in the long run they can prove beneficial. In the short run, removal of cover and disturbance of the soil may have adverse effects. There are few data which permit comparison of pretreatment and posttreatment hydrological phenomena.

One such experiment involved plowing and seeding in a big sagebrush stand in southern Idaho. Infiltration tests were made prior to and for 12 years following treatment. With few exceptions, infiltration was decreased by treatment, and in those few instances in which infiltration exceeded pretreatment rates, they were not statistically significant (Gifford, 1972). Sediment yields were, with one exception, increased from 260 to 1,200 percent. Meeuwig (1965) observed better soil properties 7 years after seeding grasses on high-altitude ranges in Utah (see Table 12.5).

Removal of piñon-juniper stands is accompanied by considerable soil disturbance depending upon the particular treatment method employed. Studies made to evaluate the effects of such treatment on a number of sites in Utah yielded inconclusive data. At 3 of 11 sites, infiltration was significantly improved by tree removal; the remainder were only slightly better or slightly poorer than untreated areas. Sedimentation was less on treated areas in one instance, but was greater in others (Gifford et al., 1970).

Wind Erosion

Wind erosion in serious form is confined to arid regions, usually below 39 cm of precipitation, the most important reasons being that here vegetation is sparse and wind velocities high. Wind velocities are greatest on level prairies. Topographic irregularities and vegetation may greatly decrease wind velocities overall, but they may intensify them locally as air currents become compressed when they flow around obstacles.

Vegetation is the most vital single element determining rate of soil blowing. It directly affects wind velocities at the soil surface (Fig. 12.9), and indirectly influences soil structure, soil organic matter, soil moisture, and atmospheric humidity. It is through these media, together with mechanical binding of the soil particles, that vegetation protects soil from blowing.

Recognizing Accelerated Erosion

No problem is more difficult for the range manager than recognizing normal or *geologic erosion* from *accelerated erosion* caused by disturbances that attend man's activities. Because of aridity and poor soils, rangelands often are too poorly vegetated to protect fully the soil from water and wind. Consequently, geologic erosion in the absence of any land use may be high. Little can be done to correct geologic erosion since it is a normal outcome of steep slopes, immature soils, and large areas of subclimax vegetation.

Accelerated erosion occurs when the stabilizing vegetation is destroyed by man and is no longer able to hold soil against the eroding forces of nature. Such erosion is an aftermath of vegetation regression. Since it is caused by man, accelerated erosion may be corrected. Corrective measures can take the form of altering land-management systems so that nature can rebuild the damaged ecosystem, although if erosion is far advanced, reoccupation by climax vegetation may be delayed decades awaiting rebuilding of the soil. In this case, it may be necessary to speed up the natural healing processes by mechanical structures and by reestablishment of vegetation.

Since changes in vegetation are the primary cause of the acceleration of erosion, vegetation also is the chief weapon in controlling ero-

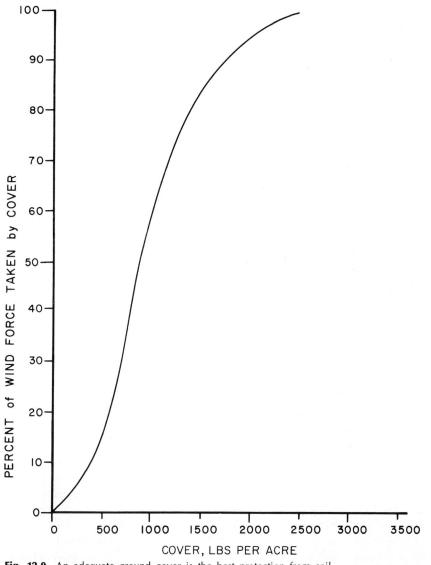

Fig. 12.9 An adequate ground cover is the best protection from soil blowing. *(From Osborn, 1956.)*

sion. The procedure of the land manager must be to determine the most advisable land use from a scientific, social, and economic viewpoint. On nontillable land it is necessary to protect the native vegetation by initiating a correct grazing method or a systematic timber-cutting program and, if the native vegetation has been destroyed, to replace it by seeding.

Mechanical means of controlling erosion and runoff are artificial and can never be considered as permanent, for they necessitate continued and expensive upkeep. Generally, a mechanical control is neither so effective nor so safe as one involving vegetation. Vegetation, because it is living, replenishes and rejuvenates itself and can continue to be effective (Fig. 12.10).

Proper grazing may be all that is needed to accomplish adequate recovery. Aldon (1963) reports reduction in sediment loads from watersheds 100 to 200 ha in size of more than 75 percent due only to control of cattle grazing and reduced utilization.

RECREATION ON RANGELANDS

In the United States, recreational use of rangelands has increased tremendously in recent years. Once this use was confined to limited acreages in more favored sites along streams and in forest settings. The primary activities, aside from hunting and fishing, were camping and picnicking. Areas once viewed as unattractive are now overrun by "rockhounds" (rock collectors), cyclists, all-terrain vehicles, and snowmobiles. No place, however remote or forbidding the terrain, is exempt. The extent to which the public interest in outdoor recreation has increased is

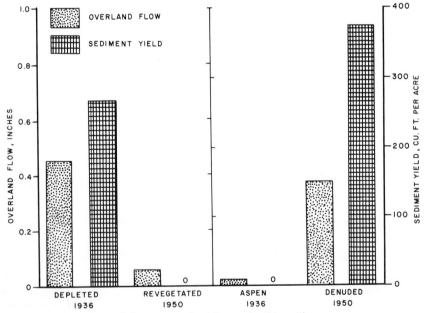

Fig. 12.10   The effects of plants on overland flow and soil loss. The area depicted on the left was revegetated; the area on the right denuded. *(From Marston, 1958.)*

typified in a sign change along a major highway in central Utah. A few years ago, there stood a sign at a road junction pointing to "Jericho," a place of importance as a rail-loading and shearing site for great numbers of sheep enroute from winter grazing grounds on the deserts of western Utah to summer ranges. That sign is now gone. In its place is one reading "Sand Dunes" pointing to an area of shifting sands near the railway siding of Jericho which now attracts great numbers of recreationists. Not all ranges are due for equally great changes in emphasis, but recreational use of rangelands is certain to continue to increase and range administration must change accordingly (Table 12.9).

The outstanding example of increased recreational use of desert areas is in southern California near to the population centers. As many as 5,000 dune buggies use these rangelands in a single weekend, and 67,000 participants and 189,000 spectators are estimated to have been attracted by competitive events involving off-road vehicles in 1972. Projected plans to control uses of this type amounted to over $3 million over a 9-year period (Development and Resources Corporation, 1971).

Recreation and Other Rangeland Uses

Different kinds of resource-based recreation are not equally compatible with other land uses. Some, such as nature study, are nonconsumptive, require no special facilities, and give rise to little impact to the resource. By contrast, maintenance of huntable animal populations, especially of big game, requires that forage be shared with livestock; wild animals

Table 12.9   Recreation Days Spent by Utah Residents in Rangeland-Oriented Activities in Utah, 1971–1972

Data from Hunt and Brown (1973)

Activity	Recreation days	Activity	Recreation days
Camping	1,756,300	Jeeping	31,300
Fishing	486,400	Tubing	27,600
Hiking	308,800	Water skiing	25,700
Skiing	299,000	Canoeing, kayaking, etc.	22,400
Horseback riding	245,300	Mountain climbing	20,100
Boating	237,000	Dune-buggy riding	13,300
Picnicking	222,900	Trapping	12,900
Trail biking	114,800	Exploring caves	4,800
Motorcycling	98,300	Snowshoeing	4,000
Rock hounding	87,900	Relic hunting	3,100
Snowmobiling	85,600	Bird watching	900
River trips	52,000		

may adversely affect other uses such as timber production; and hunters in pursuit of game deteriorate the range in off-road travel and other ways. Some forms of recreation require special facilities and elimination of other uses—for example, campgrounds and boat docks. The areas required for such facilities are usually small, and, though the effects of the use generated by them are severe, their total impact on the range is small. The most serious consequence of these types of use is from stream and lake pollution. Ski areas are extreme examples of minimal incompatibility to other rangeland uses but with a high pollution hazard, located as they most frequently are at stream sources.

Other than homesite developments, the dispersed type of recreational activities are those that create greatest problems in rangeland administration (Fig. 12.11). Since they are not site-oriented, there is little that can be done to regulate users and establish facilities to insulate the resource from their impacts. The kinds of resource-oriented recreational activities that rangelands support and their relative impact on the resources are shown in Table 12.10.

There is a certain amount of dichotomy in many of these activities. The impacts from camping are much different when camp sites are used than when camping takes place on undeveloped land at sites of the campers' choice. Horseback riding similarly differs in the degree to

Fig. 12.11 Motorcycles and other off-road vehicles provide an enormous challenge to the range manager. Shown here is part of a group of contestants and spectators preparing for a race; scores of participants and hundreds of spectators gather for such events. *(Photograph courtesy John D. Hunt.)*

Table 12.10   Popular Resource-Oriented Activities Arranged to Indicate Relationship to Rangelands and the Potential Impacts

Impacts	Primarily on rangeland	Partially on rangeland
Heavy impact	Camping*	Fishing
	Trail biking	Picnicking
	Rock hounding	Motorcycling
	Jeeping	Hunting
	Dune-buggy riding	Horse-packing trips[†]
Moderate to light impact	Hiking	Snowshoeing
	Skiing	Horseback riding
	Snowmobiling	Boating
	Mountain climbing	Water activities
	Arrowhead hunting	Nature study
	Relic hunting	Trapping

*Degree of impact varies with kind of camping—whether concentrated or dispersed.
†Hazards may be more indirect than direct.

which it affects rangelands. Not all horseback riding takes place on rangelands. When it does and involves small groups riding established roads and trails, the adverse effects are minor. Pack trips, on the other hand, which involve large groups and sizable herds of pack animals, can make tremendous demands on the limited forage crops usually available in the more primitive areas that attract this kind of use.

Indirect Effects of Recreation Use

Not all impacts from recreational use are direct; some arise from associated phenomena such as the increased incidence of fires and vandalism. This is of concern when groups of people enter the "back country" where no facilities exist and roads are lacking to service fire-protection activities. Examples of these are pack-horse and river trips.

Loss of domestic animals from theft and reduced production because of disturbance of animals become more serious when large numbers of people frequent ranges where livestock are grazed. This increased recreational use of rangelands will require attention to "people management" both by the public-land administrator and by the private landowner.

The sale of recreational home sites has an alarming impact on rangeland use in the Western United States. Workman et al. (1973) reported literally thousands of recreational home sites being sold in Utah without provisions being made for water supply, waste disposal, or access. Not only do these unplanned developments remove rangeland from production, but they indirectly affect range use through increased pollution, taxes, and other social interactions.

Recreation as a Source of Income

Some types of recreational activities may increase revenue from private rangeland. Sale of hunting privileges is a major source of income for the Western United States ranchers (Ramsey, 1965; Severson and Gartner, 1972). Similar situations exist on some of the private ranches in East Africa where charging for hunting safaris has become a profitable business.

The demand for rangeland recreation has led to the development of a great many "dude ranches" in North America where the culture of the cowboy and the beauty of the surroundings attract vacationers and form the basis of a growing tourist industry. Some of these operations are also legitimate livestock-producing ranches, and the tourist business is supplemental to other ranch income. Others have developed strictly as a tourist industry, and livestock, if present at all, are only incidental to the recreational enterprise.

INTEGRATION OF THE MULTIPLE USES

Grazing, the uses discussed here, and wildlife production (see Chapter 11) are all legitimate and important uses of rangelands. Since no single use can be maximized without affecting, and perhaps negating, the other, tradeoffs must be evaluated in some sort of optimizing process. In some cases a landowner, or country may unilaterally decide to emphasize a given product at the expense of others. However, this kind of land use is becoming rare since society demands many goods and services from land whether public or private.

Tools such as linear programming, simulation modeling, and other quantative methods of integrating many interactions are becoming useful in arriving at combinations of land uses. Today's range manager must be able to evaluate the tradeoffs among uses if rangelands are to make their maximum contribution to society.

BIBLIOGRAPHY

Aldon, Earl F. (1963): Sediment production as affected by changes in ground cover under semiarid conditions on the Rio Puerco drainage in west central New Mexico, *Trans. Am. Geophysical Union* **44:**872–873.

Ayres, Q. C. (1936): *Soil Erosion and Its Control,* McGraw-Hill, New York.

Bergerud, Arthur T., and Frank Manuel (1968): Moose damage to balsam fir–white birch forests in central Newfoundland, *Jour. Wildlife Mgt.* **32:**729–746.

Black, Hugh C., and B. T. Vladimiroff (1963): Effect of grazing on regeneration of Douglas fir in southwestern Oregon, *Proceedings of the Society of American Foresters,* Boston, Mass., pp. 69–76.

Brown, Jimmy W., and Joseph L. Schuster (1969): Effects of grazing on a hard-land site in the southern High Plains, *Jour. Range Mgt.* **22:**418–423.

Browning, Bruce M., and Earl M. Lauppe (1964): A deer study in a redwood-Douglas fir forest type, *Calif. Fish and Game* **50:**132–146.

Buckhouse, John C., and George B. Coltharp (1968): Effects of simulated graz-ing on soil moisture content in mid-elevation reseeded rangeland, *Proc. Utah Academy Sciences, Arts and Letters* **45:**211–219, part 1.

Campbell, Robert S. (1947): Forest grazing in the southern coastal plain, *Proceedings of the Society of American Foresters*, Minneapolis, Minn., pp. 262–270.

——— and Robert R. Rhodes (1944): Forest grazing in relation to beef cattle production in Louisiana, *La. Agr. Expt. Sta. Bull.* **380.**

Cheyney, E. G., and R. M. Brown (1927): The farm woodlot of southeastern Minnesota, its composition, volume, growth, value, and future possibilities, *Minn. Agr. Expt. Sta. Bull.* **241.**

Craddock, G. W., and C. K. Pearse (1938): Surface run-off and erosion on granitic mountain soils of Idaho as influenced by range cover, soil distur-bance, slope and precipitation intensity, *U.S. Dept. Agr. Circ.* **482.**

Crawford, Hewlette S., Jr. (1971): Wildlife habitat changes after intermediate cutting for even-aged oak management, *Jour. Wildlife Mgt.* **35:**275–286.

Croft, A. R., and L. V. Monninger (1953): Evapotranspiration and other water losses on some aspen forest types in relation to water available for stream flow, *Trans. Amer. Geophys. Union* **34:**563–574.

Crouch, Glenn L. (1966): Preferences of black-tailed deer for native forage and Douglas-fir seedlings, *Jour. Wildlife Mgt.* **30:**471–475.

Deam, C. C. (1923): Indiana woodlands and their management, *Ind. Dept. Con-servation Div. Forestry Bull.* **5,** Pub. 28.

Dee, Richard F., Thadis W. Box, and Ed Robertson, Jr. (1966): Influence of grass vegetation on water intake of Pullman silty clay loam, *Jour. Range Mgt.* **19:**77–79.

Denuyl, Daniel, and R. K. Day (1932): Natural regeneration of farm woods following the exclusion of livestock, *Ind. Agr. Expt. Sta. Bull.* **368.**

——— and ——— (1934): Woodland carrying capacities and grazing injury studies, *Ind. Agr. Expt. Sta. Bull.* **391.**

Development and Resources Corporation (1971): Planning guidelines for the California desert program, prepared for California State Office, Bureau of Land Management, Sacramento, Calif.

Edgerton, Paul J. (1972): Big-game use and habitat changes in a recently logged mixed conifer forest in northeastern Oregon, *Proc. Western Assoc. State Game and Fish Comm. Conference* **52:**239–246.

Flory, E. L. (1936): Comparison of the environment and some physiological responses of prairie vegetation and cultivated maize, *Ecology* **17:**67–103.

Gifford, Gerald F. (1972): Infiltration rate and sediment production trends on a plowed big sagebrush site, *Jour. Range Mgt.* **25:**53–55.

———, Gerald Williams, and George B. Coltharp (1970): Infiltration and ero-sion studies on pinyon-juniper conversion sites in southern Utah, *Jour. Range Mgt.* **23:**402–406.

Glover, Fred A. (1955): Black bear damage to redwood reproduction, *Jour. Wildlife Mgt.* **19:**437–443.

Grelen, H. E., L. B. Whitaker, and R. E. Lohrey (1972): Herbage response to precommercial thinning in direct-seeded slash pine, *Jour. Range Mgt.* **25:**435–437.

Halls, L. K., and R. Alcaniz (1971): Forage yields in an east Texas pine-hardwood forest, *Jour. Forestry* **69:**25–26.

———— and Hewlette S. Crawford, Jr. (1960): Deer-forest habitat relationships in north Arkansas, *Jour. Wildlife Mgt.* **24:**387–395.

Hanson, Clayton L., H. G. Heinemann, A. R. Kuhlman, and J. W. Neuberger (1973): Sediment yields from small rangeland watersheds in western South Dakota, *Jour. Range Mgt.* **26:**215–219.

————, Armine R. Kuhlman, Carl J. Erickson, and James K. Lewis (1970): Grazing effects on runoff and vegetation on western South Dakota rangeland, *Jour. Range Mgt.* **23:**418–420.

Harlow, R. F., and L. K. Halls (1972): Response of yellow poplar and dogwood seedlings to clipping, *Jour. Wildlife Mgt.* **36:**1076–1080.

Harper, James A., and Donald O. Swanson (1970): The use of logged timberland by Roosevelt elk in southwestern Oregon, *Proc. Western Assoc. State Game and Fish Commissioners.* **50:**318–341.

Hatton, J. H. (1920): Livestock grazing as a factor in fire protection on national forests, *U.S. Dept. Agr. Circ.* **134.**

Hedrick, D. W., and R. F. Keniston (1966): Grazing and Douglas-fir growth in the Oregon white-oak type, *Jour. Forestry* **64:**735–738.

Hibbert, Alden R. (1969): Precipitation limits the increase in streamflow after converting brush to grass in Arizona, *Trans. Amer. Geophys. Union* **50:**139 (abstract).

Hoover, M. D. (1944): Effect of removal of forest vegetation upon water yields, *Trans. Amer. Geophys. Union,* Part VI, **25:**969–975.

Hornkohl, Leon W., and Ralph A. Read (1947): Forest grazing in the Ozarks, *Proceedings of the Society of American Foresters,* Minneapolis, Minn. pp. 270–277.

Hunt, John D., and Perry J. Brown (1973): Utah resident outdoor recreation participation, 1971–1972, Report to the Utah State Department of Natural Resources (multilith).

Ingram, D. C. (1931): Vegetative changes and grazing use on Douglas fir cut-over land, *Jour. Agr. Research* **43:**387–417.

Krebill, R. G. (1972): Mortality of aspen on the Gros Ventre elk winter range, *U.S. Dept. Agr. Forest Service, Res. Paper* INT-129.

Liacos, Leonidas G. (1962a): Soil moisture depletion in the annual grass type, *Jour. Range Mgt.* **15:**67–72.

———— (1962b): Water yield as influenced by degree of grazing in the California winter grasslands, *Jour. Range Mgt.* **15:**34–42.

Linnartz, Norwin E., Chung-yun Hse, and V. L. Duvall (1966): Grazing impairs physical properties of a forest soil in central Louisiana, *Jour. Forestry,* **64:**239–243.

Lull, Howard W. (1959): Soil compaction on forest and range lands, *U.S. Dept. Agr., Forest Service Misc. Pub.* **768.**

Marston, Richard B. (1952): Ground cover requirements for storm runoff control on aspen sites in northern Utah, *Jour. Forestry* **50:**303–307.

——— (1958): Parrish canyon, Utah: a lesson in flood sources, *Jour. Soil and Water Conservation* **13:**165–167.

McConnell, Burt R., and Justin G. Smith (1970): Response of understory vegetation to ponderosa pine thinning in eastern Washington, *Jour. Range Mgt.* **23:**208–212.

Meeuwig, Richard O. (1965): Effects of seeding and grazing on infiltration capacity and soil stability of a subalpine range in central Utah, *Jour. Range Mgt.* **18:**173–180.

——— (1970): Infiltration and soil erosion as influenced by vegetation and soil in northern Utah, *Jour. Range Mgt.* **23:**185–188.

Morris, Melvin S. (1947): The grazing use of forest lands in the northern Rocky Mountain region, *Proceedings of the Society of American Foresters*, Minneapolis, Minn., pp. 303–311.

Ndawula-Senyimba, M. S., C. V. Brink, and A. McLean (1971):Moisture interception as a factor in the competitive ability of bluebunch wheatgrass, *Jour. Range Mgt.* **24:**198–203.

Osborn, Ben (1956): Cover requirements for the protection of range site and biota, *Jour. Range Mgt.* **9:**75–80.

Packer, Paul E. (1951): An approach to watershed protection criteria, *Jour. Forestry,* **49:**639–644.

——— (1963): Soil stability requirements for the Gallatin elk winter range, *Jour. Wildlife Mgt.* **27:**401–410.

Papazafiriou, Z. G., and Robert H. Burgy (1970): Effects of vegetation conversion on watershed yield, *Trans. Amer. Geophys. Union* **51:**753 (abstract).

Patton, David R., and Burd S. McGinnes (1964): Deer browse relative to age and intensity of timber harvest, *Jour. Wildlife Mgt.* **28:**458–463.

Pearson, G. A. (1931): Recovery of western yellow pine seedlings from injury by grazing animals, *Jour. Forestry* **29:**876–895.

——— (1950): Management of ponderosa pine in the southwest, U.S. Department of Agriculture, Forest Service, *Agr. Monogr.* no. 6.

Pearson, H. A., L. B. Whitaker, and V. L. Duvall (1971): Slash pine regeneration under regulated grazing, *Jour. Forestry* **69:**744–746.

Public Land Law Review Commission (1969): Study of the development, management, and use of water resources on the public lands, *U.S. Department of Commerce, Clearinghouse for Federal and Scientific Information P. B.* 18 065, Springfield, Va.

Ramsey, C. W. (1965): Potential economic returns from deer as compared with livestock in the Edwards Plateau region of Texas, *Jour. Range Mgt.* **18:**247–250.

Rauzi, Frank, and Clayton L. Hanson (1966): Water intake and runoff as affected by intensity of grazing, *Jour. Range Mgt.* **19:**351–356.

———, and Freeman M. Smith (1973): Infiltration rates: three soils with three grazing levels in northeastern Colorado, *Jour. Range Mgt.* **26:**126–129.

Rayburn, Walker, Jr., and Frederick S. Barkalow, Jr., (1973): Seasonal deer browse preference for tupelo and green ash, *Jour. Forestry* **71:**31–33.

Reid, Elbert H. (1947): Forest grazing in the Pacific Northwest, *Proceedings of the Society of American Foresters,* Minneapolis, Minn. pp. 296–302.

Reynolds, Hudson G. (1962) Effect of logging on understory vegetation and deer use in a ponderosa pine forest of Arizona, *Rocky Mountain Forest and Range Expt. Sta., Research Note* no. 80.

Roy, D. F. (1960): Deer browsing and Douglas-fir seedling growth in northwestern California, *Jour. Forestry* **58**:518–522.

Sampson, Arthur W. (1919): Effect of grazing upon aspen reproduction, *U.S. Dept. Agr. Bull.* **741.**

Severson, Keith E., and F. Robert Gartner (1972): Problems in commercial hunting systems: South Dakota and Texas compared, *Jour. Range Mgt.* **25**:342–345.

Shepherd, W. O., E. U. Dillard and H. L. Lucus (1951): Grazing and fire influences in pond pine forests, *N. C. Agr. Expt. Sta. Tech. Bull.* **97.**

Smith, Arthur D., Paul A. Lucas, Calvin O. Baker, and George W. Scotter (1972): The effects of deer and domestic livestock on aspen regeneration in Utah, *Utah Div. Wildlife Resources, Pub. No.* 72–1.

Sparhawk, W. N. (1918): Effect of grazing upon western yellow pine reproduction in central Idaho, *U.S. Dept. Agr. Bull.* **738.**

Stoeckeler, J. H., R. O. Strothmann, and L. W. Krefting (1957): Effect of deer browsing on reproduction in the northern hardwood-hemlock type in northeastern Wisconsin, *Jour. Wildlife Mgt.* **21**:75–80.

Thomas, G. W. (1969): *Progress and change in the agricultural industry,* Kendall-Hunt Publishing Co., Dubuque, Iowa.

U.S. Department of Agriculture, Southeastern Forest Experiment Station. Biennial Report for the Years 1947 and 1948, *Station Paper* 2.

Wahlenberg, W. G. (1935): Effect of fire and grazing on soil properties and the natural reproduction of longleaf pine, *Jour. Forestry* **33**:331–338.

Wallmo, Olof C., Wayne L. Regelin, and Donald W. Reichert (1972): Forage use by mule deer relative to logging in Colorado, *Jour. Wildlife Mgt.* **36**:1025–1033.

Westell, Casey E., Jr. (1954): Available browse following aspen logging in lower Michigan, *Jour. Wildlife Mgt.* **18**:266–271.

Williams, Robert E. (1952): Better management on longleaf pine forest ranges, *Jour. Range Mgt.* **5**:135–140.

Wing, Larry D., and Irven O. Buss (1970): Elephants and forests, *Wildlife Monogr.* no. 19.

Wodzicki, K. A. (1950): Introduced mammals of New Zealand, Department of Scientific and Industrial Research Bulletin no. 98, Wellington, New Zealand.

Woodward, Lowell (1943): Infiltration capacities of some plant-soil complexes on Utah watershed lands, *Trans. Amer. Geophys. Union,* part II. pp. 468–475.

Workman, John P., Donald W. MacPherson, Darwin B. Nielsen, and James J. Kennedy (1973): A taxpayer's problem—recreational subdivisions in Utah, Environment and Man Program, Utah State University Press, Logan, Utah.

Young, V. A., G. B. Doll, G. A. Harris, and J. P. Blaisdell (1942): The influence of sheep grazing on coniferous reproduction and forage on cut-over western white pine areas in northern Idaho, *Univ. Idaho Forestry Bull.* **6**, series 1.

Chapter 13

Manipulating Range Vegetation

Seldom does a range offer an array of plants in a mix that is ideal. Native ranges when first occupied by domestic livestock were in various stages of succession following fires or had been held in some seral stage by native animals. Two major alterations of the range ecosystems occurred with the appearance of modern man and his livestock: the natural fire regimes were altered, and grazing use was intensified. Consequently, successional changes have taken place, often to a less desirable and less productive stage. If rangeland is to be held at a particular stage, periodic perturbations of the ecosystem are required to counterbalance natural successional trends.

There are several ways in which vegetation can be altered: (1) with fires, (2) mechanically, (3) chemically, and (4) biologically. Each of these has its advantages and disadvantages. Each has different impacts on the species which are being controlled and on the species which are to be retained.

It should be remembered that seldom is it desirable to eliminate a species from the range even if that were possible. Most plants have some value if they are not in too great numbers. Shrubs of low-forage value may, by their presence, have beneficial effects upon other vegetation. There can be little doubt that sagebrush in the Western United States has provided protection to valuable forage plants which, otherwise, would have been eliminated by grazing. Bailey (1970) has accorded *Eleagnus commutatus* a similar role in the fescue grasslands in Alberta. Even poisonous plants may not be so serious a threat to livestock that eradication is required: many of them provide valuable forage when they are present in moderate quantities.

FIRE AS A RANGE-MANAGEMENT TOOL

Since time immemorial, fires caused by lightning and man have been one of the major forces affecting forest and grassland ecosystems. Wherever there is sufficient fuel, fires have periodically consumed trees and shrubs in those climates conducive to grasslands, thus maintaining grassland and savanna types. Sauer (1950) considered fire to have been a determinant of certain climax grassland types, and questioned the validity of a grassland climax except in certain special circumstances. Shantz (1947) estimated that one-third of the earth's natural vegetation resulted from fire.

The tall-grass prairie, generally regarded as a climax type, may owe its continuance to fire. Certainly, shrub invasion of the North American prairies has long been observed, and fire has been used to prevent it (Aldous, 1934). That this invasion may not be due to deterioration of the prairie caused by grazing is indicated by the fact that shrubs appear even under complete protection. Penfound (1964) provides evidence that invasion of *Symphoricarpos orbiculatus* follows protection from grazing. Formerly grazed prairies near Norman, Oklahoma, after 13 years of protection, had 19 times more shrub cover than did grazed prairies. Similarly, the abundance of *Juniperus virginiana* was inversely related to intensity of grazing in the Flint Hills of Kansas (Owensby et al., 1973).

Fire is a major ecological factor in maintaining tropical savannas and grasslands. Bachtelder (1967) found evidence that fire was used by primitive people at the forest-grassland interface in all tropical areas. When fire was controlled, bush types encroached on grasslands. In modern times, fire was used as a management tool in Africa (Phillips, 1936; Thompson, 1936) and in Brazil (Vincent, 1936) before its use had wide acceptance in the United States because of the influence of foresters on range management. Fire has since come to be widely used by range managers and even foresters who are questioning the desirability of excluding fire from forests in all instances. Considerable credit for this change in attitude must go to Stoddard (1931) who early promoted use of controlled fires for improving habitat for bobwhite quail in pine forests of the Southeast and demonstrated that fire could be beneficial to forest production as well.

Fire commonly is used in rangelands to (1) remove shrubby species of low-forage value and (2) to remove herbaceous vegetation that accumulates where rank-growing species are common.

Effect of Fire upon Soil

Fires were once thought to be extremely damaging to soil because of the reduction of organic matter and nitrogen due to high soil temperature. Experimental data do not bear this out. Although surface temperatures are high they are of brief duration and have little effect on the soil. Although Heyward (1938) recorded temperatures above 260°C when forests were burned, temperatures within the litter were frequently not sufficiently high to ignite organic material. Stinson and Wright (1969) observed surface temperatures, depending on the volume of fuel, of from 83 to 682°C in mixed prairie in Texas, but temperatures above 65°C seldom lasted longer than 2 to 4 minutes.

The hypotheses of soil deterioration is not supported by evidence. Annual burning in the Southeast pine forests did not result in lower organic matter (Table 13.1). Increased nitrogen content of soil and uptake of nitrogen by plants resulted from burning oak—mountain mahogany chaparral in Arizona (Mayland, 1967). No decrease in organic matter or nitrogen accompanied grassland burning in Kansas (Aldous, 1934).

Owensby and Wyrill (1973) found varying effects on the chemical properties of prairie soils in Kansas due to burning, there were more or fewer nutrients present than in unburned plots, but differences were slight. There was no indication that annual fires deteriorated the range, though some species were favored over others.

Table 13.1   Nitrogen and Organic Matter in
the Upper 6 In. of Soils from Pine Forests
Burned Annually for 8 Years Compared with
Unburned Forests

Data from Greene (1935)

Treatment	Organic matter, percent	Nitrogen, percent
Burned	4.32	0.075
Unburned	2.63	0.048

Burning Grasslands

On grass ranges, fire is used to remove the unutilized herbaceous material of the previous year in order that new growth may develop unhindered by the accumulations of dead material and be readily available to grazing animals (Fig. 13.1). Removing dead grass by fire results in more uniform utilization of the forage. This is especially important where growth is rank and coarse, typical of tropical and subtropical humid climates.

On Kansas prairies (Hensel, 1923), mean soil temperatures throughout the grazing season were higher on burned areas because of greater exposure to the sun and hence increased heat absorption. At a depth of 2.5 cm, the maximum temperature was raised as much as 6.7°C and at a 7.6-cm depth, by 2.0°C. Similar temperature differentials were observed on burned Iowa prairies (Ehrenreich, 1959). Higher temperatures existed until midsummer, though differentials were greatest during April, which resulted in more rapid growth in the early spring.

Burning grasslands generally decrease yields, but this is not always the case. Forage production may be lower on burned plots in some years and not in others (McMurphy and Anderson, 1965). Trlica and Schuster (1969) reported forage reductions of 15 to 35 percent from burning in the Texas high plains. Infrequent burning and, generally, spring burning are least harmful to subsequent forage production.

Fig. 13.1 Tall-grass prairie in North Texas. The accumulation of rank, dead grass in the background, if left standing, inhibits regrowth and hinders its use. Burning, properly timed, can improve grazing with no deterioration of the range.

Quality of forage may be increased or decreased by burning, because of changes in plant composition which vary with the time of burning. In Kansas, Aldous (1934) reported plots burned late in the fall had successional changes toward little bluestem *(Andropogon scoparius)*, and plots burned in late spring increased in coarser grasses, mostly big bluestem *(Andropogon gerardi)*. *Poa pratensis* increased on unburned plots and was virtually eliminated on burned plots.

More recently, after 30 burns over 36 years, the ungrazed plots used by Aldous were reinventoried with somewhat different results. Big bluestem had become dominant in plots burned in the fall and late spring; *Koeleria cristata* was favored by winter burning. Little bluestem was not as abundant on burned areas as on the unburned plots irrespective of the season, but it was greatly reduced by late-fall burning (McMurphy and Anderson, 1965).

Proper timing of fires can increase steer gains on prairie vegetation; best results in terms of stability of the vegetation and gains are obtained by late-spring burning (Anderson et al., 1970).

Other grassland types are less tolerant of fire. Repeated fire and overgrazing have been responsible for reducing perennial bunchgrasses in California and Washington which are replaced by annual bromegrasses. Once established, these fire-overgrazing subclimaxes, which are inferior grazing lands to the original bunchgrasses, persist. Fire is ineffective in controlling the annual species, although, properly timed fires may affect some species. Young et al. (1972) in California reduced *Bromus tectorum* by 3 years of burning, but could not affect the density of Medusae head *(Taeniatherum asperum)*.

Since livestock prefer burned range, it is essential to burn the entire range or to burn large areas and to base stocking capacity on the burned acreage; otherwise severe overgrazing will occur on the burned area to the detriment of better plants (Campbell, 1954). Utilization of *Eragrostis curvula* was increased from 8 to 52 percent by burning (Klett et al., 1971).

Burning North American Pine Forests

Burning in the southern pine region of the Eastern United States is not unlike burning in grasslands in that fires are directed to undergrowth and ordinarily do not burn the trees. Although too frequent burning is harmful to both the forest and the soil (Wahlenburg et al., 1939), controlled and regulated fire is a useful land-management tool. Probably 90 percent of the longleaf pine forest burns at least once every 3 to 4 years to improve grazing conditions. In the absence of fire, trees and shrubs increase, the composition of herbaceous cover changes, and basal area declines (Lewis and Hart, 1972). In consequence, forage production declines (Hilmon and Hughes, 1965). Although annual burning is generally not recommended, in the absence of fire, vegetation trends toward a pine-hardwood-brush mix which is undesirable to foresters, range managers, and wildlife managers alike.

Grass quality is improved in the spring following a burn, and animal gains increase markedly. Cattle were found to gain 37 percent more on annually burned longleaf pine forest than on unburned (Wahlenburg et al., 1939). Pearson et al. (1972) found increased digestibility and higher nitrogen and phosphorus contents of forage the first year after a fire in ponderosa pine forests of Arizona.

Burning to Control Shrubs

By far the most important use of fire on ranges is to control woody species. Although palatable browse species are of tremendous importance as both game and livestock feed, certain unpalatable shrubs, notably mesquite in the Southwestern United States, have increased so rapidly as to constitute serious range problems in many areas. Even moderately palatable species may become so abundant as to interfere with foraging livestock and reduce the production of forage.

Range shrubs vary greatly with respect to their tolerance to fire; hence, efforts to control them give variable results. The sclerophyllous shrubs of Mediterranean climates are judged by many ecologists to be fire types. Because of their sprouting habit, regeneration is prompt following fire and, though species composition may be affected, brush endures (Hanes and Jones, 1967). Other shrubs, sagebrush for example, are destroyed by fire and reinvasion must come by way of seed (Fig. 13.2). Many other species are intermediate to these in susceptibility. The use of fire in brushlands is highly complex and variable because of great variation among different species, field techniques, and environments.

Burning in Sclerophyll Brushlands Fire has long been regarded with disfavor in chaparral types. Forage values are low and root sprouts and fire-resistant seeds quickly replace the stand in 2 to 4 years (Sampson, 1944). In addition, soils are unprotected and torrential rains may cause great damage through floods and erosion (Adams et al., 1947).

In less xeric situations, however, fire can be effective in chaparral provided appropriate follow-up measures are taken. Love and Jones (1952) recommended a rotation involving burning, seeding to grass, careful grazing, and reburning as a means of reclaiming the more favorable brushland sites. Grazing by sheep or goats, particularly, are helpful in preventing the return of shrubs. Burning alone may not result in

Fig. 13.2 Reinvasion of brush is rapid unless proper management follows its removal. Fire and subsequent unregulated grazing has resulted in an almost pure stand of sagebrush in this range in Idaho.

greatly increased production, but the use of herbicidal sprays following burning both reduces shrubs and promotes the establishment of grass (Table 13.2).

Perhaps nowhere in the United States has the use of fire been more controversial than in California. Brush species and associations in

Table 13.2 Effects of Fire and Herbicides Followed by Seeding on Crown Cover in Chaparral Vegetation Near Dewey, Arizona

Data from Tiedemann and Schmutz (1966)

Classes of vegetation	Crown cover, percent		
	Native unburned	Burned and reseeded	Burned, reseeded and herbicide treated
Turbinella oak	24.2	19.0	3.0
Other shrubs	10.1	5.0	0.9
Halfshrubs	5.5	2.7	4.8
Grasses	0.6	3.5	23.9
Forbs	4.2	7.2	8.1
Total	44.6	37.4	40.7

California are numerous and variable. Burning itself may vary greatly in intensity and nature. These variables make any overall recommendation concerning fire impossible (Shantz, 1947), but it is widely agreed that planned burning has a place in land management in the area, provided due consideration is given to the character of the soil and vegetation and the potential of the site for production (Sampson and Burcham, 1954).

Burning Mesquite and Associated Shrubs In the early part of the century, botanists theorized that fire, formerly unchecked, had maintained Southwestern grasslands free of brush (Bentley, 1898). Grazing removed fuel until misused ranges often would not burn, and civilization brought with it fire-control efforts formerly unknown. Species known to have increased as a result include mesquite *(Prosopis juliflora)*, oak, juniper, burroweed *(Aplopappus tenuisectus)*, cholla cactus *(Opuntia fulgida)*, and snakeweed *(Gutierrezia)*. Mesquite is by far the most important and widespread. Three distinct varieties occur, and growth forms differ with habitat from many-stemmed bushes to large trees, requiring somewhat different control measures.

The role of fire in mesquite distribution is uncertain. Although some evidence tends to link mesquite spread with overgrazing, study plots in southern Arizona (Brown, 1950) showed that mesquite and other desert shrubs increased even when livestock and rodent grazing were completely eliminated (Figs. 13.3 and 13.4). Similarly, on plots open to use by cattle and rodents, mesquite numbers increased 108 percent. Where both were excluded, the increase was 129 percent (Glendening, 1952). Scifres et al. (1971), however, found evidence that a well-developed grass cover could prevent establishment of mesquite in North Texas. Nevertheless, the use of fire in control of mesquite was long vigorously opposed by many ecologists because of the fire-resistant character of the plant. Although fire often kills seedlings and small bushes, older plants sprout from the crown after fire and increase in abundance (Parker and Martin, (1952).

Even though mesquite is fire resistant, it and other shrubs can be reduced by fire, the effects lasting for 15 years (Humphrey, 1949). Late spring, the driest season of the year, gave highest shrub kills. White (1955) regarded mesquite, ocotillo *(Fouquieria splendens)*, and sotol *(Dasylirion wheeleri)* as sensitive to fire, and *Aplopappus laricifolius* as very sensitive; *Calliandra eriophylla* and *Mimosa depocaspa* were unaffected by fire. Similar differentials in susceptibility of shrub species to fire was observed in the Monte region of Argentina (Willard, 1973).

Grazing may indirectly favor mesquite spread by reducing the amount of fuel and thus the severity of fire. Where there is enough fuel,

Fig. 13.3 Santa Rita experimental range in southern Arizona in 1903. A few young plants of mesquite are visible in the stand of grama grasses to the left center of the photograph. *(Courtesy U.S. Forest Service.)*

Fig. 13.4 View of the range shown in Fig. 13.3, photographed in 1941. Mesquite has increased and grasses have decreased with a consequent reduction in grazing capacity from 40 to 50 head per section to only 12 to 15 head.

ground temperatures may be sufficiently high to kill mesquite trees
(Stinson and Wright, 1969). Since attained soil temperatures are dependent upon weather factors such as wind, relative humidity, and air
temperatures, and since age of mesquite greatly affects susceptibility to
fire, no precise fuel level nor critical temperature has yet been established. Temperatures of 425°C killed all mesquite seedlings less than a
year old, but killed only 4 percent of trees 10 years old (Wright, 1972)
making fire more effective in preventing mesquite invasion than in reducing established stands.

That introduction of livestock may contribute to mesquite increase
is supported by the fact that livestock consume the sweet, fleshy pods of
mesquite and many seeds pass through the digestive tract in viable condition at considerable distance from the parent plant (Fig. 13.5).

Burning Nonsprouting Brush   Shrubs which do not sprout from the
root such as big sagebrush *(Artemisia tridentata)* sometimes can be completely eradicated by a single fire when there is adequate ground fuel of
dry grass to carry the fire (Fig. 13.6). Although fire causes some reduction in perennial grasses, recovery is rapid without seeding. In the Snake

Fig. 13.5 Mesquite seedlings emerging from a pile of cow dung. Eight
pairs of cotyledons are visible in this photo. *(Courtesy U.S. Forest Service.)*

Fig. 13.6 Sagebrush can be completely consumed by fire when conditions are favorable, top. Bottom, boundary line between burned (left) and unburned sagebrush with excellent stand of grass. If perennial grasses are present, only protection is needed; in their absence seeding is necessary which can be done by drilling without seedbed preparation.

River Plains of Idaho, forage production was more than doubled by controlled burning followed by correct grazing (Pechanec and Stewart, 1944) (Fig. 13.7). In the absence of grass, seeding is necessary.

The success of burning sagebrush is determined by the climax status of the area. If the land normally is dominated by sagebrush, as is much of the central intermountain area, one can never hope to keep out sagebrush permanently. If the land is climax grassland, as is much of the northern intermountain area, and sagebrush has invaded as a result of improper grazing, then proper grazing after burning could be expected to keep out the sagebrush. All ranges should be protected from grazing following a burn to allow grasses time to gain in vigor; otherwise reoccupation by sagebrush may be rapid, or even worse, root-sprouting and worthless rabbitbrush *(Crysothamnus)* may take over.

The low-value burroweed in Arizona can be virtually eliminated by fire, and increased grass growth thereafter appears to cause significant reduction in new seedlings beginning growth (Humphrey, 1949). It seems possible that a fire no oftener than every 10 years will keep both burroweed and cholla cactus under control.

Nonsprouting junipers such as Utah juniper *(Juniperus osteosperma)* in the intermountain area and Ashe juniper. *(J. ashei)* in Texas burn

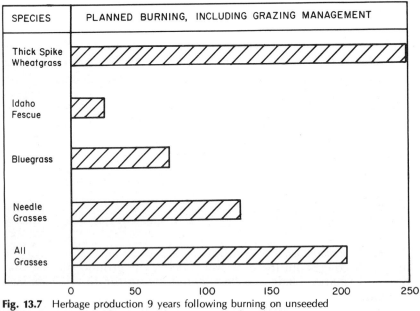

Fig. 13.7 Herbage production 9 years following burning on unseeded sagebrush range in southeastern Idaho, in percent of production on unburned area. Grazing capacity increased 83 percent. *(From Pechanec and Stewart, 1944.)*

readily and are easily killed by fire, but often ground fuel is insufficient to carry the fire from one tree to another. Where there is sufficient fuel, fire is an effective tool (Aro, 1971). A single fire has been reported to kill Ashe juniper trees in excess of 3 m in height. The burn usually is followed by dense stands of juniper seedlings, which require a second fire 10 years later to keep the range free of juniper (Wright, 1972).

Burning is often most effective when used with other forms of brush control. In dense stands of brush where there is insufficient fuel to carry a fire, it is often necessary to cut or break down the brush overstory and burn following the regrowth of herbaceous plants. Fire in combination with roller chopping, shredding, or root plowing is superior to any of the single treatments mentioned above in controlling brush in Texas brushlands (Box and White, 1969; Dodd and Holtz, 1972). Burning in East Africa is most effective in overgrazed woodland if the trees are first chopped by hand (F.A.O. 1973); the fuel from the chopped trees and any herbaceous plants that grow following bush chopping provide heat necessary to control the woody plants. Wink and Wright (1973) found that a minimum of 1,000 kg per hectare of fire fuel was required to kill seedlings of Ashe juniper and consume the trash left from bulldozing.

MECHANICAL CONTROL OF WOODY PLANTS

Mechanical control efforts have led to the adaptation or development of costly machinery, used primarily against individual trees and shrubs, although some can be mass produced. Nonsprouting species can be effectively cleared by mechanical devices alone; sprouting species require specialized equipment or companion treatments with fire, chemicals, or browsing with animals.

Among the many mechanical equipment devices used in eradicating trees and large shrubs are the following:

1. *The tree dozer,* a very large tractor bearing a push bar and a V-shaped root-cutting bar.

2. *The brush cutter or rolling chopper,* a large drum to which are affixed parallel blades. This drum can be filled with water to obtain any desired weight.

3. *The root plow,* a tractor bearing a 1- to 2-m blade, on a hydraulic lift which gouges out the root crown below the ground surface. Sometimes a U-shaped blade is pulled behind the tractor 15 to 50 cm belowground.

4. *The stinger blade,* a sharp narrow digger attached to a bulldozer and equipped with a lift device to root out small trees.

5. *The brush saw,* a circular saw operated parallel to the ground.

6. *The chain drag,* a large chain (ship-anchor chain), each end of which is attached to a track-type tractor, the two tractors being driven parallel through the brush. This relatively inexpensive method is effective only on large brush or small trees with rigid trunks and is widely used for juniper control.

7. *The brushland plow,* a heavy-duty multiple-disk plow used against low weak-stemmed shrubs such as sagebrush.

8. *The mower,* useful against small-stemmed species such as *Artemisia filifolia.*

9. *The brush beater,* a device which rapidly rotates weights or blades horizontally above the level of the ground shredding vegetation, most effectively used on small, brittle-stemmed species.

Effectiveness of Mechanical Controls

None of the devices used achieve 100 percent eradication, although all vegetation can be removed by a bulldozer blade operated below ground level. Otherwise, vigorously root-sprouting species return, and small plants survive any of the devices. Moreover, the stirring of the soil and removal of competition favors the establishment of a new stand from seed. This is not always a disadvantage for where other valuable plants are present, they too escape and produce more forage than before, a factor of special importance on ranges used by big game in winter in the Great Basin and adjoining regions. For example, piñon and juniper readily are uprooted by chaining, but bitterbrush is little harmed, and sagebrush is only reduced. Both are valuable winter forage for big game. Even the younger junipers that escape are useful as forage for big game and are valuable for cover. Vallentine (1971) discusses mechanical-control methods in detail.

Time of Treatment as a Factor in Brush Control

As with other treatment methods, proper timing is important for best results, although it may be less critical than in the case of herbicides which depend for their success on physiological activity. Sprouting shrubs are more affected by mechanical control at low stages in their food-storage cycles (See Chapter 4) and when conditions for regrowth are least favorable. Large trees can best be uprooted when the soil is moist. Smaller shrubs are more susceptible to methods that crush and pulverize the stems when they are most brittle in late season. Reinvasion may be increased, however, if treatment is done after seed formation (Table 13.3).

CHEMICAL CONTROL OF RANGE PLANTS

Chemicals have been widely used in attempts to reduce or eradicate range plants. Among the early ones were kerosene (or diesel fuel) and

Table 13.3   Effect of Time of Plowing and Seeding on Survival of Sagebrush and Establishment of Sagebrush and Wheatgrass Seedlings 1 year after Treatment

Data from Bleak and Miller (1955)

| Time of eradication | Sagebrush | | Crested wheatgrass |
	Survivors	Seedlings	Seedlings
	(No. per 100 sq ft)	(No. per 10 sq ft)	(No. per 10 sq ft)
April–May	2.26	0.33	6.24
July–August	1.28	0.01	7.77
August–September	0.38	0.01	12.23
Late September	0.54	0.04	13.10
Early October	1.24	0.20	9.19
October–November	0.88	4.09	10.42

sodium arsenite (Fisher et al., 1946). Since these had to be applied directly to individual plants, efforts were laborious and costly. In addition, the latter one was likely to be toxic to animals. The discovery of organic chemicals which disrupted, in one way or another, the normal physiological processes of the plants and which could be mechanically applied in spray form opened new horizons for manipulation of range vegetation. Herbicides can be applied more economically; are much less toxic; do not affect the stability of the soil, as do mechanical devices or fire; can be applied to steep, rocky areas; and, with some exceptions, are selective —affecting certain plants only. Their use has been remarkably successful against many species; against others they have proved of little value. There are about 150 basic chemical compounds used as herbicides (Ashton and Crafts, 1973). New compounds are constantly being formed, although environmental considerations are adding caution to this search.

Herbicides can be classified in many ways: (1) by their method of entry, (2) by their physiological action, (3) by their selectivity, (4) by their mobility in the plant, (5) by their method of application, or (6) by their chemical makeup. These are much too complicated for elaboration here. It is possible only to suggest the complexity of the subject and note instances of successful use on ranges.

Methods of Entry and Movement of Herbicides

Two major means of entry are used—through the leaves, and perhaps the stems, and through the roots. Different chemicals enter the leaf by different channels, through the stomata or through the cuticle, the latter being most important. It is thus evident that conditions that favor the application of one foliar-applied herbicide may not favor another. Agents (surfactants) thus are incorporated into foliar sprays to reduce surface tension of the liquid, aiding entry into the plant. Plants that resist entry of herbicide through the leaves, trees for example, can be treated by introducing the chemical into the stem (Leonard, 1956). Root entry is gained by the same processes as with inorganic ions, and the herbicide moves up through the xylem.

Mobility within the plant is an important factor in the effectiveness of herbicides whether absorbed through the foliage or the roots. Some foliar herbicides affect only the foliage they contact (pentachlorophenol); others move through the plant either through the phloem (2,4-D, dalapon, and amitrole) or xylem (paraquat, monuron, and simazine). The former are applied as sprays to the leaves; from there they move down into the roots. The latter are applied to the soil; from there they are carried upward into the leaves. Some herbicides move through either the phloem or the xylem depending upon where they gain entry (picloram), and can thus be applied either to the shoots or roots (National Research Council, 1968). Translocation within the plant may be affected by physiological condition of the plant so that translocation may not take place at certain times due to the absence of functional transporting tissues. Absorption may be less rapid at certain times. These and other reasons account for the greater susceptibility of a species at one phenological stage than at another.

Actions of Herbicides in Plants

Although the effects of herbicides are often clearly evident—chlorosis and morphological deformation are easily observed—the actual mechanisms of action of various herbicides within the plant are not at all clear. Research has yielded conflicting evidence leading one worker to identify a specific physiological dysfunction while others regard this to be a secondary effect. A number of mechanisms by which herbicides affect plants have been identified: hormonal imbalance and cell disfiguration, prevention of cell division, prevention of water and nutrient uptake, interference with respiration, protein degradation, prevention of the formation of nucleic acids and protein, disruption of photosynthesis, hindering formation of enzymes, and plasmolysis and desiccation.

In a review of literature on mechanisms of action, Moreland (1967) treated herbicides under three headings: (1) respiration and mitochondrial-electron transport, (2) photosynthesis and the Hill reaction, and (3) nucleic-acid metabolism and protein synthesis. Broadly these cover the biochemical responses of plants although the precise nature of the reaction within each of these may be quite different. Moreover, it is becoming clear that more than one mechanism may be involved. Thus, Ashton and Crafts (1973) attribute the effect of 2,4-D at low concentrations to breakdown of vascular tissue and starvation of the roots, and at higher concentrations to reduction in the utilization of oxygen and prevention of photosynthesis. Much more must be known about the ways in which herbicides act before they can be applied with any certainty of how a particular species of plant will respond.

Factors Affecting Usefulness of Herbicides

The most important characteristic of herbicides is the ability to affect some but not other plants. Selectivity among herbicides is achieved in different ways. Some herbicides affect only seedling plants which permits their use on established stands of perennial grasses to control annuals. Selectivity may be achieved by placement; a root-contact or root-absorbed herbicide placed in the surface layer will affect shallow-rooted species but will not affect plants rooting more deeply. Picloram applied to the soil had an adverse effect on roots of *Panicum virgatum* and *Bouteloua curtipendula,* although the latter was unaffected when the herbicide was placed at the 15-cm depth (Scifres and Halifax, 1972).

Selectivity can be increased by proper timing. Sneva and Hyder (1966) were able to achieve over 90 percent kills of sagebrush without great injury to *Purshia tridentata* by spraying prior to mid-May, before full leaf development. When spraying was delayed until the flower or bud stage, 30 percent mortality and 70 percent reduction in crown cover resulted in some years.

The most useful selective phenomen is the higher tolerance of grasses than broad-leaved plants, especially for the foliar-applied, systemic herbicides. Some herbicides exhibit no class selectivity, but differentially affect individual species of both grass and broad-leaved plants. This may be due to the ability of some plants to rapidly hydrolyze and render the particular chemical compound inert. The converse of this may occur also in which some plants appear to convert nontoxic materials into toxic substances, a process which, if widely applicable, offers possibilities not only for greater selectivity but may also make possible formulations of substances less hazardous to use.

The usefulness of herbicides depends also upon their stability and the length of time they are effective. Persistence in the soil is another factor. Some herbicides are quickly leached or rendered inert by organic matter in the soil; others remain in the soil for long periods of time. Rapidity of entry into the plant is a factor in effectiveness of foliar herbicides. Quick-entry foliar sprays are less likely to be washed from plants by subsequent rain. Rate at which a substance is moved within the plant is important also.

The variations among chemical and plant characteristics make it impossible, at present, to predict results. Their use remains a process of trial and error through experimentation. Generally the more effective a herbicide is in killing a refractory species, the more likely it is to be detrimental to some other valuable species. For example, when sufficient picloram was applied to control *Chrysothamnus parryi*, production of forbs was reduced to less than 1 percent of unsprayed areas (Paulsen and Miller, 1968). The characteristics of some of the more commonly used organic herbicides are shown in Table 13.4.

Control of Woody Plants with Chemicals

Varied results have been obtained from applications of herbicides to woody plants. Some vigorously sprouting species are affected only temporarily, and this characteristic is exploited to kill back the tops of palatable shrubs that have grown out of reach of grazing animals. The results are increased production and quality of forage. Nonsprouting species are more readily killed or thinned to desirable densities.

Control of Sprouting Shrubs   Sprouting shrubs are seldom effectively reduced by a single chemical or treatment. By using two chemicals that have different physiological effects, it is possible to get quite different results which may vary from increased to decreased responses. The effects may be greater than with either one of the two herbicides alone, equal to the sum of the effects of each applied singly, greater than the sum of the two single effects, or smaller than either one alone due to antagonism between the herbicides. There is no means of predicting results, as is illustrated by data from spraying *Quercus havardii* in Texas (Table 13.5). Stem counts gave different results than did production measurements, and the effects of soil-applied herbicides were different in the first than in the second year. Scifres and Hoffman (1972) found little difference between the canopy kill or sprout reduction on mesquite from dicamba and 2,4,5-T alone or in combination. Picloram and dicamba together were most effective.

The effect of a given herbicide or combination differs with the species (Table 13.6). Substantial reductions of *Acacia farnesiana* were obtained with any of the chemicals or combinations used, while *Berberis trifoliolata* was affected little by any of the three applications. In general, better results are obtained on sprouting species from herbicides that are applied to the roots than from foliar sprays. *Chrysothamnus nauseosus* is ineffectively controlled by foliar sprays, but high kills result when root-absorbed chemicals are used along with foliar sprays (Eckert and Evans, 1968). *Juniperus pinchoti,* a sprouting species, is unaffected by foliar sprays, is susceptible to picloram, but is not affected by some other soil-applied herbicides (Scifres, 1972b).

Table 13.4 Selected Herbicides Representative of More Common Chemical Types Grouped According to Class of Plant Principally Affected and Channel of Entry

Common name and class of plant affected	Chemical class	Effect on plant	Foliar Contact	Foliar Translocated	Roots*
Dicotelydons					
2, 4-D	Phenoxy	Hormonal imbalance		x	
		Cell deformation			
2,4,5-T	Phenoxy	Cell deformation		x	
Silvex	Phenoxy	Cell deformation		x	
Dicamba	Benzoic	Cell deformation		x	x
Diuron	Urea	Upsets photosynthesis			x
Monocotyledons					
Dalapon	Aliphatic	Affects protein metabolism		x	x
Chloropropham	Carbamate	Affects protein metabolism			x
Both Dicotelydons and Monocotyledons					
Amitrole	Triazole	Upsets photosynthesis		x	
Atrazine	Triazine	Upsets photosynthesis			x
CDAA	Amide	Affects protein metabolism			x
Dinoseb (DNBP)	Phenol	Affects protein metabolism	x		x
Diquat	Bipyridilium	Upsets photosynthesis	x		
Naptalam (NPA)	Amid	Upsets photosynthesis			x
Paraquat	Bipyridilium	Upsets photosynthesis	x	x	
Picloram	Pyridine	Growth deformation Protein		x	x
Simazine	Triazine	Upsets photosynthesis			x

*Most commonly root herbicides are taken into the plant affecting the shoot, but some operate through contact with the roots.

Control of Nonsprouting Shrubs Many nonsprouting shrubs are effectively controlled by foliar sprays. Nearly all species of sagebrush are susceptible although there are specific differences (Fig. 13.8). *Artemisia filifolia* resprouts and successive applications are required for success (Bovey, 1964). Cornelius and Graham (1958) used seven different chem-

Table 13.5   Percent Reduction in Density of Stems and Forage Production of
Quercus havardii **Caused by Herbicides Used Singly and in Combination the**
First and Second Year after Treatment

Data from Scifres (1972a)

| | Foliar spray* | | | | | | | |
| | None year after treatment | | 2,4-D year after treatment | | 2,4,5-T year after treatment | | Silvex year after treatment | |
Soil herbicide	1	2	1	2	1	2	1	2
Browse density								
None	–	–	30	32	31	26	65	62
Picloram	58	33	38	12	74	65	87	60
Dicamba	0	13	–	–	54	64	40	41
Browse production								
None	–	–	17	15	47	34	78	55
Picloram	33	0	–43	–48	86	57	30	39
Dicamba	0	50	–	–	0	48	36	25

*Applied at rate of one-half pound per acre.

Table 13.6   Percent Canopy Reduction of Several Woody Plant
Species in Texas 2 Years after Aerial Treatment with Two Herbicides
Alone and Combined

Data from Bovey et al. (1970)

Species	2,4,5-T 2 lb/ acre	Picloram 2 lb/ acre	Picloram and 2,4,5-T 1 + 1 lb/ acre
Acacia farnesiana	68	93	100
Acacia rigidula	58	84	91
Prosopis juliflora	95*	90[†]	—
Diospyros texana	33	21	26
Lycium berlandieri	15	18	24
Berberis trifoliolata	10	10	10
Whitebrush	43	99	95
Condalia	10	39	31
Celtis pallida	55	95	95

*Measured 1 1/2 years after treatment.
[†]Measured 3 years after treatment.

Fig. 13.8 Spraying with 2,4-D changed a sand-sage stand like that shown to the left to a grass stand, right, doubling the grazing capacity. *(Courtesy E. L. McIlvain, Southern Great Plains Experiment Station.)*

icals or formulations and at two different rates to three species of sagebrush, and with two exceptions, both at low rates, kills exceeded 80 percent. Complete kills of sagebrush are not unusual and kills over 90 percent are common at rates of about 2 kg per hectare (Eckert and Evans, 1968).

Production of snakeweed and *Hymenoxis cooperi* were reduced by 97 and 76 percent, respectively, by 2 years of application (Jameson, 1966).

Prickly pear cactus can be killed by esters of 2,4,5-T in oil carriers during the growing season (Young et al., 1950).

Herbicidal Control of Herbaceous Plants

There are many herbaceous plants that are poisonous, are serious competitors to other forage species, or become too abundant which may make their control desirable.

Controlling Poisonous Plants Important poisonous species of the Western United States are tall larkspur *(Delphinum,* spp.), sneezeweed *(Helenium hoopesii),* lupines *(Lupinus),* locoweeds *(Astragalus* and *Oxytropis),* water hemlock *(Cicuta),* death camass *(Zygadenus),* and halogeton *(Halogeton glomeratus).* These are varyingly susceptible to herbicides. Sneezeweed can be 90 percent eliminated with a single application of

2,4-D during prebloom stage (Doran, 1951). Reinvasion is rapid, however, unless followed by improved management.

Other poisonous plants can be controlled if not eliminated by 2,4-D. Spraying with 3 kg of 2,4-D ester killed death camas in early bud, and 2 kg killed 90 percent of the plants of lupine after spraying for 2 successive years. Locos *(Oxytropis* and *Astragalus)* are generally susceptible to 2,4-D. Single application of 1 kg per hectare gave satisfactory kills on many species, and water hemlock was killed with 2 kg (Bohmont, 1952).

Although Bohmont (1952) reported 90 percent kill of larkspur with a single application of 2,4-D, others have had much less success. Cronin and Neilson (1972) were able to approach this using 2,4,5,-T, a generally more effective herbicide, only with higher rates or repeated annual applications.

Halogeton glomeratus, an introduced species from Asia, is an example of a highly competitive and poisonous species on saline lands in the Great Basin. It can be effectively killed, but unless spraying is done early in the season, it gives only partial kills and seed is produced (Cronin, 1965).

Few poisonous plants may justify control efforts. Plants that are widely distributed are too expensive to treat, and the mortality among other associated species of value make control impractical. Moreover, most plants with poisonous properties are of low toxicity except under special circumstances or they are eaten in such small amounts on a well-vegetated range that proper management is sufficient to render them innocuous (see Chapter 10). When poisonous species come to make up the bulk of the forage in localized areas, or livestock losses are high, control can be justified.

Controlling Nonpoisonous Forbs Some highly competitive species of little or no forage value may become so abundant that only a small part of their production is utilized. An example is *Wyethia amplexicaulis,* which comes to form dense stands in foothill and mountain ranges of the Great Basin. Although of great value to big-game animals in the early spring season, it is only eaten in moderate quantities by livestock. It is highly susceptible to 2,4-D at the prebloom stage (Mueggler and Blaisdell, 1951), and the recovery of suppressed grasses following treatment is rapid.

Occasionally, highly competitive annuals may occupy denuded areas hampering seeding efforts. Tarweed *(Madia glomerata)* has occupied many high-elevation meadows and parks. It germinates very early and possesses an inhibitory substance that depresses seed germination and seedling development of grasses so that seeding efforts are nullified. Cultural-control methods are ineffective, but spraying in successive years, once before and once after seeding, has proven effective (Hull and Cox, 1968).

Time of Treatment on the Effects of Herbicides

Correct timing is even more important to success with herbicides than with other plant-control methods. Organic herbicides depend for their success on physiological processes within the plant, especially for the auxin-type herbicides (Auberg, 1964). Factors which affect physiological activity can thus have an important bearing on the results obtained and probably account for variation in results that have been reported. Even closely related species may respond quite differently (Table 13.7).

Phenological stage has been shown to be an important factor in plant susceptibility to herbicides. Many plants are most vulnerable in the prebloom stage. Others are most affected earlier in the season and still others late in the year (Fig. 13.9). It is not known whether these differences are specific or are related to environmental factors which affect the physiological processes. Few studies have dealt with specific conditions as respects soil moisture, soil temperatures, air temperatures, and other factors which might explain the results obtained. There is some indication that herbicides more effectively controlled larkspur when fertilizers were applied concurrently with the herbicides, though the data and the design of the experiment do not permit conclusive inference (Binns et al., 1971).

Dahl et al. (1971) made an exhaustive inventory of environmental factors affecting root kill on mesquite, and concluded that soil temperature was the most important factor, with phenological stage second (Fig. 13.10). Plants in full foliage with mature seed pods were most easily killed.

Table 13.7   Control of Sagebrush Sprayed with Butyl Ester of 2,4-D at Three Different Times

Data from Cornelius and Graham (1958)

	Sagebrush kill			
	May 20	June 20		July 20
Type of sagebrush	1 lb	1 lb	2 lb	1 lb
Big sagebrush	77	75	98	60
Black sagebrush	92	89	94	35
Silver sagebrush	70	81	96	38

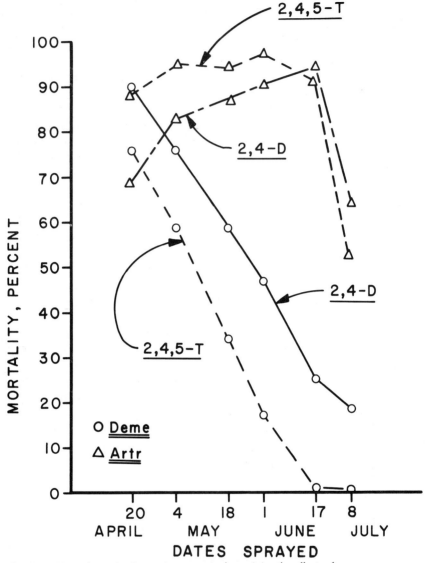

Fig. 13.9 Time of spraying is very important in determining the effects of herbicides. Big sagebrush *(Artr)* was most affected by late June applications; low larkspur *(Deme)* by spraying in April. *(Data from Hyder et al., 1956.)*

Benefits from Brush and Weed Control

Control measures, if properly done, are effective means of improving range productivity. Although the primary purpose is to increase forage yields, there are other beneficial effects in addition.

Increased Forage Production   Dramatic increases in production have resulted from reduction in shrubs due to opening of the canopy and increased moisture available to herbaceous species. Increased production of herbaceous species of nearly 300 percent was reported by Dalrymple et al. (1964), the first year following control of oak and elm woodland in Oklahoma, although most of the increase was of inferior species. Yields are usually greater in subsequent years. Increases in the same oak type in Arkansas were only 30 percent in the first year of treatment but had increased to 181 percent by the fourth year (Halls and Crawford, 1965). Widely differing results are obtained with different chemicals (Table 13.8).

Increased forage yields the first year following control of rabbitbrush amounted to 56 percent (Laycock and Phillips, 1968), and about the same increase was reported by Paulsen and Miller (1968) the second

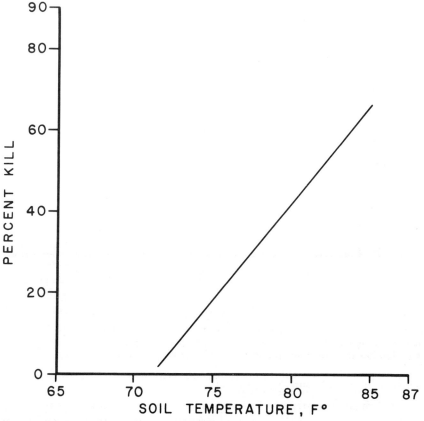

Fig. 13.10   Relationship of soil temperatures at depth of 46 cm to kill of mesquite with 2,4,5-T. *(From Dahl et al., 1971.)*

Table 13.8   Effects of Different Herbicides and Methods of Application on Production of Herbaceous Vegetation 6 Months after Application at Three Brushland Locations in Texas

Data from Bovey et al. (1972)

Herbicide	Llano		Victoria		Carlos	
	Grass	Forbs	Grass	Forbs	Grass	Forbs
	(Production, lb/acre)					
Atrazine spray	1,339*	2	2,311	3	3,814	7
Atrazine granules	620	6	2,083	79	4,456	19
Simazine spray	1,193	3	1,864	30	4,454	1
Simazine granules	1,243	173	2,129	115	4,082	62
Picloram spray	1,058	10	1,818	17	3,888	0
Picloram granules	397*	0	2,519	5	4,321	0
2,4-D spray	519	38	1,693	96	4,481	6
Control	1,086	228	1,649	165	3,267	325

*Only values significantly different, 5 percent level.

year. These comparatively small gains may substantiate Frischknecht's (1963) observations that grass yields are greater when in association with rabbitbrush.

Sagebrush control is more rewarding. Cook (1963) observed increased yields of three seeded grasses averaging 166 percent the third year after application of herbicides in northern Utah. Sprayed sagebrush ranges in Oregon produced over 200 percent more forage over a 3-year period than unsprayed range, and beef gains were increased fourfold the third year (Hyder and Sneva, 1956). (See also Table 13.9.)

How long increased yields may persist is unknown. Eckert et al. (1972) on a sagebrush range in northern Nevada, reported 400 percent increases the second year after treatment, but by 8 years increased yields diminished to 149 percent. Johnson (1969) observed no difference in forage production on sprayed and unsprayed ranges 6 years after spraying of sagebrush, and, after 17 years of protection from grazing, sagebrush was equally dense on the sprayed and unsprayed areas. The fact the observations were made on strips adjoining unsprayed areas may have hastened the invasion. Johnson and Payne (1968) thought residual mature plants rather than plants outside treated areas were most important to reinvasion of sagebrush.

Site has an important bearing on forage response to brush or tree removal. O'Rourke and Ogden (1969) found no increase in grasses at some sites in Arizona and substantial increases at others following piñon-juniper removal. Clary (1971) reported more than 200 percent increases in grass production due to juniper removal and aerial seeding.

Understory browse benefits greatly from reducing competition on piñon-juniper overstories. Bitterbrush production was increased over 300 percent and sagebrush production over 200 percent by silvicultural cutting for Christmas tree and post production (Jensen, 1972). Where piñon-juniper stands are dense and few forage species are present, seeding must accompany tree removal if production is to be increased.

Browse production can be greatly increased by fire or herbicides when applied to sprouting shrubs that have grown beyond reach. Burning reduced production of deer browse the first year following a fire in mixed pine-hardwood forests in Tennessee, but increased production by 246 and 437 percent the second and third year after the fire, respectively (Dills, 1970). Dense chamise *(Adenostema fasciculata)* brushlands can be made to produce three times as much deer forage by either fire or mechanical removal (Biswell, 1961). Production gradually declines, however, so that in 12 to 15 years brush has reoccupied the area.

Forage quality can be improved by fire. Reductions in crude fiber and increases in protein content were found in three of four woody plants observed in Maryland (DeWitt and Derby, 1955).

Top removal of brush plants by fire or mechanical means improves the palatability and use of the brush species. Although seldom eaten in the mature state, tender sprouts of *Condalia obtusifolia, Berberis trifoliolata,* and *Acacia farnesiana* became preferred forage plants of zebu steers and white-tailed deer following mechanical control in South Texas.

Nonforage Benefits from Brush Control   Control of brush has benefits other than increased forage production. Quite often major benefits are

Table 13.9 Increased Dry Forage Produced from Herbicide-Treated Gambel Oak Pastures

Date from Marquiss (1972)

Vegetation type and measurement	Increased production*				
	1966	1967	1968	1969	1970
	(Percent)				
Open grassland	60	25	20	32	64
Sagebrush grass	28	59	54	37	81
Oakbrush (beneath the canopy)	25	134	105	74	74

*Plots were treated during 1966 and again in 1969.

to be expected from improved efficiency in handling livestock. Removal of dense stands of brush lowers the cost of daily observation of grazing animals and reduces labor requirements for their handling. Removal of woody plants, especially thorny shrubs, makes the grass produced more available to grazing animals.

Mechanical control of brushland or repeated burning in tropical Africa often destroys the habitat of the tsetse fly and makes available for ranching areas previously unoccupied by domestic livestock.

Burning is useful in killing or reducing diseases and pests of both plants and animals. Fire is used in Southern pine forests to control brown-spot needle blight on pine reproduction. In Africa, grasslands are burned to kill ticks.

BIOLOGICAL MANIPULATION OF RANGE VEGETATION

The use of biological agents for plant control has been successful in some instances, but the list thus far is not impressive. Where it can be used, this means of keeping plants with undesirable characteristics in check promises long-time effects as compared to direct control methods, hence it may be less costly. As concern for the environment increases, greater emphasis will be placed on discovering effective and safe biological agents.

Biological control can be approached by (1) bringing together two unfamiliar organisms, usually by the introduction of exotic species, or (2) artificial stimulation of a native organism, by production and release. The latter is safer but more costly; the former less costly but more dangerous. Careful and complete ecological study should precede introduction of any biological agent, to be absolutely sure that the plant or animal concerned will neither displace, nor shift its activities to, desirable native species. Many unfortunate experiences have resulted from the introduction of animals into new environments. The rabbit in Australia, the starling and English sparrow in America, and the muskrat in England are examples of animals whose niche in the new land was less acceptable than that in their native land.

The following are necessary to use of a biological agent: (1) the agent must be specific to one host, or have a narrow range of alternate hosts; and (2) alternate hosts must not be economically valuable or be important to the stability of the ecosystem. Obviously, these criteria preclude elimination of a species and permit only control. The action of a biological agent may be direct or indirect. The first is accomplished by destruction of the host plant, the second by weakening it so that other pathogens attack it or making it noncompetitive with other plants (Huffaker, 1964).

Insects in Biological Control

Insects have been the agents most employed against plants. Their effects can be more readily assessed and the outcome predicted than can those of disease. Thus far, there have been two outstandingly successful examples of weed control by insects. In Australia, the prickly pear *(Opuntia)* problem has been eliminated by the introduction of moths *(Cactoblastis cactorum)* which feed upon the plant. As a result of this experience, similar efforts have been made to control prickly pears in India, Ceylon, Celebes, South Africa, and Hawaii with considerable success. Saint Johnswort *(Hypericum perforatum),* a native of Europe, has spread widely throughout Oregon and California, where it has given trouble as a stock-poisoning plant. The introduction of beetles *(Chrysolina),* from Europe and Australia, which feed exclusively on Saint Johnswort, has effectively controlled this troublesome plant (Table 13.10).

Other attempts have given little or only partial success. Seed-eating insects were imported into Hawaii and elsewhere to control lantana *(Lantana camara)* but without notable success. Attempts to control gorse *(Ulex europeus)* in New Zealand and in the United States with weevils have not given adequate control.

Puncture vine *(Tribulus terrestris)* has been found to be vulnerable to the seed-eating weevil of *Microlarinus lareynii* which is native to the Mediterranean and India (Daniels and Wise, 1962). The cinnabar moth was introduced into California to control tansy ragwort *(Senecio jacobeaeae)* but did not prevent its spread despite the fact that it found

Table 13.10 Vegetational Composition on Sheep-Grazed Range during Biological Control of Klamath Weed from 1948 to 1953

Data from Murphy (1955)

	1948	1949	1950	1951	1953
			(Percent)		
Klamath weed	40	38	T	0	T
Annualgrasses	28	40	50	43	50
Forbs	33	23	50	58	50

the plant a suitable host. A second insect *(Hylemya senciella)* whose larvae feed upon the seeds of ragwort later was released in Oregon and California to aid the foliage-eating moth (Anonymous, 1967). A possible agent for mesquite control is the twig girdler *(Oncideres rhodosticta)* which eats the foliage and girdles smaller stems. A native species, it would require mass rearing (Ueckert et al., 1971). A summary of successes attained with insects is shown in Table 13.11.

In addition to these, Holloway (1964) identified 20 other weeds that were under investigation in the United States.

Grazing Animals as Agents for Vegetation Manipulation

Manipulating vegetation does not always involve worthless or dangerous plants. Often times, plants which have forage value may become so abundant that they limit production of better associated species resulting in either less or poorer quality forage than otherwise would be possible. In these situations grazing animals may prove the most effective means of altering the composition of forage. Often they can be used in conjunction with other devices, such as fire or chemicals.

The effectiveness of domestic livestock as agents for manipulating vegetation rests upon four conditions: (1) effective control over livestock, (2) acceptance of the target plant as forage by livestock, (3) presence of other forage species which can replace the target species or a site favorable to seeding, and (4) differential susceptibility of plants to grazing at some time of the year.

Table 13.11   Success of Biological Agents in Control of Undesirable Plants Throughout the World

From National Research Council (1968)

Plant	Degree of success		
	Complete	Substantial	Partial
	(Number of instances)		
Opuntia spp.	3	10	7
Hypericum perforatum	1	1	6
Clidemia hirta	1		1
Leptospermum scoparium	1		
Cordia macrostachya		1	1
Linaria vulgaris			2
Lantana camara		2	9
Eupatorium adenophorum	1		
Emex spp.		2	
Tribulus spp.	2		

Control of Unwanted Shrubs Livestock are most successfully used against sprouting shrubs in connection with other control methods —fire, herbicides, or mechanical methods. Sprouts are more palatable than mature plants and are reached readily by grazing animals. Lay (1967) observed twice as heavy utilization by deer on shrubs following fire in a Southern pine forest. Moreover, the sprouting plant is weakened physiologically and maximum impact can be delivered at this time.

Cattle have not been particularly effective when used against brigalow *(Acacia harpophyla)* in eastern Australia; sheep have proven more successful (Johnson, 1964). The following sequence of treatment is used: mechanical control, seeding, burning to clear the area of debris, and heavy stocking ("flogging") with sheep.

Sheep grazing in the fall of the year can slow the invasion of sagebrush plants into crested wheatgrass stands. Control can be accomplished only if sagebrush has not become too dense, 14 or more plants per 10 sq m (Frischknecht and Harris, 1973).

The Use of Goats for Brush Control Goats are used in many countries as agents of brush control, although Malechek and Leinweber (1972) did not consider browse in Texas a first-choice item of forage. The goat's reputation comes from its ability to accept browse more readily than other livestock in the absence of other vegetation (Fig. 13.11). Huss (1972) suggests that goats require much the same assortment of forage classes as do other livestock for good gains (Table 13.12). DuToit (1973), in South Africa, gives evidence that goats can effectively hold *Acacia karroo* in check following other control methods under continuous grazing without affecting production from herbaceous vegetation. Despite the increased production of herbs, goat gains declined in response to the decline in acacia yields. Magee (1957) regards goats as both effective and profitable agents in brush suppression in Texas.

When goats are used as agents of brush control, animal production usually suffers. This condition occurs because it is generally necessary to stock at high intensities for short periods of time to ensure that brush plants are killed. If Angora nannies are used, kid crops drop and production of mohair is lowered, hence wether goats, usually of a hair variety, are recommended.

Goats as well as cattle appear destructive to desirable species and thus promote the less desirable species, both shrubs and grass, if allowed to graze Israelean brushlands immediately after a fire (Naveh, 1972). Once the better species have attained a height of 2 m, goats can be useful in keeping undergrowth down. Cassady (1972) found goats to be an effective means of control of brush in hand-cleared woodland areas in Kenya.

Fig. 13.11 Goats can be effective in holding shrubs in check. In the absence of forage on the ground, they can climb into small trees such as this *Argania spinosa,* Morocco. *(Photograph courtesy, H. N. Le Houérou.)*

Table 13.12 First-Year Effects of Roller Cutting and Burning after Roller Cutting upon Goat Production and Vegetation, Montemorelos, N. L., Mexico

Data from Huss (1972)

Treatment	Shrub cover, percent*	Forb cover, percent†	Grass production, kg per ha*	Daily goat gains, gm†
Original vegetation	86	12	542	191.7
Roller cutting	62	20	1,146	252.8
Burned after roller cutting	37	24	1,628	250.2

*All treatments significantly different.
†Control only significantly different.

Manipulation of Big-Game Ranges for Wildlife Although unrestricted livestock use can be highly competitive with big game, benefits may derive from controlling numbers and time of use by livestock. It has been shown that livestock can use big game winter ranges heavily in spring and early summer without greatly affecting subsequent production of shrubs, since at this season they subsist primarily upon herbaceous forage.

In northern Utah, livestock grazing prior to about July 1 results in only moderate use of the browse species, upon which game animals depend, and most of the annual-twig growth is made after that date so that forage production from them is affected little. If grazing is permitted through summer into fall, half of the production of bitterbrush may be removed (Smith and Doell, 1968). Summer sheep use is even more detrimental, since utilization of bitterbrush is as high as 80 percent in midsummer, but they, too, can be permitted to graze big-game ranges early without greatly reducing annual browse production (Jensen et al., 1972). This use of the shrubs may be compensated for by increased browse production due to reduced competition from herbaceous plants. For example, increased twig growth on bitterbrush of 40 to 60 percent resulted from keeping herbaceous plants adjacent to bitterbrush plants clipped throughout the summer (Smith and Doell, 1968). Similar benefits may result from livestock grazing, but this remains to be demonstrated.

The use of cattle to break down shrubs that have grown beyond reach of smaller animals has been recommended as a means of improving winter forage for deer.

The use of wildlife and livestock combinations to control woody plants has been demonstrated in Kenya. Here eland and giraffe were used in an improved cattle-ranching operation to keep brush from reinvading pastures following clearing (Skovlin, 1971). Although wild animals offer a wide variety of diets that could be used to manipulate vegetation, they seldom have the flexibility of diet required to survive the heavy stocking rates usually necessary for brush control.

The extent to which wildlife or livestock alter the ecosystem depends upon the particular role of each as consumers and the land-use objective of the manager. Alternating combinations and numbers or species of animals will force plant succession in different directions, which are predictable if the grazing habits of animals are fully known.

Much more needs to be known of the limitations and possibilities offered through the use of the grazing animal as a tool in range manipulation, but considering the possibilities, and in view of the knowledge of successional impacts of grazing animals this phase of range management seems badly neglected.

ENVIRONMENTAL CONSIDERATIONS

Experience has demonstrated that range vegetation can be manipulated so as to provide more forage for grazing animals. Far too often in the past this was done without knowledge of or regard to environmental degradation. Most often a single purpose was pursued—the production of forage for livestock, although it was argued sometimes, with little documentation, that watershed conditions were improved. Important wildlife habitat often was destroyed.

Even from the standpoint of livestock forage, adverse effects may follow the use of herbicides. This makes it important to consider associated vegetation when herbicide treatments are contemplated. Inevitably others besides the target species will be affected and some may well be important forage producers. Spraying to control sagebrush severely affected six associated shrubs that were valuable forage plants although all but one recovered through sprouting. Among the associated herbaceous plants 12 were unharmed, 10 lightly affected, 3 moderately affected, and 10 suffered heavy mortality. Several important forage species were among those most injured (Blaisdell and Mueggler, 1956).

With the enactment of the Environmental Protection Act (83 Stat. 852) in the United States, which requires public review of federal action programs, this course is no longer possible. Nor is it sufficient to ensure that other resource values, soil, watershed and wildlife needs are protected. Even altering the landscape by clear-cutting forests, or chaining piñon-juniper stands are viewed with wide public disfavor. The use of chemical agents is opposed by increasingly large numbers of people for the foregoing reasons and also because of the threat of pollution and the possible chain effects they may produce.

Accordingly, managers of public ranges in the United States will be required to plan better than they have done in the past. Instead of bulldozing piñon-juniper areas clear of all vegetation, less effective measures such as chaining, which leave associated plants that have value as food and cover for wildlife, will more often be used. Moreover, greater care in the configuration and design of projects to lessen the unsightliness of treated areas will be a necessity.

Some beginnings have been made toward this by removing unwanted brush in undulating strips rather than along straight lines thereby improving the appearance as well as retaining better game-habitat conditions. Research to evaluate these and other secondary effects of range-manipulation techniques is badly needed.

It is important in any assessment of range-rehabilitation programs to distinguish between short-run and long-run effects. What may appear immediately after treatment to be unaccepted may in a short time be more acceptable than before. Vogl and Beck (1970) found white-tailed deer more than twice as abundant 8 years following a fire in an oak-pine forest in Wisconsin than in the unburned forest. Nature under protection does not provide the most productive and useful vegetation. Climax vegetation, particularly forest types, are little suited to wildlife and, often, are not the most productive of timber. Many of our most valuable timber types are not climax but subclimax types. Man's hand in directing succession is no less natural than catastrophic and uncontrolled determinants, such as fire and disease which, since time immemorial, have fashioned the "natural" vegetation. In order to duplicate historic forces, drastic alterations are sometimes necessary to maintain an ecosystem.

BIBLIOGRAPHY

Adams, Frank, et al. (1947): Hydrologic aspects of burning brush and woodland-grass ranges in California, California Division of Forestry, January.

Aldous, A. E. (1934): Effect of burning on Kansas bluestem pastures, *Kan. Agr. Expt. Sta. Tech. Bull.* **38.**

Allred, B. W. (1949): Distribution and control of several woody plants in Texas and Oklahoma, *Jour. Range Mgt.* **2:**17–29.

Anderson, Kling L., Ed F. Smith, and Clenton E. Owensby (1970): Burning bluestem range, *Jour. Range Mgt.* **23:**81–92.

Anonymous (1967): Second insect for range weed control, *Western Livestock Jour.,* August, p. 240.

Aro, Richard S. (1971): Evaluation of pinyon-juniper conversion to grassland, *Jour. Range Mgt.* **24:**188–197.

Ashton, Floyd M., and Alden S. Crafts (1973): *Mode of Action of Herbicides,* Wiley, New York.

Auberg, Ewert (1964): Susceptibility: Factors in the plant modifying the response of a given species to treatment, in L. J. Audus (ed.), *The Physiology and Biochemistry of Herbicides,* Academic Press, London and New York, pp. 401–422.

Bachtelder, R. B. (1967): Spatial and temporal patterns of fire in the tropical world, *Proc. Tall Timbers Fire Ecol. Conference* **6:**171–208.

Bailey, Arthur W. (1970): Barrier effect of the shrub *Elaeagnus commutata* on grazing cattle and forage production in central Alberta, *Jour. Range Mgt.* **23:**248–251.

Bentley, H. L. (1898): A report upon the grasses and forage plants of central Texas.

Binns, Wayne, Lynn F. James, and A. Earl Johnson (1971): Control of larkspur with herbicides plus nitrogen fertilizer, *Jour. Range Mgt.* **24:**110–113.

Biswell, H. H. (1961): Manipulation of chamise brush for deer range improvement, *Calif. Fish and Game* **47:***125*–144.

Blaisdell, James P., and Walter F. Mueggler (1956): Effect of 2, 4-D on forbs and shrubs associated with big sagebrush, *Jour. Range Mgt.* **9:**38–40.

Bleak, A. T., and Warren G. Miller (1955): Sagebrush seedling production as related to time of mechanical eradication, *Jour. Range Mgt.* **8:**66–69.

Bohmont, Dale W. (1952): Chemical control of poisonous range plants, *Wyo. Agr. Expt. Sta. Bull.* **313.**

Bovey, R. W. (1964): Aerial application of herbicides for control of sand sagebrush, *Jour. Range Mgt.* **17:**253–256.

———, J. R. Baur, and H. L. Morton (1970): Control of huisache and associated woody species in south Texas, *Jour. Range Mgt.* **23:**47–50.

———, R. E. Meyer, and H. L. Morton (1972): Herbage production following brush control with herbicides in Texas, *Jour. Range Mgt.* **25:**136–142.

Box, Thadis W., and Richard S. White (1969): Fall and winter burning of south Texas brush ranges, *Jour. Range Mgt.* **22:**373–376.

British Weed Control Council (1965): E. K. Woodford and S. A. Evans (eds.), *Weed Control Handbook,* Blackwell Scientific Publications, Oxford.

Britton, Carlton M., and Henry A. Wright (1971): Correlation of weather and fuel variables to mesquite damage by fire, *Jour. Range Mgt.* **24:**136–141.

Brown, Albert L. (1950): Shrub invasion of southern Arizona desert grassland, *Jour. Range Mgt.* **3:**172–177.

Buehring, Normie, P. W. Santelmann, and Harry M. Elwell (1971): Responses of eastern red cedar to control procedures, *Jour. Range Mgt.* **24:**378–382.

Burkhardt, J. Wayne, and E. W. Tisdale (1969): Nature and successional status of western juniper vegetation in Idaho, *Jour. Range Mgt.* **22:**264–270.

Cable, Dwight R. (1965): Damage to mesquite, Lehmann lovegrass, and black grama by a hot June fire, *Jour. Range Mgt.* **18:**326–329.

Campbell, R. S. (1954): Fire in relation to forest grazing, Food and Agriculture Organization, United Nations (mimeo.).

Cassady, John (1972): Bush control at Buchuma station, Final Report FAO Range management Project, Kenya, Rome (mimeo.).

Clary, Warren P. (1971): Effects of Utah juniper removal on herbage yields from Springerville soils, *Jour. Range Mgt.* **24:**373–378.

Cook, C. Wayne (1963): Herbicide control of sagebrush on seeded foothill ranges in Utah, *Jour. Range Mgt.* **16:**190–195.

——— (1965): Grass seedling response to halogeton competition, *Jour. Range Mgt.* **18:**317–321.

——— and Clifford E. Lewis (1963): Competition between big sagebrush and seeded grasses on foothill ranges in Utah, *Jour. Range Mgt.* **16:**245–250.

——— and L. A. Stoddart (1953): The halogeton problem in Utah, *Utah State Agr. Expt. Sta. Bull.* **364.**

Cornelius, Donald R., and Charles A. Graham (1958): Sagebrush control with 2, 4-D, *Jour. Range Mgt.* **11:**122–125.

Cronin, Eugene H. (1965): Ecological and physiological factors influencing chemical control of *Halogeton glomeratus, U.S. Dept. Agr. Tech. Bull.* **1325.**

———, and D. B. Nielsen (1972): Controlling tall larkspur on snowdrift areas in the subalpine zone, *Jour. Range Mgt.* **25:**213–216.

Dahl, B. E., R. B. Wadley, M. R. George, and J. L. Talbot (1971): Influence of site on mesquite mortality from 2, 4, 5-T, *Jour. Range Mgt.* **24:**210–215.

Dalrymple, R. L., D. D. Dwyer, and P. W. Santelmann (1964): Vegetational responses following winged elm and oak control in Oklahoma, *Jour. Range Mgt.* **17:**249–253.

Daniels, Norris E., and Allen F. Wise (1962): Survival and spread of the puncturevine seed weevil in Texas, *Tex. Agr. Expt. Sta. M.P.* **827.**

DeWitt, James B., and James V. Derby, Jr. (1955): Changes in nutritive value of browse plants following forest fires, *Jour. Wildlife Mgt.* **19:**65–70.

Dills, Gary G. (1970): Effects of prescribed burning on deer browse, *Jour. Wildlife Mgt.* **34:**540–545.

Dodd, J. D., and S. T. Holtz (1972): Integration of burning with mechanical manipulation of south Texas grassland, *Jour. Range Mgt.* **25:**130–136.

Doran, Clyde W. (1951): Control of orange sneezeweed with 2, 4-D, *Jour. Range Mgt.* **4:**11–15.

DuToit, P. F. (1973): The goat as a factor in bush control, *The Angora Goat and Mohair Jour.* **15:**49–51, 57.

Eckert, Richard E., Jr., Allen D. Bruner, and Gerard J. Klomp (1972): Response of understory species following herbicidal control of low sagebrush, *Jour. Range Mgt.* **25:**280–285.

———— and Raymond A. Evans (1968): Chemical control of low sagebrush and associated green rabbitbrush, *Jour. Range Mgt.* **21:**325–328.

Ehrenreich, John H. (1959): Effect of burning and clipping on growth of native prairie in Iowa, *Jour. Range Mgt.* **12:**133–137.

Food and Agricultural Organization (1973): F.A.O. Range management project in Kenya, Final Report, Rome (mimeo.).

Fisher, C. E. (1950): The mesquite problem in the Southwest, *Jour. Range Mgt.* **3:**60–70.

————, Jess L. Fults, and Henry Hopp (1946): Factors affecting action of oils and water-soluble chemicals in mesquite eradication, *Ecol. Monogr.* **16:**109–126.

Fisser, Herbert G. (1968): Soil moisture and temperature changes following sagebrush control, *Jour. Range Mgt.* **21:**283–287.

Frischknecht, Neil C. (1963): Contrasting effects of big sagebrush and rubber rabbitbrush on production of crested wheatgrass, *Jour. Range Mgt.* **16:**70–74.

———— and Lorin E. Harris (1973): Sheep can control sagebrush on seeded range if—, *Utah Science* **34:**27–30.

Glendening, George E. (1952): Some quantitative data on the increase of mesquite and cactus on a desert grassland range in southern Arizona, *Ecology* **33:**319–328.

Graves, James E., and Wilfred E. McMurphy (1969): Burning and fertilization for range improvement in central Oklahoma, *Jour. Range Mgt.* **22:**165–168.

Greene, S. W. (1935): Relation between winter grass fires and cattle grazing in the longleaf pine belt, *Jour. Forestry* **33:**338–341.

Halls, L. K., and H. S. Crawford (1965): Vegetation response to an Ozark woodland spraying, *Jour. Range Mgt.* **18:**338–340.

Hanes, Ted L., and Harold W. Jones (1967): Post fire chaparral succession in southern California, *Ecology* **48**:259–264.

Harniss, Roy O., and Robert B. Murray (1973): 30 years of vegetal change following burning of sagebrush-grass ranges, *Jour. Range Mgt.* **26**:322–325.

Hensel, R. L. (1923): Effect of burning on vegetation in Kansas pastures, *Jour. Agr. Research* **23**:631–643.

Heyward, F. (1938): Soil temperature during forest fires in the longleaf pine region, *Jour. Forestry* **36**:478–491.

Hilmon, J. B., and R. H. Hughes (1965): Fire and forage in the wiregrass type, *Jour. Range Mgt.* **18**:251–254.

Holloway, J. K. (1964): Projects in biological control of weeds, in Paul DeBach and Evert I. Schlinger (eds.), *Biological Control of Insect Pests and Weeds,* Reinhold, New York, pp. 650–670.

Huffaker, C. B. (1964): Fundamentals of biological weed control, in Paul De-Bach and Evert I. Schlinger (eds.), *Biological Control of Insect Pests and Weeds,* Reinhold, New York, pp. 631–649.

Hull, A. C., Jr., and Hallie Cox (1968): Spraying and seeding high elevation tarweed rangelands, *Jour. Range Mgt.* **21**:140–144.

Humphrey, R. R. (1949): Fire as a means of controlling velvet mesquite, burroweed, and cholla on southern Arizona ranges, *Jour. Range Mgt.* **2**:175–182.

Huss, Donald L. (1972): Goat response to use of shrubs as forage, in Wildland shrubs—their biology and utilization, *U.S. Dept. Agr. Forest Service, Gen. Tech. Rep.* INT-1, 1972, pp. 331–338.

Hyder, Donald N., and Forrest A. Sneva (1956): Herbage response to sagebrush spraying, *Jour. Range Mgt.* **9**:34–38.

———, ———, and Lyle D. Calvin (1956): Chemical control of sagebrush larkspur, *Jour. Range Mgt.* **9**:184–186.

Isaac, L. A., and H. G. Hopkins (1937): The forest soil of the Douglas fir region, and changes wrought upon it by logging and slash burning, *Ecology* **18**:264–279.

Jameson, Donald A. (1966): Competition in a blue grama-broom snakeweed-actinea community and response to selective herbicides, *Jour. Range Mgt.* **19**:121–124.

Jensen, Charles H., Arthur D. Smith, and George W. Scotter (1972): Guidelines for grazing sheep on rangelands used by big game in winter, *Jour. Range Mgt.* **25**:346–352.

Jensen, Neil E. (1972): Pinyon-juniper woodland management for multiple use benefits, *Jour. Range Mgt.* **25**:231–234.

Johnson, James R., and Gene F. Payne (1968): Sagebrush reinvasion as affected by some environmental influences, *Jour. Range Mgt.* **21**:209–213.

Johnson, R. W. (1964): Ecology and control of brigalow in Queensland, Queensland Department of Primary Industries, Australia.

Johnson, W. M. (1969): Life expectancy of a sagebrush control in central Wyoming, *Jour. Range Mgt.* **22**:177–182.

Klett, W. Ellis, Dale Hollingsworth, and Joseph L. Schuster (1971): Increasing utilization of weeping lovegrass by burning, *Jour. Range Mgt.* **24**:22–24.

Lay, Daniel W. (1967): Browse palatability and the effects of prescribed burning in southern pine forests, *Jour. Forestry* **65**:826–828.

Laycock, William A., and Thomas A. Phillips (1968): Long-term effects of 2, 4-D on lanceleaf rabbitbrush and associated species, *Jour. Range Mgt.* **21**:90–93.

Leonard, Oliver A. (1956): Effect on blue oak (*Quercus douglasii*) of 2, 4-D and 2, 4, 5-T concentrates applied to cuts in trunks, *Jour. Range Mgt.* **9**:15–19.

Lewis, Clifford E., and Richard H. Hart (1972): Some herbage responses to fire on pine-wiregrass range, *Jour. Range Mgt.* **25**:209–212.

Lillie, D. T., George E. Glendening, and C. P. Pase (1964): Sprout growth of shrub live oak as influenced by season of burning and chemical treatments, *Jour. Range Mgt.* **17**:69–72.

Love, R. Merton, and Burle J. Jones (1952): Improving California brush ranges, *Calif. Agr. Expt. Sta. Circ.* **371**, rev.

Lyon, L. Jack, and Walter F. Mueggler (1968): Herbicide treatment of north Idaho browse evaluated six years later, *Jour. Wildlife Mgt.* **32**:538–541.

Magee, A. C. (1957): Goats pay for clearing Grand Prairie rangelands, *Tex. Agri. Expt. Sta. M.P.* **206.**

Malechek, John C., and C. L. Leinweber (1972): Forage selectivity by goats on lightly and heavily grazed ranges, *Jour. Range Mgt.* **25**:105–111.

Marquiss, Robert W. (1972): Soil moisture, forage, and beef production benefits from gambel oak control in southwestern Colorado, *Jour. Range Mgt.* **25**:146–150.

Mayland, H. F. (1967): Nitrogen availability on fall-burned oak-mountain mahogany chaparral, *Jour. Range Mgt.* **20**:33–35.

McCarty, M. K., and C. J. Scifres (1972): Herbicidal control of western ragweed in Nebraska pastures, *Jour. Range Mgt.* **25**:290–292.

McCorkle, C. O., Jr., A. H. Murphy, Lynn Rader, and D. D. Caton (1964): Cost of tree removal through chemicals, *Jour. Range Mgt.* **17**:242–244.

McKell, C. M., Victor W. Brown, Charles F. Walker, and R. M. Love (1965): Species composition changes in seeded grasslands converted from chaparral, *Jour. Range Mgt.* **18**:321–326.

McMurphy, Wilfred E., and Kling L. Anderson (1965): Burning Flint Hills range, *Jour. Range Mgt.* **18**:265–269.

Moreland, Donald E. (1967): Mechanisms of action of herbicides, *Plant Physiology* **18**:365–386.

Mueggler, Walter F., and James P. Blaisdell (1951): Replacing wyethia with desirable forage species, *Jour. Range Mgt.* **4**:143–150.

Murphy, Alfred H. (1955): Vegetational changes during biological control of Klamath weed, *Jour. Range Mgt.* **8**:76–79.

National Research Council (1968): Principles of Plant and Animal Pest Control, vol. 2, Weed Control, National Academy of Science Publication 1597, Washington.

Naveh, A. (1972): The role of shrubs and shrub ecosystems in present and future Mediterranean land use, in Wildland shrubs—their biology and utilization, *U.S. Dept. Agr. Forest Service Gen. Tech. Rep.* INT-1, 1972, pp. 414–427.

O'Rourke, J. T., and P. R. Ogden (1969): Vegetative response following pinyon-juniper control in Arizona, *Jour. Range Mgt.* **22**:416–418.

Owensby, Clenton E., Kenneth R. Blan, B. J. Eaton, and O. G. Russ (1973): Evaluation of eastern red cedar infestation in the northern Kansas Flint Hill, *Jour. Range Mgt.* **26**:256–260.

—— and John Bruce Wyrill, III (1973): Effects of range burning on Kansas Flint Hills soils, *Jour. Range Mgt.* **26**:185–188.

Parker, Kenneth W. (1943): Control of mesquite on southwestern ranges, *U.S. Dept. Agr. Leaflet* **234.**

—— and S. Clark Martin (1952): The mesquite problem on southern Arizona ranges, *U.S. Dept. Agr. Circ.* **908.**

Paulsen, Harold A., Jr., and John C. Miller (1968): Control of Parry rabbit brush on mountain grasslands of western Colorado, *Jour. Range Mgt.* **21**:175–177.

Pearson, H. A., J. R. Davis, and G. H. Schubert (1972): Effects of wildfire on timber and forage production in Arizona, *Jour. Range Mgt.* **25**:250–253.

Pechanec, Joseph F., and George Stewart (1944): Sagebrush burning—good and bad, *U.S. Dept. Agr. Farmers' Bull.* **1948.**

Penfound, William T. (1964): The relation of grazing to plant succession in the tall grass prairie, *Jour. Range Mgt.* **17**:256–260.

Phillips, J. (1936): Fire in vegetation: A bad master, a good servant and a national problem, *Jour. S. African Botany* **2**:35–45.

Powell, Jeff, and Thadis W. Box (1967): Brush management influences preference values of south Texas woody species for deer and cattle, *Jour. Range Mgt.* **19**:212–214.

Sampson, Arthur W. (1944): Plant succession on burned chaparral lands in northern California, *Calif. Agr. Expt. Sta. Bull.* **685.**

—— and L. T. Burcham (1954): Costs and returns of controlled brush burning for range improvement in northern California, *Div. of Forestry, Range Improvement Studies* no. 1.

Sauer, Carl O. (1950): Grassland climax, fire, and man, *Jour. Range Mgt.* **3**:16–21.

Scifres, C. J. (1972a): Herbicide interactions in control of sand shinnery oak, *Jour. Range Mgt.* **25**:386–389.

—— (1972b): Redberry juniper control with soil-applied herbicides, *Jour. Range Mgt.* **25**:308–310.

——, J. H. Brock, and R. R. Hahn (1971): Influence of secondary succession on honey mesquite invasion in north Texas, *Jour. Range Mgt.* **24**:206–210.

—— and J. C. Halifax (1972): Root production of seedling grasses in soil containing picloram, *Jour. Range Mgt.* **25**:44–46.

—— and G. O. Hoffman (1972): Comparative susceptibility of honey mesquite to dicamba and 2, 4, 5-T, *Jour. Range Mgt.* **25**:143–146.

Shantz, H. L. (1947): The use of fire as a tool in the management of the brush ranges of California, California Division of Forestry, January.

Skovlin, Jon M. (1971): Ranching in East Africa: a case study, *Jour. Range Mgt.* **24**:263–270.

Smith, Arthur D., and Dean D. Doell (1968): Guides to allocating forage between cattle and big game on winter range, *Utah State Div. Fish and Game*, pub. no. 68–11.

Smith, Dixie R. (1969): Is deferment always needed after chemical control of sagebrush, *Jour. Range Mgt.* **22**:261–263.

Sneva, Forrest A., and D. N. Hyder (1966): Control of big sagebrush associated with bitterbrush in ponderosa pine, *Jour. Forestry* **64**:677–680.

Stinson, Kenneth J., and Henry A. Wright (1969): Temperatures of headfires in the southern mixed prairie of Texas, *Jour. Range Mgt.* **22**:169–174.

Stoddard, Herbert L. (1931): *The Bob-White Quail, Its Habits, Preservation, and Increase,* Scribner, New York.

Tabler, Ronald D. (1968): Soil moisture response to spraying big sagebrush with 2, 4-D, *Jour. Range Mgt.* **21**:12–15.

Thompson, W. R. (1936): Veld burning: its history and importance in South Africa, *Pretoria Univ. Ser.* 1, 31.

Tiedemann, Arthur R., and Ervin M. Schmutz (1966): Shrub control and reseeding effects on the oak chaparral of Arizona, *Jour. Range Mgt.* **19**:191–195.

Trlica, M. J., Jr., and J. L. Schuster (1969): Effects of fire on grasses of the Texas High Plains, *Jour. Range Mgt.* **22**:329–333.

Ueckert, Darrell N., Kenith L. Polk, and Charles R. Ward (1971): Mesquite twig girdler: a possible means of mesquite control, *Jour. Range Mgt.* **24**:116–118.

Vallentine, John F. (1971): *Range Development and Improvements,* Brigham Young University Press, Provo, Utah.

Van Overbeek, J. (1964): Survey of mechanisms of herbicide action, in L. J. Audus (ed.), *The Physiology and Biochemistry of Herbicides,* Academic Press, London and New York, pp. 387–400.

Vincent, C. (1936): The burning of grassland (trans. from the Portuguese), *Rev. Indus. Anim.* **4**:286–299.

Vogl, Richard J., and Alan M. Beck (1970): Response of white-tailed deer to a Wisconsin wildfire, *Amer. Midland Naturalist* **84**:270–273.

Wahlenberg, W. G., S. W. Greene, and H. R. Reed (1939): Effects of fire and cattle grazing on longleaf pine lands, as studied at McNeill, Miss., *U.S. Dept. Agr. Tech. Bull.* **683**.

White, Larry D. (1955): Effects of wildfire on several desert grassland shrub species, *Jour. Range Mgt.* **22**:284–285.

Willard, E. Earl (1973): Effect of wildfires on woody species in the Monte region of Argentina, *Jour. Range Mgt.* **26**:97–100.

Wink, Robert L., and Henry A. Wright (1973): Effects of fire on an ashe juniper community, *Jour. Range Mgt.* **26**:326–329.

Wright, Henry A. (1972): Shrub response to fire, in "Wildland Shrubs—their Biology and Utilization," *U.S. Forest Service Gen. Tech. Rep. INT-1, 1972,* pp. 204–217.

Young, James A., Raymond A. Evans, and John Robison (1972): Influence of repeated annual burning on a medusahead community, *Jour. Range Mgt.* **25**:372–375.

Young, Vernon A., et al. (1948): Brush problems on Texas ranges, *Tex. Agr. Expt. Sta. Misc. Pub.* **21**.

———, et al. (1950): Recent developments in the chemical control of brush on Texas ranges, *Tex. Agr. Expt. Sta. Bull.* **721**.

Chapter 14

Range Improvements for Increasing Forage Production

Native ranges do not always provide the maximum usable forage of which they are capable. Furthermore, heavy and indiscriminate grazing or other factors may have so modified the natural vegetation that only low-value species remain. In other cases brush fires occurring on ranges where herbaceous plants are few in number leave an area without sufficient vegetative cover to protect it from erosion. For these and other reasons improvement practices have been developed to increase forage production without waiting for nature to restore the area to its potential.

Before the turn of the century, range-improvement practices were proposed, along with livestock management, to restore impaired Texas ranges to full capabilities (Smith, 1898). These included scattering grass seed, loosening the soil by means of harrows, and cutting furrows to intercept grass seed that might be blown away by wind. Kennedy and Doten (1901) recognized the possibility of range improvement using native or foreign plants from similar climates, but expressed the view that proper management was the better course. With respect to seeding, they observed that "the average man is likely to think there is some cure-all plant somewhere which will grow everywhere and do anything just like a patent medicine at one dollar per bottle which will really cure all diseases." Their pessimism was not altogether unfounded, for means of artificially improving some rangelands still have not been found. Great strides, however, have been made in finding species and methods for restoring many deteriorated ranges.

The most commonly employed means of improving ranges other than those discussed in Chapter 13 are (1) seeding, (2) fertilizing, and (3) mechanical treatments designed to conserve precipitation and increase production.

RANGE SEEDING

A number of techniques are employed in seeding rangelands, most of which require some form of soil preparation to make a suitable seedbed and reduce competition. Since this may cause some deterioration in soil structure and permeability and loss of existing plants, the advantages of seeding should be carefully weighed. Existing vegetation should be appraised to determine the relative merit of artificial planting compared with natural revegetation. Further, possibilities of changing managment of the area should be studied to determine how the misuse which made revegetation necessary can be corrected. Then, and only then, should artificial seeding be planned. The suitability of the area as to climate and soil should be carefully considered in terms of prior experience on similar sites. Careful, scientific seeding has wonderful possibilities, but it is a major undertaking which should be attempted only after proper planning. Seeding is costly and may not always be profitable in terms of forage produced (Plath, 1954).

As early as 1895, the federal government began grass-planting experiments in the Western United States, but 1,500 trials brought largely failure, and the optimism over range seeding waned. Again in 1907, the U.S. Forest Service began a large-scale seeding program, but 500 tests in 11 states gave only about 16 percent success, even on favorable locations. Failures were due primarily to the fact that seed was available only for cultivated forage plants, which were adapted to more humid climates than most range sites, and to the inadequate measures of site preparation that were used.

The need for range seeding was intensified by the droughts in the early 1930s which led to widespread range abuse and the abandonment of extensive acreages of rangelands. Since 1935, rapid progress has been made. Development of new machinery and planting methods, along with promising new species of grass, has paved the way for success.

The following objectives may justify seeding: (1) to revegetate barren areas such as abandoned croplands, (2) to replace vegetation destroyed by fire, (3) to expand the grazing season, (4) to improve the quantity and quality of forage, (5) to reestablish native forage plants which would not naturally become established, and (6) to protect an area from erosion.

Seeding Denuded Ranges or Abandoned Croplands

The decision to seed abandoned croplands is easiest to make. The fact that crops were once grown is evidence that these areas are productive which almost ensures success provided proper methods are employed and species suited to the site are used. Natural revegetation is a slow process requiring 25 to 50 years under favorable conditions for climax vegetation to reoccupy the site. The alternative is to permit invasion by annual grasses and weeds which are usually inferior as forage and which compete with better perennial species, thus delaying their reestablishment. If seeding is done prior to this invasion, seed may often be drilled without seedbed preparation, although usually cultivation improves success (Table 14.1).

Once highly competitive annuals become established, site preparation treatments, such as plowing or disking, are required. Such treatments are costly and the seedbed may be so loose as to hinder establishment of the seeded species. Chemical soil sterilants have been proposed to obviate these difficulties. *Bromus tectorum* can be effectively controlled in this way and the establishment of seeded species improved (Eckert and Evans, 1967), although residual chemicals in the soil may adversely affect establishment of some seeded species (Eckert et al., 1972).

Depleted range sites bear some similarity to abandoned croplands,

Table 14.1 Number of Seeded Grass Plants Established 2 Years after Planting, Using Three Methods of Seedbed Preparation

Data from Bement et al. (1965)

Species	Summer fallow	Spring cultivated	Direct
	(Number per meter of drill row)		
Crested wheatgrass	2.56	1.71	0.56
Russian wildrye	1.31	0.72	0.07
Blue grama	0.39	0.36	0.30
Side-oats grama	0.52	0.13	0.10
Mean	1.18	0.72	0.26

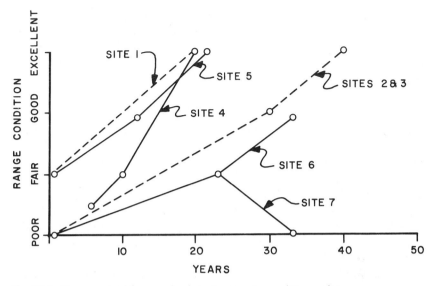

Fig. 14.1 Time required for rangelands to improve in condition under complete protection, British Columbia. Site 7 retrogressed because of rodent activity. *(Data from McLean and Tisdale, 1972.)*

although they present some additional problems. Recovery under protection is slow, an estimated 20 to 40 years in southern British Columbia (Fig. 14.1). Depleted ranges usually have present low-value vegetation which must be eliminated. Loss of topsoil may have reduced their productive capacity, and rough topography may make the use of equipment more difficult and costly.

Seeding Following Fires

Range fires frequently encompass large areas leaving them bare and subject to erosion. Where perennial grasses are present they may recover adequately to make seeding unnecessary. Often, however, only annual grasses and shrubs were present prior to fire and natural recovery is slow. An example is found in the intermountain region on sagebrush sites. Fires consume all or almost all of the sagebrush leaving the area clean of obstacles to seeding equipment. Drilling without further treatment has given excellent results (Fig. 14.2). Kay and Street (1961) were able to get satisfactory establishment of grasses by drilling into sprayed sagebrush stands in northeastern California. In some cases aerial seeding has proved successful. Only the more mesic sites with loose, uncompacted soils are likely to give success from aerial seeding, for best results are obtained when seed is covered from 6 to 12 mm (Franzke and Hume, 1942).

Fig. 14.2 *This sagebrush range was accidentally burned in late fall. One week later it was drilled to crested wheatgrass with no other treatment. This excellent stand resulted after 1 year. Unburned brush is seen in the background.*

Seeding to Expand the Grazing Season

Native vegetation often may provide green forage for a limited part of the year. In areas where warm-season grasses predominate, such as in the plains and prairies of North America and in the pampas of South America, growth does not begin early in the spring but is delayed until the arrival of hot weather and summer rains. In the northern plains, seedings of crested wheatgrass, a cool-season grass, provide green forage well before native plants begin growth thus reducing the period livestock must be fed and lengthening the period during which cattle can make weight gains. Provided soil moisture is available, the length of the growing season for cool-season grasses may be 2 to 3 months longer than for warm-season grasses (Conard and Youngman, 1965).

Comparisons of gains of steer and lengths of grazing season were made at Lincoln, Nebraska, on prairie species as compared to pastures seeded with introduced species (Table 14.2). The cool-season grasses did not extend the grazing season in the fall because of inadequate moisture, but they could be grazed a month earlier in the spring, thus lengthening the grazing season. Miles (1954) used crested wheatgrass pastures for spring and summer grazing and native range in fall and winter in Montana. By combining native range, and ranges seeded to different exotic species, animal production was increased 25 percent (Currie, 1969).

On tropical monsoonal ranges in Australia the native sorghums become leached in the dry season. Pastures planted to legumes, which retain their protein content when dry, provide a longer grazing season and better quality forage.

Seeding to Improve Quantity or Quality of Forage

Often, even when native forage is present, it is neither as abundant nor of as good quality as may be desired. Two options are available: (1) cultivating the soil, destroying all or most of the vegetation present, or (2) *interseeding* the existing stand with better forage plants. The latter entails cutting furrows or otherwise destroying the existing vegetation in strips spaced across the area at fixed intervals into which the desired plants are sown.

The choice between these depends upon the existing and potential productivity of the area. Seeding is costly and not always effective. Normally, native ranges in reasonably good condition should not be plowed to make way for introduced species, although Rumsey (1971) reported that seedings gave greater production than native range in excellent condition in eastern Idaho. These results may be biased by the fact that seeded ranges are invariably the best sites, chosen for their greater promise of success.

Interseeding has several advantages over complete tillage: (1) there is less disturbance to the site, (2) the species introduced can be those that

Table 14.2   Animal Production and Grazing Season from Cool-Season and Warm-Season Pastures, Lincoln, Nebraska

Data from Conard and Youngman (1965)

Year	Kind of pasture	From	To	Days	Steer-days of grazing per acre	Gains in body weight per acre, lb
1956	Cool-season	May 21–Aug. 19		90	54	57
	Warm-season	June 18–Oct. 2		106	121	140
1957	Cool-season	May 13–Oct. 9		149	135	134
	Warm-season	June 19–Oct. 9		112	137	133

Table 14.3   Mean Production for 3 Years Following Treatment from Native and Interseeded Species in Montana

Data adopted from Houston and Adams (1971)

Treatment	Fall 1965		Spring 1966		Fall 1966	
	Seeded species	Native species	Seeded species	Native species	Seeded species	Native species
	(Lb/acre, air dry)					
Control		845		845		845
Furrowed only		1,163		797		1,053
Green needlegrass	190	1,407				
Thickspike wheatgrass	120	807				
Prairie clover			30	1,237		
Rambler alfalfa			53	1,097		
Orenberg alfalfa*					405	685

*Two-year mean.

complement existing forage, as when cool-season grasses are sown into warm-season stands, (3) forage production remains high during the treatment period, (4) the introduction of legumes may result in higher production from the existing species, and (5) it is less costly than complete cultivation. Since specialized mechanical equipment is involved, interseeding is best used on ranges of moderate topography, free of large rocks and brush.

Interseeding has been successful in Texas (Robertson and Box, 1969) and elsewhere throughout the Great Plains. The results of some interseedings suggested that the effects of furrows on native vegetation may be more important than are the introduced species in improving yields (Table 14.3). After three growing seasons, the yields from plots furrowed in the fall were equally productive to those seeded with grasses. Only one of the three legumes seeded contributed substantially to forage yields; however, in the mountain ranges of central Utah, legume-grass mixtures were superior to grass alone, the increased production coming from the legumes (Bleak, 1968). There was no indication in either case that legumes increased the growth of associated grasses, a response observed in legume-grass seedings in the Canadian plains (Kilcher and. Heinrichs, 1968).

Replacing Native Plants Which Would Not Naturally Become Established

Often valuable native plants have been eliminated and in the absence of a natural source of seed they must be reestablished artificially. Shrubs especially fit into this category. Because of low motility of the seed of most shrubs, rapid reoccupancy of abandoned croplands and burns cannot be expected.

Seeding of shrubs requires adherence to the same requirements as for herbaceous plants, a suitable seedbed, freedom from competition, and a period of protection while they become established. There are additional problems connected with browse plantings. Seeds are larger and often difficult to handle by mechanical means because of accessory structures such as plumed awns, hulls, or pulpy fruits. Because browse seed is not readily harvested by ordinary mechanical devices as are grasses, seed must be hand-collected which limits the amounts available and makes it costly, though there is some promise that vacuum-cleanerlike machines can be developed to facilitate seed collection (Nord, 1963). Germination percentages are often low and treating seed to break dormancy may be necessary. Because of their large size, browse seed is attractive to rodents, and they may consume much of the seed (Table 14.4). This necessitates protective measures, either repellents or rodenticides.

Among the many browse species tested in the intermountain area, bitterbrush (*Purshia tridentata*) has been most successfully used; since it presents fewer obstacles to collection, processing, and the establishment than others. A number of other native shrubs have been successfully reproduced, though with greater difficulty. Plummer et al. (1968) have summarized the characteristics and methods of handling a number of native and introduced shrub species.

Table 14.4   Percentage of Bitterbrush Seed Spots Disturbed by Mice at Various Intervals of Time after Planting

Data from Holmgren and Basile (1959)

Date of planting	Time elapsed between seeding date and observation date			
	2 weeks	7 weeks	3 months	5 months
	(Percent)			
Oct. 6, 1955	55	—	66	67
Oct. 29, 1956	29	88	—	88

Criteria for Selecting Seeding Sites

Not all rangelands can be successfully seeded at the present state of the art. The appearance and vigor of vegetation growing on an area is the best guide to selection for seeding, since plants integrate many ecological factors—moisture, soil, climate, etc. Sites with healthy, lush growth of plants, even though only weedy species, are the ones for reseeding. Shown et al. (1969), for example, found that success in seeding sagebrush varied directly with vigor index of big sage (*Artemisia tridentata*); black sage (*A. nova*) sites were less successful. If *Sarcobatus* or *Atriplex confertifolia* were present poor grass establishment resulted, and these native shrubs became firmly established. In the intermountain region of the Western United States, 25 cm of precipitation are thought necessary for success. Areas which support big sagebrush, piñon pine and juniper, oak brush, aspen, or meadow types at higher elevations have proved suitable for seeding. Seeding in salt-desert sites has not been successful in trials made in Wyoming (Hull, 1963b) nor in western Utah and Nevada (Bleak et al., 1965).

Results have generally been good throughout the Great Plains both with native and some introduced species. Efforts to seed in the low deserts of southern Arizona and New Mexico have generally given poor results, although areas which supported desert-grassland types have given satisfactory results (Reynolds et al., 1949).

In Mexico and other arid portions of North America, seedings of native grasses have been successful in the deep soils where water accumulates, but the uplands and the extreme desert areas are difficult to revegetate (Dwyer, 1969). Reseeding abandoned farmlands in Zacatecas, Mexico, has been successful when native grasses are seeded on a properly prepared seedbed (Gonzalez, 1973).

In Australia, seeding in the arid rangeland area has not been extensively practiced. In the higher rainfall area—50 cm or more—improved mixes of clover and grass are successful when accompanied by phosphorus fertilizer.

Mixed versus Pure Stands

There are certain advantages and disadvantages to either pure stands or mixtures. Mixtures of plants appear to have several advantages over pure stands: (1) All areas have variable conditions of soil, moisture, and slope. In mixtures, each species produces abundantly on the site more nearly supplying its needs. (2) Different rooting habits may result in more efficient use of soil moisture and nutrients from various soil depths. (3) Seasonal forage production is likely to be more uniform because periods of lush growth and dormancy vary in different species.

(4) A mixed diet is likely to be more desirable to the animals and produce greater gains, especially when browse plants and legumes are included. (5) Mixtures may have greater longevity, those species being better suited to a particular site replacing the less-suited species as they disappear from the stand. (6) Some plants of the mixtures may have favorable influences on others. A notable example is the nitrification accomplished by legumes, which increases vigor of grass growth and also increases its protein content. Nitrogen fixation is not limited to legumes. A number of rosaceous shrubs, such as *Ceanothus betuloides,* possess this ability (Vlamis et al., 1964). Nichols and Johnson (1969) reported a 50 percent increase in grass production from including sweet clover in the seed mix.

Many believe that production from mixed stands is greater than from monocultures, although this has not been well documented, and the evidence is contradictory (Bleak, 1968; Rumsey, 1971). Theoretically, mixtures should be more productive due to rooting and drawing moisture from different depths, and to a longer combined growing season, thus utilizing more of the precipitation that falls. For example, crested wheatgrass makes its growth in the spring and becomes dormant throughout the summer despite the presence of soil moisture. Other species continue to grow during the warm summer months. Frischknecht (1963) attributes benefits from a low-value shrub (*Chrysothamus nauseosus*) to the more succulent forage produced near rabbitbrush plants. Pure stands are managed more easily because the growth requirements of only one species need be considered. Mixtures inevitably require greater management skills, since they are subject to compositional changes resulting from differential utilization by animals and tolerance to grazing. In mixtures, highly palatable species may be killed out, and low-palatability species may be underutilized. Grasses that mature or reach maximum palatability or nutritive value at different seasons might best be planted alone so that they can be managed according to individual requirements, although the quality of forage provided may be poorer and the period of use shorter than from mixed stands.

An even more serious shortcoming of monocultures may be their greater susceptibility to insects and disease. Many stands of crested wheatgrass in Utah have been made almost valueless as forage due to the depredation of the black-grass bug. (See Fig. 11.7.)

Management Requirements of Seeded Ranges

Many of the early seedings were short lived because of poor management. Protection from grazing is imperative to permit seedlings to become well rooted and established. Often two or three growing seasons may be required. Unfortunately, this is not always possible, as for instance on big-game ranges. Once established, seedings should be grazed properly, otherwise they will deteriorate and be invaded by low-value plants. Introduced plants are, after all, not miracle plants that possess greatly superior resistance to use than native plants. Properly managed seedings may last many years, but even managed seedings cannot be maintained indefinitely. Pure seedings are disclimax communities, hence nature proceeds to reestablish the climax. Grazing management can slow but may not prevent invasions of climax plants. In Utah seeded ranges were grazed by cattle at different intensities. Records over a 12-year period showed that sagebrush increased irrespective of the intensity of utilization, although much more sagebrush was present under heavy use (Fig. 14.3).

Seeding for Erosion Control

At times the most compelling need is to halt erosion and stabilize the soil

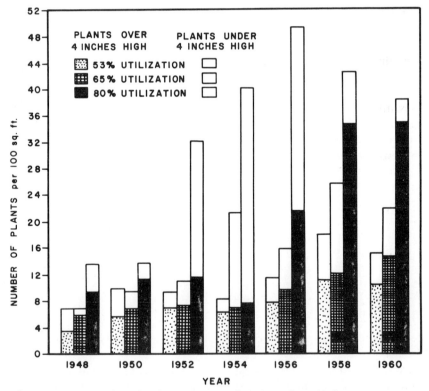

Fig. 14.3 Invasion of sagebrush into grass seedings in northern Utah is affected by intensity of grazing. After 12 years, a great many more sagebrush plants were present under heavy use. *(From Frischknecht and Harris, 1958.)*

either against water or wind. In such instances, consideration need not be given to forage quality but to ease of establishment, hardiness, and soil-protecting qualities. Sod-forming grasses and layering shrubs are best suited for this purpose, since once they are established they most effectively protect the soil and are long-lived. The latter characteristic is especially important at high elevations where the growing season is short and production of seed uncertain.

In special situations such as along highway cut-and-fill banks, unpalatability is desirable; otherwise, game animals are attracted to the seeding at considerable peril to the animals and motorists. Low inflammability is desirable also to reduce fire hazards.

FERTILIZATION OF RANGES

In recent years a great deal of attention has been directed to the possibilities of increased forage production by fertilization. More readily available and cheaper fertilizers (comparatively) and the high costs of seeding have sparked this interest. Additionally, the steady increase in consumption of red meat has encouraged greater meat production.

Fertilization has advantages over other means of range improvement. It requires no highly specialized equipment; costs are less than for seeding; and a period of nonuse is unnecessary. Fertilization is undertaken primarily to increase forage production, but other benefits may be forthcoming as well. Among these are (1) a more varied forage mix, (2) more palatable and nutritious forage, (3) a longer grazing season, (4) better distribution of animals, and (5) better seedling establishment.

Increased Production from Fertilization

Fertilizers have been shown to increase forage yields tremendously under some conditions but very little under others. Increased production varied from 4 to 250 percent (Table 14.5). The results obtained are affected by (1) the kind of fertilizer applied, (2) the rate of application, (3) the kind and fertility of the soil, (4) the kind of vegetation, (5) the amount of precipitation, and (6) the time of application.

Kind of Fertilizer Although animal manures and mulches have proved effective in increasing range production, it is unlikely that they will be used widely. Commercial fertilizers, because of ease of handling and ready availability, are most adapted to range situations.

The most widely applied elements are nitrogen, phosphorus, sulfur, and potassium. Others have been tested, but their use is less general being associated with specific regional soil deficiencies such as in the case of calcium in warmer humid climates where leaching is excessive.

Table 14.5 First-Year Increases in Forage Production from Applying Nitrogen to Rangelands at Different Rates

Area	Vegetation	Rate of application, kg/ha (Percent)						Source
		25–45	55–85	90–120	135–165	180–300	300+	
Northwest								
British Columbia	*Agropyron inerme*	32	58	75	99	91	111	Mason and Miltimore (1964)
	Cool-season grasses		85			99		Mason and Miltimore (1969)
California	Annualgrass-forbs			188				Walker and Williams (1963)
Great Basin								
Oregon	*Juncus, Carex*	14		17	143			Rumburg (1972)
Southern Utah	Subalpine meadow		55					Bowns (1972)
California	*Agropyron, Poa*		53					Wilson et al. (1966)
Southwest								
Arizona	Desert grassland	66	114	117				Holt and Wilson (1961)
New Mexico	*Hilaria mutica*	37	60	32				Herbel (1963)
New Mexico	Desert grassland	92		148				Dwyer (1971)
Northern Plains								
Montana	*Stipa viridula*		76	64	150			White et al. (1972)
North Dakota	Mid-grasses	4	47	70				Goetz (1969)
	Short grasses	22	61					Goetz (1969)
North Dakota	Native grasslands	21	67	92	100	170	179	Power and Alessi (1971)
	Bromus inermis		138	250				Power et al. (1972)
	Agropyron desertorum			161	187			Power and Alessi (1970)
Central Plains								
Wyoming	Warm-season grasses	19	18					Rauzi et al. (1968)
	Cool-season grasses	42	63					Rauzi et al. (1968)
Wyoming	Cool- and warm-season grasses	72		128		180		Cosper et al. (1967)
Nebraska	Warm-season grasses	49	56					Burzlaff et al. (1968)
Nebraska	Tall grasses	34		72				Rehm et al. (1972)
Coastal Prairie								
Texas	Grasses			7			36	Drawe and Box (1969)

Of these, nitrogen has given the greatest and most universal response in western North America, but it is not effective everywhere; Mason and Miltimore (1969) reported increased production on some sites in British Columbia but not on others. Where it is effective, production is directly correlated with the rate of application up to about 135 kg per hectare (120 lb per acre) (Fig. 14.4). Beyond that level, increased production is usually not experienced, though there are exceptions. The source of nitrogen is unlikely to be a factor influencing increased production (Power and Alessi, 1970).

Phosphorus has not usually given increased production under Western range conditions. Some increases were reported in meadows in Oregon (Cooper and Sawyer, 1955); and, in California where legumes comprised from 16 to 60 percent of the vegetation, yields were increased by the application of superphosphate by 75 to over 500 percent (Table 14.6). In both cases the increased production was provided by legumes. Some studies indicate that under certain conditions grasses may respond to phosphorus (Power and Alessi, 1970). Phosphates are effective in increasing forage yield and quality in Florida (Lewis, 1970).

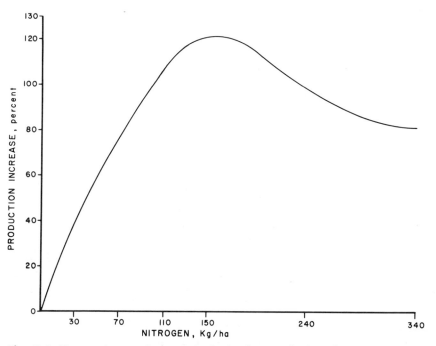

Fig. 14.4 First-year increase in forage production from application of increasing rates of nitrogen to rangelands in the Western United States. *(Data from many sources.)*

Table 14.6   Effect of Phosphorous Fertilizers on Forage Production on Annualgrass-Clover Range in California

Data from Williams et al. (1956)

Fertilizer and date	Phosphorus applied (lb/acre)	Harvest date	
		May 1953	May 1954
		(lb/acre, dry weight)	
Fall 1951			
None	0	1,625	1,480
Superphosphate	200	3,820	3,955
Fall 1952			
None	0	755	1,300
Superphosphate and treble superphosphate	150	3,185	3,620

Although phosphorus alone is generally ineffective in increasing yields, nitrogen and phosphorus in combination at high levels of application have given higher yields than nitrogen alone (Table 14.7). In-

Table 14.7   Herbage Production Attributable to Addition of Phosphorus to Applications of Nitrogen as Percent of Production when Nitrogen Only Was Applied

Data from various sources

Nitrogen, kg/ha	Phosphorus, kg/ha					
	22	34	45	67	90	180
34		117				
45					109	104
					115	101
67				98		
				109		
				107		
				104		
				101		
90*	125		133		112	126
					125	106
143*		143			154	137
					145	136

*Combinations including 90 kg or more of nitrogen were significant at 5 percent level in all or some of the years going to make up the averages shown.

creased production occurred only at nitrogen applications of 90 kg or more per hectare. The level of phosphorus was seemingly unimportant in increasing yields.

Sulfur, whether applied as elemental sulfur, as gypsum ($CaSO_4$), or as superphosphate, has been shown to stimulate forage production on granitic soils in the annual-range type of California. Legumes often provide most of the increase. Grasses respond much less, possibly because of increased nitrogen fixation by the legumes (Bentley and Green, 1954), although Bentley et al. (1958) found year-to-year differences with grasses yielding most forage in some years and legumes in others.

As with phosphorus, greatest yields came when nitrogen and sulfur are combined; from 26 to 43 percent more forage was produced than when nitrogen was applied alone at the same rate (Walker and Williams, 1963). In pot trials of soils from a pine-grass range in British Columbia, little increased production from sulfur alone was observed; but when nitrogen and sulfur were applied together, yields sometimes were increased (Freyman and van Ryswyk, 1969). The authors believed that sulfur enabled plants to utilize nitrogen more efficiently; recovery of nitrogen was almost twice as great when sulfur was added.

Potassium generally does not increase yields on Western ranges in the United States, although Retzer (1954) reported response by plants growing on granite-type soils in Colorado. In the Southeastern United States, some tree species responded to calcium (Ward and Bowersox, 1970) and Bermuda grass gave greater yields from potassium-nitrogen mixes than from nitrogen alone (Cook and Baird, 1967). In New Zealand, where cobalt-deficient soils are found, cobaltized superphosphate has given good results (Philipp, 1959).

Effect of Available Moisture on Response to Nitrogen   Precipitation greatly affects the response of ranges to fertilization. In years of low precipitation little increase in production occurs (Luebs et al., 1971); growth is limited by lack of moisture. In years of high precipitation, especially if it comes during the growing season, mineral nutrients may limit growth, in which case fertilization increases yields (Fig. 14.5). Williams et al. (1964) found no benefits from application of sulfur in years when precipitation was low. Similarly, tobosa (*Hilaria mutica*) responded to nitrogen only when the area received floodwaters (Herbel, 1963). Stroehlein et al. (1968) observed greater response when nitrogen was applied at the start of the summer rainy period.

Power (1967) traced nitrogen applied as fertilizer on irrigated and unirrigated plots. Under irrigation (ample moisture) most of the nitrogen was recovered from the tops and roots and only 5 to 10 percent

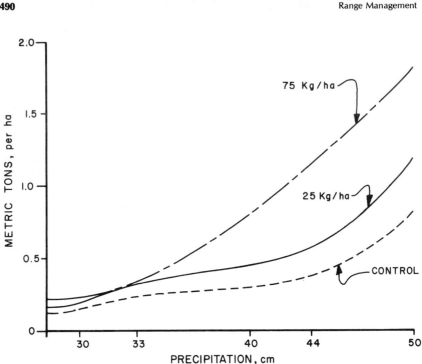

Fig. 14.5 Increased production resulting from two levels of nitrogen, under different amounts of precipitation, Saskatchewan, Canada. Fertilization is ineffective when moisture is limited. *(Data from Kilcher, 1958.)*

was fixed in the soil. Without irrigation the amount fixed in the soil equaled or exceeded that recovered by plants. He observed that the primary effect of water was to hasten the absorption of nitrogen. Wight and Black (1972) claimed increased efficiency in use of precipitation due to fertilizers, but since no measurement of actual water use was made, the term may be misleading. More simply the problem is one in which plant nutrients are limiting at high moisture levels and water is limiting moisture at low levels.

Effect of Fertilization on the Length of the Growing Season A little-documented response to fertilization, but one supported by observation, is that of the effect on the length of the growing season. Humid fertilized pastures remain green longer in the fall than those unfertilized. Under range conditions where moisture may be limiting, the effects may be less pronounced. Desert grasslands in Arizona remained green 1 week longer when fertilizers were applied (Holt and Wilson, 1961), and fertilized *Stipa viridula* matured 4 weeks later than when unfertilized (White et al., 1972).

Persistence of Fertilizers   Phosphorus is readily fixed in the soil and is little subject to leaching. It is thus possible to fertilize heavily with expectation of carryover effects for several years. By contrast, under California conditions, sulfur is readily leached and that unutilized the year of application is not available in subsequent years. Three-fourths of the sulfur applied may be lost through percolation in the first year (McKell and Williams, 1960).

Nitrogen is subject to leaching by water (Fig. 14.6), and that lost to the atmosphere in the form of gas may be as much as 20 percent (Power, 1972). Leaching is likely to vary greatly depending primarily upon the amount and distribution of precipitation. High precipitation, especially if it comes during the nongrowing season, results in high percolation which carries with it much of the nitrogen. Summer rainfall in semiarid areas seldom results in deep percolation and leaching may be of little importance, although surface runoff may cause losses. That some ecosystems may have a high capacity to retain nitrogen in the system and give increased yields for several years is indicated by results from North

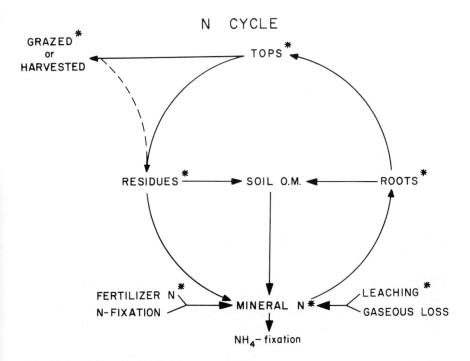

Fig. 14.6 Nitrogen cycle showing sources of nitrogen and avenues where losses occur. Items starred can be determined; those unstarred can only be estimated. *(From Powers, 1972.)*

Table 14.8   Dry-Matter Production by Native Grass as Affected by Rate and Timing of Nitrogen Fertilization

Data from Power and Alessi (1971)

Nitrogen rate	Years applied	Years						
		1963	1964	1965	1966	1967	1968	Total
Kg/ha/year		(Dry matter, kg/ha)						
0		670	580	550	270	220	770	3,060
34	1	780	570	400	280	250	700	2,980
135	1	1,340	1,520	900	340	300	950	5,350
540	1	1,870	2,940	2,320	900	840	2,220	11,090
11	3	490	570	640	280	250	870	3,100
45	3	830	1,330	1,440	430	370	930	5,330
180	3	1,740	2,790	2,710	810	1,100	2,230	11,410
6	6	640	730	660	390	290	1,010	3,720
22	6	660	930	950	450	520	1,700	5,210
90	6	1,160	2,430	2,480	900	1,150	2,700	10,820
LSD, 0.05		280	310	460	220	200	380	1,180

Dakota where different rates of nitrogen were applied under three schedules—all at once, one-third each year for 3 years, and one-sixth each year for 6 years (Table 14.8). At the higher rates of application, 6-year yields differed very little irrespective of how fertilizer was applied which led the authors to conclude that it was possible to get the same production with a single heavy application as from lighter yearly applications. There were some species changes that may have affected production apart from direct fertilizer effects. Mason and Miltimore (1972), however, report carryover effects from nitrogen application up to 10 years.

Effect of Fertilizers on Vegetative Composition

Not all plants respond alike to fertilization. Grasses and shrubs generally respond to nitrogen, though the response is different among species. Legumes, because of the fact that they fix nitrogen, do not respond to nitrogen but make dramatic responses to phosphorus. Because of these differential species responses, the possibility exists of changing the forage mix in stands of native vegetation provided the responses are known. As an example, alfalfa seeded with grasses was virtually eliminated when nitrogen alone was applied at rates of 168 kg per hectare (Cooke et al., 1968).

Rate of application also affects different plants differently (Table 14.9). *Festuca* produced most at the low level of nitrogen, wheatgrass and bluegrass at intermediate rates, and cheatgrass yielded most at the higher rates. With no fertilization, cheatgrass contributed less than 10 percent of the forage; at high rates it contributed more than half. Similar differential responses between Harding grass (*Phalaris*) and annual bromes to different rates of nitrogen have been reported, but annual-grasses were favored at low fertilizer rates, and the perennial grasses were favored at high rates (Martin et al., 1964).

In mixtures of grasses and legumes, nitrogen can be expected to increase the grass component and phosphorus will increase the legumes (Fig. 14.7). Where sulfur is effective, legumes are favored by its use (McKell et al., 1971). As between other forbs and grasses and among individual grass species, the effects of nitrogen fertilization are uncertain. Cosper et al. (1967) found that the effects on intermediate grasses, short grasses, and other species (mostly forbs) differed with the area, the time of application, the rate of application, and the condition and composition of the range. Moreover, first-year effects differed from those observed the second year.

On the Great Plains, cool-season grasses generally respond better to nitrogen than warm-season grasses (Launchbaugh, 1962; Rauzi et al., 1968), but response may differ among species. *Agropyrons* increased and *Stipa comata* and *Koeleria* decreased in cover from either nitrogen alone or with phosphorus (Johnston et al., 1967). In Oklahoma, production from warm-season grasses was not greatly increased by applications of nitrogen alone or nitrogen plus superphosphate, but there was considerably greater production from forbs (Huffine and Elder, 1960).

Table 14.9 Oven-Dry Herbage Produced by Four Grass Species at Different Rates of Nitrogen Application, Hooper, Washington

Data from Patterson and Youngman (1960)

Nitrogen, lb/acre	Agropyron inerme	Festuca idahoensis	Poa secunda	Bromus tectorum	Total
	(Lb/acre)				
0	605	44	110	80	840
20	664	401	290	540	1,700
40	781	118	495	1,043	2,438
60	419	50	319	1,470	2,259
80	408	27	292	1,740	2,468
Average	275	88	302	975	1,941

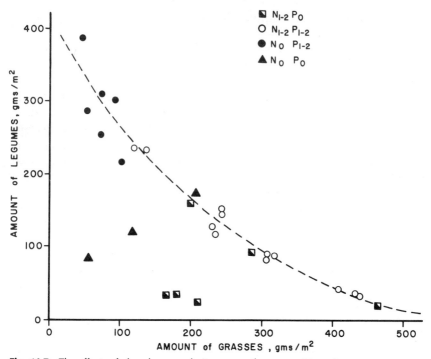

Fig. 14.7 The effects of phosphorus and nitrogen on the composition of an annual grass-legume mixture in Israel. Nitrogen favors grass; phosphorus favors legumes. *(From Ofer and Seligman, 1969.)*

Dolan and Taylor (1972) reduced clubmoss (*Selaginella densa*), a plant of no forage value that often makes up as much as 80 percent of the ground cover in the northern Great Plains, equally well with applications of manure as with intense mechanical treatments.

Effects of Fertilization on Quality of Forage

One of the objectives sought through fertilization is to improve the nutrient quality of the forage, especially crude protein (nitrogen) and phosphorus contents. That concentration of these and other elements can be altered by the use of fertilizers, especially in soils where certain elements are deficient, has been demonstrated. In addition, it is believed by some that digestibility of forage can be increased by fertilization.

Effects of Fertilization on Nitrogen Content of Forage Results obtained under range conditions are variable. Mason and Miltimore (1964) show steady increases in plant nitrogen in *Agropyron inerme* as nitrogen applications were gradually increased from 28 to 500 kg per hectare (Fig.

14.8). In contrast to this, Klipple and Retzer (1959) found no significant increase in protein from ammonium nitrate, but both cattle manure and phosphorus resulted in higher protein percentages in mature blue grama. Somewhat different results were obtained with cool-season grasses in South Dakota when water was applied to fertilized plots. Nitrogen content of forage was increased by nitrogenous fertilizers on both watered and unwatered plots. Only on the unwatered plots did phosphorus, alone or with nitrogen, increase nitrogen contents of forage (Cosper and Thomas, 1961). Luebs et al. (1971) report increases in nitrogen content of forage by application of nitrogen and phosphorus together on some sites in California but not in others.

The nitrogen content of forage increased almost immediately upon application of fertilizer to meadows up to mid-July, but the magnitude of the initial increase became less as the season advanced (Rumburg, 1972). By mid-July, the effect of the May application was gone; effects of the later applications were still evident on September 1.

There is little question that nitrogen content of forage, and presumably protein content, can be increased by fertilization, but the results are sometimes small and erratic, probably because of differences in the amounts and kind of fertilizer used, the stage of growth, the soil condition, the availability of moisture, and the kinds of vegetation. Thus, it is not possible to predict average outcome under a given set of conditions. Considering the influence on yield, even small changes in protein content can result in large increases in crude-protein production. As an

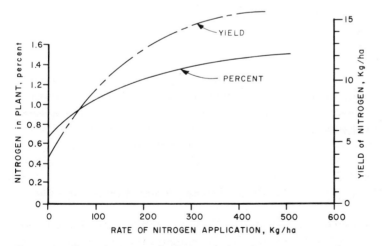

Fig. 14.8 Effect of nitrogen fertilizers on the nitrogen content of *Agropyron inerme* and on total yield of nitrogen per hectare. *(Data from Mason and Miltimore, 1964.)*

example, increasing percent protein in the forage by 2 1/2 times increased the crude-protein production nearly six times (Drawe and Box, 1969). Nitrate concentrations in *Atriplex polycarpa* similarly were increased by increasing nitrogen concentrations of the substrate (Chatterton et al., 1971).

The effect of increased yield and quality of forage can greatly increase livestock production. Dwyer and Schickendanz (1971) found animal gains were doubled on desert grasslands due to fertilization.

Poisoning from Too-High Nitrogen Contents   An adverse side effect may result from using nitrogenous fertilizers. Where a high proportion of the nitrogen present in the forage occurs as nitrates (NO_3) rather than as crude protein, so-called nitrate poisoning may result. Actually, the adverse effects come not from the nitrate, but from the intermediate reduction product nitrite (NO_2). Where there is an excess of nitrate, the reduction to ammonia is hampered and nitrite accumulates in the bloodstream affecting hemoglobin and the ability of the blood to carry oxygen (Dodge, 1970).

Few researchers have made determinations of the form in which nitrogen occurs in range forage, reporting either total nitrogen or crude protein calculated from total nitrogen. Ryan et al. (1972) differentiated nitrate nitrogen from total nitrogen in four grasses at different levels of fertilization and found that nitrate concentrations increased as fertilizer was increased. At low rates nitrate levels were judged safe, but when 540 kg per hectare of fertilizer was applied, the percentages of nitrate nitrogen present at times exceeded the "safe level" (0.15 percent). A possible solution where heavy fertilization is contemplated would be to supply the fertilizer in increments throughout the growing season.

Effect of Fertilizers on Phosphorus Content of Forage   Phosphorus content of plant tissue can also be increased by fertilization, although the results obtained vary greatly. Where the soils are deficient in phosphorus, phosphate fertilizers increase forage phosphorus, the amount of increase at a given place depending upon the amount of fertilizer used, soil-moisture levels, and the stage of growth.

During early growth stages low levels of phosphate application increased the phosphorus content of rough fescue *(Festuca scabrella);* at later stages, only high levels were effective, although they persisted even to the weathered stage (Johnston et al., 1968). Whether the low response is a function of stage growth or availability of moisture is uncertain, for Cosper and Thomas (1961) found that when water was applied to fertilized prairie vegetation in western South Dakota, phosphorus in the forage was greater than on comparable unwatered plots.

Plant phosphorus percentages may be depressed by nitrogenous fertilizers. Both commercial fertilizer and manure depressed phosphorus percentages when applied without phosphorus (Klipple and Retzer, 1959). Even when phosphorus and nitrogen are applied in combination, phosphorus percentage may be lower than where phosphorus is applied alone at equivalent rates (Cosper and Thomas, 1961).

Effect of Fertilization on Digestibility   There is little evidence regarding the effect of fertilization on digestibility. Nitrogen and nitrogen-phosphorus combinations, but not phosphorus alone, have been reported to decrease cellulose percentages and increase digestibility of rough fescue (Johnston et al., 1968). Contrary to this, Cooke et al. (1968) cite evidence that indicates a reduction in digestibility as a result of fertilization. Considering such factors as stage of maturity and rankness of growth, which affect chemical composition, and the problems inherent in sampling vegetation and in measuring digestibility, documentation may prove extremely difficult.

Effect of Fertilization on Plant Palatability   There is evidence that grazing animals prefer fertilized over unfertilized plants, although what gives rise to this preference is not certain (see Palatability and Preference, Chapter 6). Deer preferred fertilized shrubs even though the differences in calcium and phosphorus contents were minor (Ward and Bowersox, 1970). Deer use of Douglas fir seedlings was greater when they were fertilized with nitrogen alone or in combination, but not when sulfur or phosphorus was applied (Oh et al., 1970). Jackrabbits have been observed to congregate on plots to which fertilizers were applied (Table 14.10). Erratic results were obtained by Gibbens and Pieper (1962); deer showed a preference for manzanita *(Arctostaphylos mariposa)*

Table 14.10   White-tailed Jackrabbit Distribution in Winter in Relation to Kind and Level of Fertilizer

Data from Johnston et al. (1967)

Fertilizer	Rate of application			
	None	Low	Medium	High
	(Pellets per sq m)			
Phosphorus	4.0	4.2	10.6	4.1
Nitrogen	4.0	5.7	6.2	8.2
Nitrogen and phosphorus	4.0	17.1	26.7	39.4

when it was fertilized with sulfur, only slight preference when nitrogen was used, and no preference for plants fertilized with phosphorus. Merino sheep in Austrialia, however, showed a preference for the mature herbage fertilized with phosphorus when the phosphorus content of forage was below 0.10 to 0.15 percent (Ozanne and Howes, 1971). Utilization was increased by phosphorus fertilization in Florida (Lewis, 1970).

The preference shown by herbivores for fertilized vegetation has led to proposals for using fertilization as a tool to obtain better animal distribution. Brown and Mandery (1962) proposed to minimize elk- and deer-damage problems by this means. On annual ranges in California, fertilizing the drier slopes with sulfur induced greater utilization there and lessened the pressure on the more moist swales (Green et al., 1958).

Experience on northern Utah ranges provided less favorable indications. Cattle did not of themselves locate and use the nitrogen-fertilized portions of the range but made greater use of them when they were driven there (Hooper et al., 1969). Smith and Lang (1958) reported significantly greater use by cattle due to fertilization with urea in Wyoming.

Fertilization and Seedling Establishment

The period between germination and establishment is a critical one for a plant. Unless root development can proceed rapidly enough to maintain contact with moist soil there is little chance that the seedling plant will survive. Fertilization has been tried as a means of stimulating growth and, thus, survival.

Pot trials have demonstrated that growth of seedlings can be affected by adding nutrients. Thus, phosphorus was found to stimulate development of crested wheatgrass and Russian wild rye seedlings in the very early stages, whereas, nitrogen gave no response prior to about 25 days,[1] even though mature grasses seldom respond to phosphorus.

Under field conditions fertilization has not always resulted in increased establishment. Gates (1962) was unable to increase seedling emergence and survival on high-altitude ranges in northern Idaho nor in the greenhouse. Often, applying fertilizer at the time of seeding results in poorer establishment as was found with 20 warm-season prairie grasses in Nebraska (Warnes and Newell, 1969) and in the case of Ceanothus cuneatus seedlings (Gibbens and Pieper, 1962). In both these instances weed growth was stimulated, and the poor survival was attri-

[1] R.B.D. Whalley, personal communication.

buted to competition from weeds. This is supported by the results obtained by Gartner et al. (1957), wherein seedlings of *Ceanothus cuneatus* produced much smaller root systems growing in competition with ryegrass *(Lolium perenne)* even though top growth was increased by fertilization. Similarly, fertilization failed to aid in establishment of sideoats grama *(Bouteloua curtipendula)* in western Texas (Robertson and Box, 1969). McKell (1972) suggests using slow-release fertilizers as a means of overcoming the adverse effects on seedlings during the early, critical stages.

In addition to the increased competition from associated plants, fertilization may have an adverse effect on the competitive ability of the seedling. Top growth often is increased and root growth reduced, thus increasing water loss and reducing the capacity of the roots to supply it. Ayeke and McKell (1969) found root growth of perennial ryegrass and smilo in the first 5 days after germination to be reduced by increasing concentrations of nitrogen and phosphorus combined. Growth of either species was unaffected by nitrogen alone at any concentration. Similar effects, however, may not be found once the seedling reaches the photosynthetic stage.

CONSERVING WATER TO INCREASE RANGE PRODUCTION

The most universal factor limiting production is lack of adequate soil moisture. Precipitation is low all or parts of the year and frequently comes at such high rates that much of it runs off (see Chapter 12). Often soils are heavy and impenetrable. Under these conditions any mechanical modification of range sites that will improve infiltration into the soil will reduce soil-moisture stress and increase production. "Pitting," "chiseling," furrowing (contour or otherwise), terraces, and water spreading have all been employed to reduce surface runoff and increase storage of soil moisture.

Pitting and Chiseling

Pitting involves the creation of small basins to catch and hold precipitation. Various mechanical devices may be used in their construction from bulldozers to eccentric disks either in established stands of vegetation or in conjunction with seeding operations (Fig. 14.9).

In southern Arizona, pits have proved successful in increasing penetration of rainfall (Table 14.11). Pits of intermediate size (2 1/2 × 3 m and 15 cm deep) proved more effective than smaller or larger ones in

Fig. 14.9 Rangeland pitted with an eccentric disk to promote infiltration and increase forage production. *(Photograph by O. K. Barnes.)*

establishment of and production from seedlings (Slayback and Renney, 1972). Increases in grazing capacity of 33 percent have been noted on short-grass plains from pitting (Barnes, 1950). Pitting significantly increased soil moisture availability on Montana ranges (Houston, 1965).

Chiseling is used on heavy clay soils and where hardpans form beneath the soil surface. Not only is water penetration slow on these

Table 14.11   Rainfall and Moisture Penetration in and Adjacent to Intermediate-Sized Pits on the Santa Rita Experimental Range, 1963

Data from Slayback and Cable (1970)

	Period ending					
	7/16	7/27	8/1	8/8	8/16	8/25
	(In.)					
Rain during period	1.50	0.90	2.00	0.68	0.25	0.67
Moisture penetration:						
Sandy loam—flat*	5	9	22	10	17	22
—basin	24	26	34	27	28	28
Loam —flat	7	9	11	7	17	15
—basin	15	15	27	27	27	23
Clay loam —flat	2	8	15	5	5	6
—basin	11	12	19	16	15	12

*Penetration measured 2 to 3 ft from pits.

areas, but plants find it difficult to establish themselves in the compacted soils. Fryrear and McCully (1972) found the roots of *Bouteloua curtipendula* seedlings were unable to penetrate compacted soils; chiseling is necessary for effective seedling establishment. Road rippers or special machines with strong teeth designed to break through the compacted layers are used in order to permit infiltration. Experience indicates that chiseling is ineffective unless the soil surface receives considerable disturbance (Barnes and Nelson, 1945).

Water Spreading

Of all the mechanical treatments, water-spreading structures are most beneficial to forage production especially on the Great Plains and in the Southwest. They consist of dams and dikes which intercept surface runoff and convey it out of natural drainage areas at low gradients across the land surface where it can be absorbed. The aim should be to bring the maximum land under the influence of the dikes by so spacing and locating them that flooding will affect the greatest possible area.

Spreading water has so greatly increased grass growth on range areas in New Mexico that hay has been harvested from lands formerly of low production even for grazing. Forage increases of 350 percent have been noted on areas in Montana as a result of spreader construction. Spreaders are perhaps fully effective only on heavy soils; results in the Southwest on light soils showed no increased forage production (Valentine, 1947). Increased herbage production due to spreading of 62 and 353 percent were reported in Montana on *Agropyron smithii* and *Bouteloua gracilis* ranges, respectively (Houston, 1960).

Miller et al. (1969) summarized the results obtained from water spreading in the Western United States. These showed forage production of from 1,500 to 8,600 kg per hectare. The following conclusions were drawn: (1) spreaders were successful only if at least one flooding occurred each year, (2) forage production was less when water was ponded and could not drain, and (3) the moisture retention capacity of the A and B horizons were more important than soil texture on the amount of forage produced.

Similar systems of water management are used by African pastoralists for crop production and for range-forage supplementation. Dams and brushwood deflectors in the ephemeral streams of Arabia and the Sahelian Zone divert water from the wadis and onto "run-on" areas. These small areas, periodically irrigated, may produce almost as much forage as the vast area of surrounding rangeland. Although water spreading dates back to prebiblical times, it is still one of the most promising methods of increasing forage production in the arid zone.

Terraces

The primary aim of a terrace is to retain water on the land and prevent erosion, although vegetation also is benefited. Terraces should be level or have a very slight gradient conducting the water away from the natural drainage. On ranges, expensive terracing is resorted to only when the terrace cost is exceeded by the property damage being done on lower-lying lands by the runoff or when a jeopardized city-water supply or irrigation system demands immediate attention. Occasionally, however, an inexpensive terrace is advisable, often alternating with a number of furrows.

The size or capacity of the terrace varies with the topography and the amount and distribution of precipitation to which past weather records show the area to be subjected. A coefficient of runoff must be computed for each slope, and the needed terrace capacity calculated allowing a margin of safety of 10 to 25 percent. The coefficient will vary considerably depending primarily upon soil, vegetation, and topography.

Italian workers originated the terracing of nontillable lands for erosion control, but they have since been modified and widely used by government agencies in the United States. Most commonly, they are resorted to on areas of steep topography of high erosion-damage potential (Bailey and Croft, 1937).

Contour Furrows

Because of the high cost of terracing, increasing interest has developed in the use of contour furrows for rangeland. They serve an important role because they are inexpensive yet effective in increasing production on low-value ranges. The furrow differs from the terrace in being smaller and less elaborately designed. Contour furrows are plowed or listed strips placed close together and generally not smoothed after plowing (Fig. 14.10).

It has been observed that rainfall may percolate into the ground 15 to 45 cm deeper on furrowed land (Semple, 1937). Grasses have become reestablished in 1 to 3 years after furrowing in the southern plains (McCorkle and Dale, 1941). Branson et al. (1962) found greater infiltration rates and increased soil moisture storage on heavy soils in Montana that had been furrowed and reseeded 10 years earlier. There was no way that furrowing effects could be separated from seeding effects.

Tests on ranges in Texas (Langley and Fisher, 1939) in which the land was listed to a depth of 7 to 8 cm showed grass yield to increase as much as 3.9 times as a result of increased soil moisture and depth of penetration. Native grass increased in both ground cover and produc-

Fig. 14.10 Contour furrows in southeastern Idaho, with baffles to prevent drainage of the entire trench in the event of failure at any point. Evidence of accumulation of water can be seen in the mud cracks at the bottom of the furrow.

tion. An increase of 79 percent in the weight of root material resulted from list furrowing (Dickson et al., 1940).

Branson et al. (1966) reviewed the responses of range vegetation to various soil treatments. Contour furrows, broad-based ones particularly, were superior to pitting or other similar soil-modification treatments, and the results were greater on fine-textured soils.

Contour furrows and ridges usually are not desirable on the southern Great Plains on loose, sandy soils, on rough broken lands, or on steep slopes. They are impractical when large quantities of sand and silt accumulate in them.

CHEMICAL CURING OF RANGE FORAGE

The growing period, during which herbaceous plants provide highly nutritious forage, occupies only part of the year, and during the remainder, unless evergreen browse is plentiful, range forage is of low nutritive value, especially in humid areas and on annualgrass ranges. Herbicides which arrest growth without killing the plants have been advocated to avoid this loss of nutriments.

Results reported thus far indicate that chemical curing is feasible, although in what areas and on what plants is yet to be determined. On Oregon ranges, primarily on wheatgrasses, results using paraquat indicate that summer declines in protein levels can be arrested (Fig. 14.11). The occurrence of summer and fall rains reduced mineral contents of the cured forage, especially calcium, phosphorus, and potassium. Curing bleached the treated forage with consequent loss of carotene (Sneva,

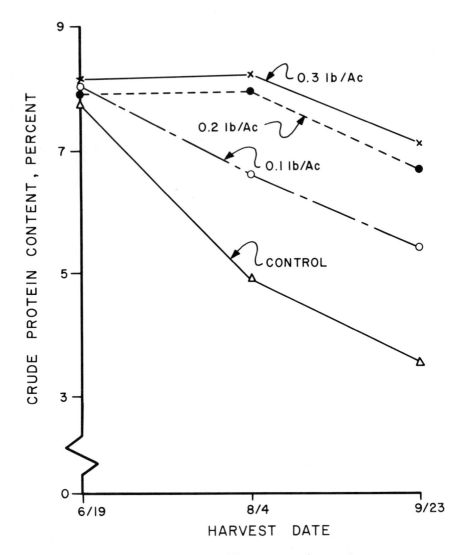

Fig. 14.11 Effect of applying paraquat at different rates on the protein content of crested wheatgrass *(Agropyron desertorum)*. Nutrient quality of the forage was increased, but yields were substantially reduced in wet seasons. *(From Sneva, 1967.)*

1967). Paraquat plus nitrogen increased nitrogen yields in forage even more (Sneva, 1973).

On annual range paraquat and two other similar herbicides produced similar effects on protein content to those reported by Sneva, but on an area where drought terminated growth naturally coincident with spraying, no effects were evident (Kay and Torrell, 1970). Plant composition was changed by spraying; grasses were reduced and legumes increased. Sheep made better gains on treated than on untreated forage, but less forage was produced under the spray treatment.

Although the results of these tests indicate that curing standing forage crops may be practical, much more will have to be known about the effects in other range environments before the usefulness of this practice is established.

Comparative Costs of Range Improvements

Physical and biological factors dictate the kinds of improvement that are required and that can be used, but within these there are options of methods that vary greatly in costs and effectiveness. All are expensive though they vary greatly from place to place. The probable returns should be carefully weighed against the costs involved before one decides on a particular method. Table 14.12 shows estimated costs for some common improvement practices. Since some of these must be combined (e.g., seeding and juniper removal), the total costs are greater than for any single practice shown. Combinations that were considered necessary resulted in costs from $16 to $34 per acre.

Table 14.12   Comparative Costs of Some Range-Improvement Practices in the Southern Rocky Mountain and Southern Intermountain Areas, 1970

Data from Anderson et al. (1973)

Practice	Cost per acre
Bulldozing juniper	$20.30
Chaining juniper (one way)	5.80
Drill seeding*	4.38
Aerial seeding	3.34
Deferred grazing	.57
Fencing	6.52

*Includes seedbed preparation.

BIBLIOGRAPHY

Anderson, Jay C., J'wayne McArthur, and Darwin B. Nielson (1973): Feasibility of range improvement on the rangelands of the Four Corners Economic Development Region, Report to the Four Corners Regional Commission, *N. Mex. State Univ. Special Rep.* no. 23.

Ayeke, Cyril Alende, and Cyrus M. McKell (1969): Early seedling growth of Italian ryegrass and smilo as affected by nutrition, *Jour. Range Mgt.* **22:**29–32.

Bailey, R. W., and A. R. Croft (1937): Contour trenches control floods and erosion on range lands, *Emergency Conservation Work Forestry Pub.* **4.**

Barnes, O. K. (1950): Mechanical treatments on Wyoming range land, *Jour. Range Mgt.* **3:**198–203.

—— and A. L. Nelson (1945): Mechanical treatments for increasing the grazing capacity of shortgrass range, *Wyo. Agr. Expt. Sta. Bull.* **273.**

Bement, R. E., et al. (1965): Seeding of abandoned croplands in the central Great Plains, *Jour. Range Mgt.* **18:**53–59.

Bentley, J. R., and L. R. Green (1954): Stimulation of native annual clovers through application of sulfur on California foothill range, *Jour. Range Mgt.* **7:**25–30.

——, ——, and K. A. Wagnon (1958): Herbage production and grazing capacity on annual-plant range pastures fertilized with sulfur, *Jour. Range Mgt.* **11:**133–140.

Bleak, A. T. (1968): Growth and yield of legumes in mixtures with grasses on a mountain range, *Jour. Range Mgt.* **21:**259–261.

——, N. C. Frischknecht, A. Perry Plummer, and R. E. Eckert, Jr. (1965): Problems in artificial and natural revegetation of the arid shadscale vegetation zone of Utah and Nevada, *Jour. Range Mgt.* **18:**59–65.

Bowns, James E. (1972): Low level nitrogen and phosphorus fertilization on high elevation ranges, *Jour. Range Mgt.* **25:**273–276.

Branson, F. A., R. F. Miller, and I. S. McQueen (1962): Effects of contour furrowing, grazing intensities and soils on infiltration rates, soil moisture and vegetation near Fort Peck, Montana, *Jour. Range Mgt.* **15:**151–158.

——, ——, and —— (1966): Contour furrowing, pitting, and ripping on rangelands of the western United States, *Jour. Range Mgt.* **19:**182–190.

Brown, E. Reade, and John H. Mandery (1962): Planting and fertilization as a possible means of controlling distribution of big game animals, *Jour. Forest.* **60:**33–35.

Burzlaff, D. F., G. W. Fick, and L. R. Rittenhouse (1968): Effect of nitrogen fertilization on certain factors of a western Nebraska range ecosystem, *Jour. Range Mgt.* **21:**21–24.

Chatterton, N. Jerry, J. R. Goodin, and Cameron Duncan (1971): Nitrogen metabolism in *Atriplex polycarpa* as affected by substrate nitrogen and NaCl salinity, *Agron. Jour.* **63:**271–274.

Conard, E. C., and Vern E. Youngman (1965): Soil moisture conditions under pastures of cool-season and warm-season grasses, *Jour. Range Mgt.* **18:**74–78.

Cook, E. D., and R. W. Baird (1967): Coastal Bermudagrass fertilization, *Blackland Expt. Sta. Texas Agr. Expt. Sta. M.P.* **849.**

Cooke, D. A., S. E. Beacom, and W. K. Dawley (1968): Response of six-year-old grass-alfalfa pastures to nitrogen fertilizer in north-eastern Saskatchewan, *Can. Jour. Plant Sci.* **48:**167–173.

Cooper, Clee S., and W. A. Sawyer (1955): Fertilization of mountain meadows in eastern Oregon, *Jour. Range Mgt.* **8:**20–22.

Cosper, H. R., and J. R. Thomas (1961): Influence of supplemental run-off water and fertilizer on production and chemical composition of native forage, *Jour. Range Mgt.* **14:**292–297.

———, ———, and A. Y. Alsayegh (1967): Fertilization and its effect on range improvement in the northern Great Plains, *Jour. Range Mgt.* **20:**216–222.

Currie, Pat O. (1969): Use seeded ranges in your management, *Jour. Range Mgt.* **22:**432–434.

Dickson, R. E., B. C. Langley, and C. E. Fisher (1940): Water and soil conservation experiments at Spur, Texas, *Tex. Agr. Expt. Sta. Bull.* **587.**

Dodge, Marvin (1970): Nitrate poisoning, fire retardants and fertilizers—any connection?, *Jour. Range Mgt.* **23:**244–247.

Dolan, John J., and John E. Taylor (1972): Residual effects of range renovation on dense clubmoss and associated vegetation, *Jour. Range Mgt.* **25:**32–37.

Drawe, D. Lynn, and Thadis W. Box (1969): High rates of nitrogen fertilization influence coastal prairie range, *Jour. Range Mgt.* **22:**32–36.

Dwyer, Don D. (1969): Practicas de mejoramiento en Agostaderos de zona arida y semi-arida, in *Simposio Internacional Sobre al Aumento de la Producion de Alimentos en Tonas Aridas,* Texas Tech. University Press, Lubbock, Tex., pp. 149–154.

——— (1971a): Nitrogen fertilization of blue grama range in the foothills of south-central New Mexico, *N. Mex. Agr. Expt. Sta. Bull.* **585.**

———, and Jerry G. Schickendanz (1971b): Vegetation and cattle responses to nitrogen-fertilized rangeland in south-central New Mexico, *N. Mex. Agr. Expt. Sta. Res. Rep.* **215.**

Eckert, Richard E., Jr., and Raymond A. Evans (1967): A chemical-fallow technique for control of downy brome and establishment of perennial grasses on rangeland, *Jour. Range Mgt.* **20:**35–41.

———, Gerard J. Klomp, Raymond A. Evans, and James A. Young (1972): Establishment of perennial wheatgrasses in relation to atrazine residue in the seedbed, *Jour. Range Mgt.* **25:**219–224.

Franzke, C. J., and A. N. Hume (1942): Regrassing areas in South Dakota, *S. Dak. Agr. Expt. Sta. Bull.* **361.**

Freeman, Barry N. and Robert R. Humphrey (1956): The effects of nitrates and phosphates upon forage production of a southern Arizona desert grassland range, *Jour. Range Mgt.* **9:**176–180.

Freyman, S., and A. L. van Ryswyk (1969): Effect of fertilizer on pinegrass in southern British Columbia, *Jour. Range Mgt.* **22:**390–395.

Frischknecht, Neil C. (1963): Contrasting effects of big sagebrush and rubber rabbitbrush on production of crested wheatgrass, *Jour. Range Mgt.* **16:**70–74.

—— and Lorin E. Harris (1968): Grazing intensities and systems on crested wheatgrass in central Utah: Response of vegetation and cattle, *U.S. Dept. of Agr., Forest Service, Tech. Bull.* **1388.**

Fryrear, D. W., and W. G. McCully (1972): Development of grass root systems as influenced by soil compaction, *Jour. Range Mgt.* **25:**254–257.

Gartner, F. R., A. M. Schultz, and H. H. Biswell (1957): Ryegrass and brush seedling competition for nitrogen on two soil types, *Jour. Range Mgt.* **10:**213–220.

Gates, Dillard H. (1962): Revegetation of a high-altitude, barren slope in northern Idaho, *Jour. Range Mgt.* **15:**314–318.

Gibbens, R. P., and R. D. Pieper (1962): The response of browse plants to fertilization, *Calif. Fish and Game* **48:**268–281.

Goetz, Harold (1969): Composition and yields of native grassland sites fertilized at different rates of nitrogen, *Jour. Range Mgt.* **22:**384–390.

Gonzales, Martin H. (1973): Plan Zacatecas, Proceedings of El Hombre in Las Americas, AAAS Symposium, Mexico D.F.

Green, L. R., K. A. Wagnon, and J. R. Bentley (1958): Diet and grazing habits of steers on foothill ranges fertilized with sulfur, *Jour. Range Mgt.* **11:**221–227.

Herbel, Carlton H. (1963): Fertilizing tobosa on flood plains in the semi-desert grassland, *Jour. Range Mgt.* **16:**133–138.

Holmgren, Ralph C., and Joseph V. Basile (1959): Improving southern Idaho deer winter ranges by artificial revegetation, *Idaho Dept. Fish and Game, Wildlife Bulletin* 3.

Holt, Gary A., and David G. Wilson (1961): The effect of commercial fertilizers on forage production and utilization on a desert grassland site, *Jour. Range Mgt.* **14:**252–256.

Hooper, Jack F., John P. Workman, Jim B. Grumbles, and C. Wayne Cook (1969): Improved livestock distribution with fertilizer—a preliminary economic evaluation, *Jour. Range Mgt.* **22:**108–110.

Houston, Walter R. (1960): Effects of water spreading on range vegetation in eastern Montana, *Jour. Range Mgt.* **13:**289–293.

—— (1965): Soil moisture response to range improvement in the northern Great Plains, *Jour. Range Mgt.* **18:**25–30.

—— and Robert E. Adams (1971): Interseeding for range improvement in the northern Great Plains, *Jour. Range Mgt.* **24:**457–461.

Hubbard, W. A., and J. L. Mason (1967): Residual effects of ammonium nitrate and ammonium phosphate on some native ranges of British Columbia, *Jour. Range Mgt.* **20:**1–5.

Huffine, Wayne W., and W. C. Elder (1960): Effect of fertilization on native grass pastures in Oklahoma, *Jour. Range Mgt.* **13:**34–36.

Hull, A. C., Jr., (1963a): Fertilization of seeded grasses on mountainous rangelands in northeastern Utah and southeastern Idaho, *Jour. Range Mgt.* **16:**306–310.

—— (1963b): Seeding salt-desert shrub ranges in western Wyoming, *Jour. Range Mgt.* **16:**253–258.

Johnston, A., L. M. Bezeau, A. D. Smith, and L. E. Lutwick (1968): Nutritive value and digestibility of fertilized rough fescue, *Can. Jour. Plant Sci.* **48:**351–355.

———, S. Smoliak, A. D. Smith, and L. E. Lutwick (1967): Improvement of southeastern Alberta range with fertilizers, *Can. Jour. Plant Sci.* **47:**671–678.

Kay, Burgess L., and James E. Street (1961): Drilling wheatgrass into sprayed sagebrush in northeastern California, *Jour. Range Mgt.* **14:**271–273.

——— and Donald T. Torrell (1970): Curing standing range forage with herbicides, *Jour. Range Mgt.* **23:**34–41.

Kennedy, P. Beveridge, and Samuel B. Doten (1901): A preliminary report on the summer ranges of western Nevada sheep, *Nev. State Univ. Agr. Expt. Sta. Bull.* **51.**

Kilcher, Mark R. (1958): Fertilizer effects on hay production of three cultivated grasses in southern Saskatchewan, *Jour. Range Mgt.* **11:**231–234.

——— and D. H. Heinrichs (1968): Seeding rambler alfalfa with dryland pasture grasses, *Jour. Range Mgt.* **21:**248–249.

Klipple, G. E., and John L. Retzer (1959): Response of native vegetation of the central Great Plains to applications of corral manure and commercial fertilizer, *Jour. Range Mgt.* **12:**239–243.

Langley, B.C., and C. E. Fisher (1939): Some effects of contour listing on native grass pastures, *Jour. Am. Soc. Agron.* **31:**972–981.

Launchbaugh, J. L. (1962): Soil fertility investigations and effects of commercial fertilizers on reseeded vegetation in west-central Kansas, *Jour. Range Mgt.* **15:**27–34.

Lavin, Fred (1967): Fall fertilization of intermediate wheatgrass in the southwestern ponderosa pine zone, *Jour. Range Mgt.* **20:**16–21.

Lewis, Clifford E. (1970): Responses to chopping and rock phosphate on south Florida ranges, *Jour. Range Mgt.* **23:**276–282.

Luebs, R. E., A. E. Laag, and M. J. Brown (1971): Effect of site and rainfall on annual range response to nitrogen and phosphorus, *Jour. Range Mgt.* **24:**366–370.

Martin, W. E., Cecil Pierce, and V. P. Osterli (1964): Differential nitrogen response of annual and perennial grasses, *Jour. Range Mgt.* **17:**67–68.

Mason, J. L., and J. E. Miltimore (1964): Effect of nitrogen content of beardless wheatgrass on yield response to nitrogen fertilization, *Jour. Range Mgt.* **17:**145–147.

——— and ——— (1969): Yield increases from nitrogen on native range in southern British Columbia, *Jour. Range Mgt.* **22:**128–131.

——— and ——— (1972): Ten year yield response of beardless wheatgrass from a single nitrogen application, *Jour. Range Mgt.* **25:**269–272.

McCorkle, J. S., and T. Dale (1941): Conservation practices for the range lands of the southern Great Plains, U.S. Department of Agriculture, Soil Conservation Service, Washington.

McKell, C. M. (1972): Seedling vigor and seedling establishment, in V. B. Younger and C. M. McKell (ed.), *The Biology and Utilization of Grasses*, Academic Press, New York, pp. 74–89.

——, R. Derwyn B. Whalley, and W. A. Williams (1971): Competition for sulfur by three annual-range species in relation to temperature, *Ecology* **52:**664–668.

—— and William A. Williams (1960): A lysimeter study of sulfur fertilization of an annual-range soil, *Jour. Range Mgt.* **13:**113–117.

McLean, A., and E. W. Tisdale (1972): Recovery rate of depleted range sites under protection from grazing, *Jour. Range Mgt.* **25:**178–184.

Miles, Arthur D. (1954): Improved pasture for spring and summer, range for fall and winter, *Jour. Range Mgt.* **7:**149–152.

Miller, R. F., et al. (1969): An evaluation of range floodwater spreaders, *Jour. Range Mgt.* **22:**246–257.

Nichols, James T., and James R. Johnson (1969): Range productivity as influenced by biennial sweetclover in western South Dakota, *Jour. Range Mgt.* **22:**342–347.

—— and Wilfred E. McMurphy (1969): Range recovery and production as influenced by nitrogen and 2,4-D treatments, *Jour. Range Mgt.* **22:**116–119.

Nord, Eamor C. (1963): Bitterbrush seed harvesting: when, where, and how, *Jour. Range Mgt.* **16:**258–261.

Ofer, Yitzchak, and No'am G. Seligman (1969): Fertilization of annual range in northern Israel, *Jour. Range Mgt.* **22:**337–341.

Oh, John H., Milton B. Jones, William M. Longhurst, and Guy C. Connolly (1970): Deer browsing and rumen microbial fermentation of douglas-fir as affected by fertilization and growth stage, *Forest Sci.* **16:**21–27.

Ozanne, P. G., and K. M. W. Howes, (1971): The effects of grazing on the phosphorus requirement of an annual pasture, *Austral. Jour. Agr. Res.* **22:**81–92.

Patterson, J. K., and V. E. Youngman (1960): Can fertilizers effectively increase our range land production?, *Jour. Range Mgt.* **13:**255–257.

Philipp, Perry F. (1959): The economics of grassland development and improvement in New Zealand, *Jour. Range Mgt.* **12:**170–175.

Plath, C. V. (1954): Reseed now?, *Jour. Range Mgt.* **7:**215–217.

Plummer, A. Perry, Donald R. Christensen, and Stephen B. Monson (1968): Restoring big-game range in Utah, *Utah Div. Fish and Game, Pub.* 68–3.

Power, J. F. (1967): The effect of moisture on fertilizer nitrogen immobilization in grasslands, *Soil Sci. Soc. Am. Proc.* **31:**223–226.

—— (1972): Fate of fertilizer nitrogen applied to a northern Great Plains rangeland ecosystem, *Jour. Range Mgt.* **25:**367–371.

—— and J. Alessi (1970): Effects of nitrogen source and phosphorus on crested wheatgrass growth and water use, *Jour. Range Mgt.* **23:**175–178.

—— and —— (1971): Nitrogen fertilization of semi-arid grasslands: Plant growth and soil mineral N levels, *Agron. Jour.* **63:**277–280.

——, ——, G. A. Reichman, and D. L. Grunes (1972): Effect of nitrogen source on corn and bromegrass production, soil pH, and inorganic soil nitrogen, *Agron. Jour.* **64:**341–344.

Rauzi, Frank, Robert L. Lang, and L. I. Painter (1968): Effects of nitrogen fertilization on native rangelands, *Jour. Range Mgt.* **21:**287–291.

Rehm, G. W., W. J. Moline, and E. J. Schwartz (1972): Response of a seeded mixture of warm-season prairie grasses to fertilization, *Jour. Range Mgt.* **25**:452–456.

Retzer, John L. (1954): Fertilization of some range soils in the Rocky Mountains, *Jour. Range Mgt.* **7**:69–73.

Reynolds, H. G., F. Lavin, and H. W. Springfield (1949): A preliminary guide for range seeding in Arizona and New Mexico, *Southwestern Forest and Range Expt. Sta. Research Rep.* **7.**

Robertson, Truman E., Jr., and Thadis W. Box (1969): Interseeding sideoats grama on the Texas High Plains, *Jour. Range Mgt.* **22**:243–245.

Rogler, George A., and Russell J. Lorenz (1957): Nitrogen fertilization of northern Great Plains rangelands, *Jour. Range Mgt.* **10**:156–160.

Rumburg, C. B. (1972): Yield and N accumulation of meadow forage fertilized at advancing maturity with N, *Agron. Jour.* **64**:187–189.

Rumsey, Walter B. (1971): Range seedings versus climax vegetation on three sites in Idaho, *Jour. Range Mgt.* **24**:447–450.

Ryan, M, W. F. Wedin, and W. B. Bryan (1972): Nitrate-N levels of perennial grasses as affected by time and level of nitrogen application, *Agron. Jour.* **64**:165–168.

Semple, A. T. (1937): Following contour furrows across the United States, *Soil Conservation* **2**:134–138.

Shown, L. M., R. F. Miller, and F. A. Branson (1969): Sagebrush conversion to grassland as affected by precipitation, soil, and cultural practices, *Jour. Range Mgt.* **22**:303–311.

Slayback, Robert D., and Dwight R. Cable (1970): Larger pits aid reseeding of semidesert rangeland, *Jour. Range Mgt.* **23**:333–335.

———— and Clinton W. Renney (1972): Intermediate pits reduce gamble in range seeding in the Southwest, *Jour. Range Mgt.* **25**:224–227.

Smith, D. R., and R. L. Lang (1958): The effect of nitrogenous fertilizers on cattle distribution on mountain ranges, *Jour. Range Mgt.* **11**:248–249.

Smith, Jared G. (1898): Experiments in range improvements, *U.S. Dept. of Agr., Div. of Agrostology Circ.* **8.**

Sneva, Forrest A. (1967): Chemical curing of range grasses with paraquat, *Jour. Range Mgt.* **20**:389–394.

———— (1973): Nitrogen and paraquat saves range forage for fall grazing, *Jour. Range Mgt.* **26**:294–295.

Stroehlein, J. L., P. R. Ogden, and Bahe Billy (1968): Time of fertilizer application on desert grasslands, *Jour. Range Mgt.* **21**:86–89.

Valentine, K. A. (1947): Effect of water-retaining and water-spreading structures in revegetating semi-desert range land, *N. Mex. Agr. Expt. Sta. Bull.* **341.**

Vlamis, J., A. M. Schultz, and H. H. Biswell (1964): Nitrogen fixation by root nodules of western mountain mahogany, *Jour. Range Mgt.* **17**:73–74.

Walker, Charles F., and William A. Williams (1963): Responses of annual-type range vegetation to sulfur fertilization, *Jour. Range Mgt.* **16**:64–69.

Ward, W. W., and T. W. Bowersox (1970): Upland oak response to fertilization with nitrogen, phosphorus, and calcium, *Forest Sci.* **16**:113–120.

Warnes, D. D., and L. C. Newell (1969): Establishment and yield responses of warm-season grass strains to fertilization, *Jour. Range Mgt.* **22:**235–240.

White, Larry M., Clee S. Cooper, and Jarvis H. Brown (1972): Nitrogen fertilization and clipping effects on green needlegrass *(Stipa viridula* Trin.): I. Development, growth, yield, and quality, *Agron. Jour.* **64:**328–331.

Wight, J. Ross, and A. L. Black (1972): Energy fixation and precipitation-use efficiency in a fertilized rangeland ecosystem of the northern Great Plains, *Jour. Range Mgt.* **25:**376–380.

Williams, William A., R. Merton Love, and John P. Conard (1956): Range improvement in California by seeding annual clovers, fertilization and grazing management, *Jour. Range Mgt.* **9:**28–33.

———, Cyrus M. McKell, and Jack N. Reppert (1964): Sulfur fertilization of an annual-range soil during years of below-normal rainfall, *Jour. Range Mgt.* **17:**1–5.

Wilson, A. M., G. A. Harris, and D. H. Gates (1966): Fertilization of mixed cheatgrass-bluebunch wheatgrass stands, *Jour. Range Mgt.* **19:**:134–137.

Index

513